PREFACE

With Hints for Using the Indexes

IT seems strange compiling a composite index system for a work with indexes already available at the end of volumes I–V and volume VII. Yet by combining them it has been possible to provide a thread of continuity through time as modern technology has evolved. Additionally this revision of the indexes provides opportunities for introducing some improvements in style and layout which were not so necessary for the indexes in their separate volumes.

It is only proper to recognize the painstaking work provided by my predecessors as indexers of each volume, upon which the consolidated indexes are founded. They are: Mr P. G. Burbridge for volumes I and II; Miss M. A. Hennings for volumes III, IV, and V; and Mr J. Tudge for combined indexes to volumes VI and VII.

It also appropriate to thank my helpers who made this work possible. They are: Miss D. Bragg; Mrs Nina Kurkjian; Mrs Kay Smith; and Mrs Mary Symons. I also have to express my appreciation to Dr Trevor I. Williams, the remaining Editor of the history, and Mr Ivon Asquith of the OUP for their patience during the extended time this work took for its completion. At first I thought the task would be straightforward; but when long strings of page numbers occurred in the combined version a great amount of re-indexing became necessary. With the job now done, however, I wish users some happy hunting and fruitful answers in this comprehensive *History of Technology*.

Hints for users

It may seem unusual to offer suggestions to potential readers of the history on the use of the indexes. After all, most of us with sufficient background of learning to make the content of the *History of Technology* worth study will be no strangers to indexes. Yet there are methods for using indexes to better advantage and the few suggestions given here may stimulate users to create their own search strategies. In these 'hints' further details given about the four parts of this volume are followed by some brief notes on search strategy.

PART 1—It summarizes the lists of contents for all the seven volumes of the text and is useful for tracing the development of main subject areas through the ages.

PART 2—The Index of Names lists the names of inventors, discoverers, and

achievers referred to in the text. Their dates are not included although such information may be found in the indexes to each volume. For the ease of a search, many references to separate individuals are subdivided by subjects. For information on the literature, users should consult the appropriate entries cited in the bibliographies at the end of chapters. Although the names listed in the Part 2 index may be a starting-point for a search, the destination might well be a reference in a bibliography.

PART 3—The Index of Place Names lists towns and cities and some countries cited in the text. As with Part 2, this index provides further subdivisions by subjects for places with more than a few page references. Some towns and cities are also listed in the Part 4 index with reference to Town planning. Although some countries are listed briefly in Part 3, more detail on them is provided in Part 4 where they are treated as subjects.

PART 4—The Index of Subjects, including countries, provides access to the subject content of the text. Many cross-references are provided to assist users with their search.

Notes—Most users will find the correct reference with little difficulty. Sometimes, however, the information sought is not to be found immediately, so some alternative strategy becomes necessary for its ultimate location.

Users seeking information can apply several methods depending on the sort of information required and the purpose of such a search. For example, the Index of Names will suffice for a search on the work of Leonardo da Vinci; similarly the Index of Place-names would be consulted for reported happenings in Marseilles. When a researcher has a fairly precise search objective, then these two indexes will provide both volume and page references quickly.

However, some investigators may not know exactly what they want, but vaguely have a wish to search within a certain broad subject area and then let inspiration do the rest. For example, a search might begin on the effect of mechanization, lead through aspects of 'Mass production' (for which there is a subject heading) and finish with the evolution of the motor car. In such cases a user may expedite the process by compiling a list of five or six related subjects that seem relevant. A browse through the subject index using the chosen subjects as a starting-point will help the investigator define the objective more closely. This is where the cross-reference system can be used constructively to link up with subject material not previously considered. Finally, sub-entry subjects of reasonable importance should not be ignored as they are usually indexed again as main subjects. For example

under Drainage, 'pumps' offers only three page references; yet under the main heading 'Pumps' some sixty-three references are cited in thirty-one subject subheadings.

Hove   RICHARD RAPER
February, 1983

CONTENTS

PART 1   LIST OF CONTENTS FOR VOLUMES I–VII   1

PART 2   INDEX OF NAMES   13

PART 3   INDEX OF PLACE-NAMES   73

PART 4   INDEX OF SUBJECTS   127

PART 1

LIST OF CONTENTS FOR
VOLUMES I–VII

LIST OF CONTENTS FOR
VOLUMES I–VII

VOLUME I

FROM EARLY TIMES TO FALL OF ANCIENT EMPIRES

CHRONOLOGICAL TABLES by F. E. ZEUNER   xlviii

MAPS   lvii

1. SKILL AS A HUMAN POSSESSION by KENNETH P. OAKLEY   1
2. EARLY FORMS OF SOCIETY by V. GORDON CHILDE   38
3. DISCOVERY, INVENTION, AND DIFFUSSION by H. S. HARRISON   58
4. SPEECH AND LANGUAGE by A. SOMMERFELT   85
5. PRIMITIVE TIME-RECKONING by E. R. LEACH   110
6. WORKING STONE, BONE, AND WOOD by L. S. B. LEAKEY   128
7. GRAPHIC AND PLASTIC ARTS by L. S. B. LEAKEY   144
8. FORAGING, HUNTING, AND FISHING by DARYLL FORDE   154
9. ROTARY MOTION by V. GORDON CHILDE   187
10. FIRE-MAKING, FUEL, AND LIGHTING by H. S. HARRISON   216
11. CHEMICAL, CULINARY, AND COSMETIC ARTS by R. J. FORBES   238
12. BUILDING IN WATTLE, WOOD, AND TURF by JOHN BRADFORD   299
13. DOMESTICATION OF ANIMALS by F. E. ZEUNER   327
14. CULTIVATION OF PLANTS by F. E. ZEUNER   353
15. POTTERY by the late SIR LINDSAY SCOTT, K.B.E.   376
16. TEXTILES, BASKETRY, AND MATS by GRACE M. CROWFOOT   413
    with *A Note on the Materials of Ancient Textiles and Baskets* by JULIUS GRANT   447
    and *Rope-making* by K. R. GILBERT   451
17. BUILDING IN BRICK AND STONE by SETON LLOYD, O.B.E.   456
    with *A Note on Stonehenge* by R. H. G. THOMSON   490
18. DIFFERENTIATION OF NON-METALLIC TOOLS by S. M. COLE   495
19. WATER-SUPPLY, IRRIGATION, AND AGRICULTURE by M. S. DROWER, M.B.E.   520
20. MINING AND QUARRYING by the late C. N. BROMEHEAD   558
21. EXTRACTING, SMELTING, AND ALLOYING by R. J. FORBES   572

22. METAL IMPLEMENTS AND WEAPONS by H. H. COGHLAN   600

23. FINE METAL-WORK by HERBERT MARYON, O.B.E., and H. J. PLENDERLEITH   623

24. FINE IVORY-WORK by R. D. BARNETT   663

25. FINE WOOD-WORK by CYRIL ALDRED   684

26. LAND TRANSPORT WITHOUT WHEELS. ROADS AND BRIDGES by S. M. COLE   704

27. WHEELED VEHICLES by V. GORDON CHILDE   716

28. BOATS AND SHIPS by ADRIAN DIGBY   730

29. RECORDING AND WRITING by S. H. HOOKE   744

30. MEASURES AND WEIGHTS by F. G. SKINNER, O.B.E.   774

31. ANCIENT MATHEMATICS AND ASTRONOMY by O. NEUGEBAUER   785

VOLUME II

THE MEDITERRANEAN CIVILIZATIONS AND THE MIDDLE AGES
c.700 B.C.–c.A.D. 1500

HISTORICAL NOTES by A. R. HALL   li

1. MINING AND QUARRYING TO THE SEVENTEENTH CENTURY by the late C. N. BROMEHEAD   I

2. METALLURGY by R. J. FORBES   41

3. AGRICULTURAL IMPLEMENTS by E. M. JOPE   81

4. FOOD AND DRINK by R. J. FORBES   103

5. LEATHER by JOHN W. WATERER   147
     with *A Note on Parchment* by H. SAXL   187

6. SPINNING AND WEAVING by R. PATTERSON   191

7. PART I. FURNITURE: TO THE END OF THE ROMAN EMPIRE by CYRIL ALDRED   221
     PART II. FURNITURE: POST-ROMAN by R. W. SYMONDS   240

8. PART I. CERAMICS: FROM c.700 B.C. TO THE FALL OF THE ROMAN EMPIRE by GISELA M. A. RICHTER   259
     PART II. CERAMICS: MEDIEVAL by E. M. JOPE   284

9. GLASS AND GLAZES by D. B. HARDEN, O.B.E.   311

10. PRE-SCIENTIFIC INDUSTRIAL CHEMISTRY by the late F. SHERWOOD TAYLOR and CHARLES SINGER   347
     with *A Note on Military Pyrotechnics* by A. R. HALL   374

11. THE MEDIEVAL ARTISAN by R. H. G. THOMSON   383

12. BUILDING CONSTRUCTION by MARTIN S. BRIGGS   397

13. FINE METAL-WORK by HERBERT MARYON, O.B.E.   449
     with *A Note on Stamping of Coins and Other Objects* by PHILIP GRIERSON   485

14. ROADS AND LAND TRAVEL, WITH A SECTION ON HARBOURS, DOCKS, AND LIGHTHOUSES by R. G. GOODCHILD and R. J. FORBES   493

15. VEHICLES AND HARNESS by E. M. JOPE   537

16. SHIPBUILDING by T. C. LETHBRIDGE   563

17. POWER by R. J. FORBES   589
    with *A Note on Windmills* by REX WAILES   623

18. MACHINES by BERTRAND GILLE   629
    with *A Note on Ancient Cranes* by A. G. DRACHMANN   658

19. HYDRAULIC ENGINEERING AND SANITATION by R. J. FORBES   663

20. MILITARY TECHNOLOGY by A. R. HALL   695

21. ALCHEMICAL EQUIPMENT by E. J. HOLMYARD   731

EPILOGUE: EAST AND WEST IN RETROSPECT by CHARLES SINGER   753

VOLUME III

FROM THE RENAISSANCE TO THE INDUSTRIAL REVOLUTION
*c.*1500–*c.*1750

1. FOOD AND DRINK by R. J. FORBES   I

2. METALLURGY AND ASSAYING by CYRIL STANLEY SMITH and R. J. FORBES   27

3. COAL MINING AND UTILIZATION by J. U. NEF   72

4. WINDMILLS by REX WAILES   89

5. TRADESMEN'S TOOLS *c.*1500–1850 by R. A. SALAMAN   110
    with *A Note on the Wheelwright's Trade* by J. GERAINT JENKINS and R. A. SALAMAN   123
    and *A Note on Coopering* by J. GERAINT JENKINS and R. A. SALAMAN   128

6. FARM-TOOLS, VEHICLES, AND HARNESS 1500–1900 by OLGA BEAUMONT and J. GERAINT JENKINS   134

7. SPINNING AND WEAVING by R. PATTERSON   151
    with *A Note on Knitting and Knitted Fabrics* by JAMES NORBURY   181

8. FIGURED FABRICS by the late J. F. FLANAGAN   187

9. GLASS by R. J. CHARLESTON and L. M. ANGUS-BUTTERWORTH   206

10. BUILDING CONSTRUCTION by MARTIN S. BRIGGS   245

11. TOWN-PLANNING FROM THE ANCIENT WORLD TO THE RENAISSANCE by MARTIN S. BRIGGS   269

12. LAND DRAINAGE AND RECLAMATION by L. E. HARRIS   300

13. MACHINES AND MECHANISMS by A. P. USHER   324

14. MILITARY TECHNOLOGY by A. R. HALL   347

15. PRINTING by MICHAEL CLAPHAM   377
    with *A Note on Technical Advances in the Manufacture of Paper before the*
    Nineteenth Century by JOHN OVERTON   411

16. BRIDGES by S. B. HAMILTON, O.B.E.   417

17. CANALS AND RIVER NAVIGATIONS BEFORE 1750 by A. W. SKEMPTON   438

18. SHIPS AND SHIPBUILDING by G. P. B. NAISH   471

19. CARTOGRAPHY, SURVEY, AND NAVIGATION TO 1400 by CHARLES SINGER,
    DEREK J. PRICE, and E. G. R. TAYLOR   501

20. CARTOGRAPHY, SURVEY, AND NAVIGATION 1400–1750 by E. G. R. TAYLOR   530

21. THE CALENDAR by SIR HAROLD SPENCER JONES, K.B.E.   558

22. PRECISION INSTRUMENTS TO 1500 by DEREK J. PRICE   582
    with a section (VII) on Hero's instruments by A. G. DRACHMANN

23. THE MANUFACTURE OF SCIENTIFIC INSTRUMENTS FROM *c*.1500 TO *c*.1700 by
    DEREK J. PRICE   620

24. MECHANICAL TIMEKEEPERS by H. ALAN LLOYD, M.B.E.   648

25. INVENTION IN CHEMICAL INDUSTRIES by F. W. GIBBS   676

EPILOGUE: THE RISE OF THE WEST by A. R. HALL   709

VOLUME IV

THE INDUSTRIAL REVOLUTION
c.1750 TO *c*.1850

 1. PART I. AGRICULTURE: FARM IMPLEMENTS by OLGA BEAUMONT and
    J. W. Y. HIGGS   1

    PART II. AGRICULTURE: FARM TECHNIQUES OF FARMING by G. E. FUSSELL   13

 2. FISH PRESERVATION by C. L. CUTTING   44
    with *A Note on Whaling* by L. HARRISON MATTHEWS   55

 3. METAL AND COAL MINING 1750–1875 by the late J. A. S. RITSON, O.B.E.   64

 4. PART I. EXTRACTION AND PRODUCTION OF METALS: IRON AND STEEL by
    H. R. SCHUBERT   99

    PART II. EXTRACTION AND PRODUCTION OF METALS: NON-FERROUS
    METALS by F. W. GIBBS   118

 5. POWER TO 1850 by R. J. FORBES   148

 6. THE STEAM-ENGINE TO 1830 by the late H. W. DICKINSON   168

 7. WATERMILLS *c*.1500–*c*.1850 by A. STOWERS   199

 8. PART I. THE CHEMICAL INDUSTRY: DEVELOPMENTS IN CHEMICAL THEORY
    AND PRACTICE by E. J. HOLMYARD   214

    PART II. THE CHEMICAL INDUSTRY: INTERACTION WITH THE INDUSTRIAL
    REVOLUTION by A. and N. L. CLOW   230

9. GAS FOR LIGHT AND HEAT by SIR ARTHUR ELTON, Bt.   258

10. PART I. THE TEXTILE INDUSTRY: MACHINERY FOR COTTON, FLAX, WOOL,
    1760–1850 by JULIA DE L. MANN   277

    PART II. THE TEXTILE INDUSTRY: SILK PRODUCTION AND MANUFACTURE,
    1750–1900 by W. ENGLISH   308

11. CERAMICS FROM THE FIFTEENTH CENTURY TO THE RISE OF THE STAFFORDSHIRE
    POTTERIES by A. and N. L. CLOW   328

12. GLASS by L. M. ANGUS-BUTTERWORTH   358

13. PRECISION MECHANICS by MAURICE DAUMAS   379

14. MACHINE-TOOLS by K. R. GILBERT   417

15. BUILDING AND CIVIL ENGINEERING CONSTRUCTION by S. B. HAMILTON, O.B.E.   442

16. PART I. SANITARY ENGINEERING: WATER-SUPPLY by J. KENNARD   489

    PART II. SANITARY ENGINEERING: SANITATION by J. RAWLINSON, C.B.E.   504

17. ROADS TO c.1900 by R. J. FORBES   520

18. PART I. CANALS: INLAND WATERWAYS OUTSIDE BRITAIN by ROGER
    PILKINGTON   548

    PART II. CANALS: INLAND WATERWAYS OF THE BRITISH ISLES by CHARLES
    HADFIELD   563

19. SHIP-BUILDING by GEORGE NAISH   574

20. CARTOGRAPHY by R. A. SKELTON   596

21. DREDGING by G. DOORMAN   629

22. TELEGRAPHY by G. R. M. GARRATT   644

23. THE BEGINNING OF THE CHANGE FROM CRAFT MYSTERY TO SCIENCE AS A
    BASIS FOR TECHNOLOGY by A. R. J. P. UBBELOHDE   663

VOLUME V

THE LATE NINETEENTH CENTURY
c.1850 TO c.1900

PART I. PRIMARY PRODUCTION

1. GROWTH OF FOOD PRODUCTION by G. E. FUSSELL   1

2. MANAGEMENT AND PRESERVATION OF FOOD by T. N. MORRIS   26

3. THE STEEL INDUSTRY by H. R. SCHUBERT   53

4. NEW EXTRACTION PROCESSES FOR METALS by R. CHADWICK   72

5. PETROLEUM by R. J. FORBES   102

PART II. PRIME MOVERS

6. THE STATIONARY STEAM-ENGINE 1830–1900 by A. STOWERS   124

7. THE MARINE STEAM-ENGINE by H. PHILIP SPRATT   141

8. INTERNAL COMBUSTION ENGINES by D. C. FIELD   157

PART III. THE RISE OF THE ELECTRICAL INDUSTRY

9. THE GENERATION OF ELECTRICITY by C. MACKECHNIE JARVIS   177

10. THE DISTRIBUTION AND UTILIZATION OF ELECTRICITY by C. MACKECHNIE JARVIS   208

PART IV. THE CHEMICAL INDUSTRY

11. HEAVY CHEMICALS by TREVOR I. WILLIAMS   235

12. DYESTUFFS IN THE NINETEENTH CENTURY by E. J. HOLMYARD   257

13. EXPLOSIVES by J. MCGRATH   284

14. FINE CHEMICALS by ARTHUR W. SLATER   299

PART V. TRANSPORT

15. THE DEVELOPMENT OF RAILWAY ENGINEERING by C. HAMILTON ELLIS   322

16. SHIP-BUILDING by A. M. ROBB   350

17. AERONAUTICS by PETER W. BROOKS   391

18. MECHANICAL ROAD-VEHICLES by D. C. FIELD   414

19. CARTOGRAPHY AND AIDS TO NAVIGATION by D. H. FRYER, O.B.E.   438

PART VI. CIVIL ENGINEERING

20. BUILDING MATERIALS AND TECHNIQUES by S. B. HAMILTON, O.B.E.   466

21. BRIDGES AND TUNNELS by H. SHIRLEY SMITH, O.B.E.   499

22. HYDRAULIC ENGINEERING by J. ALLEN   522

23. WATER-SUPPLY by F. E. BRUCE   552

PART VII. MANUFACTURE

24. PART I. THE TEXTILE INDUSTRY: WOVEN FABRICS by D. A. FARNIE   569
PART II. THE TEXTILE INDUSTRY: HOSIERY AND LACE by F. A. WELLS, O.B.E.   595

25. THE WORKING OF METALS by R. CHADWICK   605

26. MACHINE-TOOLS by D. F. GALLOWAY   636

27. CERAMICS by IREEN JAMESON   658

28. GLASS TECHNOLOGY by R. W. DOUGLAS   671

29. Printing and Related Trades by W. Turner Berry   683

30. Part I. The Photographic Arts: Photography by Helmut and Alison Gernsheim   716

    Part II. The Photographic Arts: Cinematography by Anthony R. Michaelis   734

31. Production and Utilization of Rubber by S. S. Pickles   752

PART VIII. THE THRESHOLD OF THE TWENTIETH CENTURY

32. Education for an Age of Technology by Sir Eric Ashby   776

33. Technology and Industrial Organization by Charles Wilson   799

34. Technology and its Social Consequences by Sir Alexander Fleck, K.B.E.   814

VOLUME VI

THE TWENTIETH CENTURY
c.1900 TO c.1950

PART I

1. The Setting in World History by T. K. Derry, O.B.E.   1

2. The Sources of Innovation by David Sawers   27

3. The Economics of Technological Development by F. R. Bradbury   48

4. Management by Glenn Porter   77

5. Trade Unions by Harold Pollins   93

6. The Role of Government by Alexander King   113

7. Education in Industrialized Societies by David Layton   138

8. The Fossil Fuels by Arthur J. Taylor   172

9. Natural Sources of Power by Lord Wilson of High Wray

    Part I: Water Power   195

    Part II: Other Natural Sources of Power   214

10. Atomic Energy by Lord Hinton of Bankside

    Part I: The Early History   223

    Part II: Uranium Enrichment   228

    Part III: The Development of Nuclear Reactors   236

    Part IV: Chemical Processes for Atomic Energy   256

11. The Development of the Nuclear Weapon by E. F. Newley, C.B.E   268

12. Electricity by Brian Bowers   284

13. AGRICULTURE by LYNNETTE J. PEEL
    PART I: ANIMAL PRODUCTION   298
    PART II: FOOD AND INDUSTRIAL CROPS   318

14. FISHING AND WHALING by G. H. O. BURGESS and J. J. WATERMAN   342

15. COAL-MINING by SIR ANDREW BRYAN   359

16. PRODUCTION OF PETROLEUM OIL AND NATURAL GAS by H. R. TAINSH and S. E. CHURCHFIELD   376

17. METAL MINING by JOHN TEMPLE   410

18. THE UTILIZATION OF METALS by W. O. ALEXANDER   427

19. IRON AND STEEL by M. L. PEARL   462

20. THE CHEMICAL INDUSTRY: A GENERAL SURVEY by L. F. HABER   499

21. THE CHEMICAL INDUSTRY by FRANK GREENAWAY, R. G. W. ANDERSON, SUSAN E. MESSHAM, ANN M. NEMARK, and D. A. ROBINSON
    PART I: HEAVY INORGANIC CHEMICALS   514
    PART II: RAW MATERIALS FOR ORGANIC CHEMICALS (INCLUDING EXPLOSIVES)   532
    PART III: POLYMERS, DYES, AND PIGMENTS   551

22. GLASS MANUFACTURE by R. W. DOUGLAS   570

23. PAINT by HENRY BRUNNER   590

24. PAPER by ERIC HAYLOCK, J. P.   607

25. CERAMICS by W. F. FORD   622

26. THE TEXTILES INDUSTRIES: A GENERAL SURVEY by D. T. JENKINS   638

27. TEXTILE MANUFACTURE by C. S. WHEWELL   648

28. CLOTHING MANUFACTURE by H. C. CARR   675

VOLUME VII

THE TWENTIETH CENTURY
c.1900 TO c.1950

PART II

29. THE DEVELOPMENT OF THE WORLD TRANSPORT MARKET by K. M. GWILLIAM   691

30. ROAD VEHICLES by LORD MONTAGU OF BEAULIEU   711

31. SHIPS AND SHIPBUILDING by BRIAN BAXTER   736

32. RAILWAYS by O. S. NOCK   762

33. AIRCRAFT AND THEIR OPERATION by PETER W. BROOKS   789

34. Navigational Equipment by wg. cdr. e. w. anderson 837

35. Space Technology by anthony r. michaelis 857

36. Civil Engineering by p. s. a. berridge
    Part I: Roads, Bridges, and Tunnels 871
    Part II: Land Reclamation, Canals, Ports, and Harbours 907

37. Building and Architecture by r. j. mainstone 930

38. Town Planning by sir colin buchanan 958

39. Engineering: The Scientific Basis by aubrey f. burstall 983

40. The Internal Combustion Engine by lynwood bryant 997

41. The Steam Turbine by allen smith 1025

42. Machine Tools by robert s. woodbury 1035

43. Production Engineering by f. koenigsberger 1046

44. Fluid Power by lord wilson of high wray 1060

45. The Generation, Distribution, and Utilization of Electricity by brian bowers 1068

46. Electronic Engineering by j. r. tillman 1091

47. Domestic Appliances by g. b. l. wilson 1126

48. Computers by tom margerison 1150

49. Instruments by frank greenaway 1204

50. Electrical Communication by d. g. tucker 1220

51. Printing by james moran 1268

52. Photography by helmut gernsheim 1281

53. Cinematography by david b. thomas 1302

54. Medical Technology by audrey b. davis 1317

55. Water Supply and Waste Disposal by f. e. bruce
    Part I: Water Supply 1363
    Part II: Sewerage and Sewage Disposal 1382

56. Food Technology by j. b. m. coppock
    Part I: The Scientific Background 1399
    Part II: Food Processing and Distribution 1411

57. Technology of the Ocean Depths by t. f. gaskell 1446

58. Technology and the Quality of Life by trevor i. williams 1467

PART 2
INDEX OF NAMES

INDEX OF NAMES

Notes

1. This index consolidates separate volume indexes at the end of volumes I–V and VII and the Index of Companies in volume V.

2. Although most dates and occupations have been omitted here, this information is obtained by referring to the separate indexes at the ends of volumes I to V.

Abbe, Ernst, V 673, 682; VII 1216, 1354
Abbott, Edwin, V 790
— W., V 769
Abel, Sir Frederick A., V 290, 292, 295
Abercrombie, P., VII 968
Abo, C. von, V 502
Abū Bakr Muḥammad ibn Zakariyya al-Rāzī, II 736–8
Abu Manṣūr Ḥaidar, *see* Ḥaidar . . .
Abū Manṣūr Muwaffak, II 737
Abulcacim Abnacahm, III 616 and n.
Abu'l-Qāsim of Kashan, II 302
Accum, Friedrich Christian, chemist
  foods, adulteration treatise, III 22; V 834; VII 1402
  gas-lighting, IV 268–70, 272
Achard, Franz, IV 28
Acheson, E. J., VII 1042
Ackerman, A. S. E., VI 221
— Rudolf, IV 268
Adād al-Daula, Caliph, III 612
Adadnirari II, I 534
Adalhard, II 140
Adam, Robert, IV 473–4
Adams, George, III 556, 645–6; IV 390, 602; V 457
— L. A., VI 588
— Robert, III 481
— W. (*fl* 1880), V 584
— William (1829–1904), V 334–5
— William Bridges, V 331, 335
Adamson, Robert, V 723
Addison, Joseph, IV 668
Ader, C., V 409
Aerial Transit Company, V 408
Aeschylus, II 168, 233
Æthelwulf, II 481
Agas, Ralfe, III 317
Agatharchides, I 581
Agnes, St, II 461
Agricola, Georgius (Georg Bauer)
  assaying, III 59, 62, 63, 65

balances, metallurgical, II 746
bearing-metals, III 41
bitumen, I 252; V 103
chain-dredgers, IV 638
chalk as fertilizer, II 34–5
chemical processes, III 677
De re metallica, II 13–15; III 27–8, 324; IV 233
fire-setting, mining, II 8 n.
glass-furnaces, III 210, 212–14, 216, 217, 700
glass-making, II 315; III 206, 208, 209, 224
iron smelting, II 68
lead purity, II 44, 45
machines, III 324–32
metals, description, III 29–30, 691
mining practices, II 13–23, 655; III 82
ore-furnaces, III 49, 50, 51, 52
potter's kick-wheel, II 290
publications, III 27–8
railways, V 322
succinnic acid, description, II 741
sulphur manufacture, II 379, 380
sulphuric acid, IV 242
treadmills, IV 671
ventilation, mines, IV 93
water-wheel, IV 181
— Gnaesus Julius, II 404
Agrippa, Cornelius, of Nettesheim, III 676–7
— Marcus Vipsanius, II 671; III 504
Ahiram, I 763
Aḥmad ibn Mu'tasim, II 635
Aiken, H. H., engineer
  computer designs, VII 1154, 1184–7, 1189
  drum information storage, VII 1195
Airy, Sir George, IV 399; V 442, 443
Akers, E., VII 1324
Akhenaten, I 690
Aktiengesellschaft Grusonwerk, V 764
Alan of Walsingham, II 442
Al-Battānī, III 590
Albers, H. C., IV 624
Albert, K., VI 596

Albert, Prince Consort, V 828, 782, 785
— I, Prince of Monaco, VII 1447, 1448
— the Great, V 394
— of Saxony, V 394
Alberti, L. B., III 250, 429, 446–7
Albertus Magnus, II 64
Al-Bīrūnī, II 737
Albright, Arthur, V 252–3
— W. F., I 762, 763, 766
Albright & Wilson, V 253, 832
Albucasis of the Latins, see Khalaf . . .
Alcock, J., VII 806, 1218
Aldhelm, St, II 32
Al-Dimashqī, II 615, 616
Alembert, Jean le Rond d', IV 417, 674
Alessio Piemontese, see Alexis
Alexander VI, III 288
— VII, III 289
— the Great, Macedonian King
  cavalry, use, II 697–8
  chronology, I tables E, F; II 772
  cotton imports, II 199
  machinery, military, II 700–1, 716
  roads, I 713
  steel from India, II 57
  sugar-cane, India, I 371
  town-planning, III 275
  world map, III 505
Alexanderson, E. F. W., VII 1235
Alexis (Alessio Piemontese), III 43, 679, 703
Al-Farghānī (Alfraganus), III 522
Al-Fazārī, III 609
Alfonso X, the Wise, II 12; III 526, 648–9
Alfort, d', IV 32
Alfraganus of the Latins, see Al-Farghānī
Alfred, English king, II 580–1; III 518
Al-Ḥākim, III 584
Al-Ḥasan al-Rammāḥ, II 379
Alhazen (Ibn al-Haitham), III 230; V 716, 735
Al-Jazarī, Muḥammad ibn Ibrahīm, II 614
Al-Kāshī, III 617
Al-Khwarīzmī, III 608
Allan, Alexander, V 324
Allen, B. T., VI 38
— Elias, III 630–1
— J., V 522; VI 195
— (Alleyn), John, III 687
Alley & MacLellan, V 134
Allgood, Edward, III 689
— Thomas, III 688
Allgoods, the, III 697
Allsopp & Sons, V 273
All-Ma'mūn, II 735; III 521, 584
Al-Mas'ūdī, II 616
Aloysius Lilius, III 579
Alpini, Prospero of Padua, III 5
Al-Qazwīnī, II 597, 614

Alstroemer, IV 26
Aluminium Company, V 249, 250
Aluminium Crown Metal Company, V 91
Al-'Urdī, III 584
Al-Zarqālī, III 584–5, 616
Amant, III 672, 674
Amasis II, I 262
Ambler, G. H., VI 474
Ambrose, St, III 202
Amenemhet I, I 537, table F
— II, I 657
Amenhetep I, I 776, table F
— II, Egyptian king
  copper, casting method, I 627
  robe design, I 440
  toilet-casket, veneered, I 686, 687, 693
— III, Egyptian king
  architecture, I 483
  capacity measure, I 782, 784
  chronology, I table F
  ointments, use of, I 286
  timber trade, I 685 n.
  vineyard gift, I 283
  water clock, III 601
Amenhotep, see Amenhetep
Amenophis, I 540
Amman, Jost, wood engraver
  engraving-wheel, III 223
  printing press, III 389, 399
  trades, sixteenth century, III 116
  typefounder at work, III 391, 392
  type-metal, III 43–4, 412
  type-storage, III 388, 389, 394
  wheelwright's shop (woodcut), II 552
Ammannati, Bartolomeo, III 430
Ammianus Marcellinus, II 713, 723
Amos (Easton and), IV 195
Ampère, André Marie, physicist
  commutator suggestion, V 180
  electricity, discoveries, IV 649, 653
  electromagnetism, V 178, 655, 679
Anderson, John, IV 445; V 780
— William, V 566, 567
Andersson, A. G., VI 474
Andrade, E. N. da C., VII 984
Andrée, S. A., V 400
Andreolis of Gubbio, III 678
Andreossy, François, III 464–6
Andrews, E. B., V 112
Androuin, J., VII 1052
Angelus, Jacobus, III 530
Angilcus, Robertus, III 537
Anglo-American Brush Electric Light Corporation,
  V 194, 198, 216, 217
Anglo-Swiss Condensed Milk Company, V 35
Anisson-Duperron, E. A. J., V 691
Anna Comnena, II 30, 376

Anschütz, Ottomar, V 743; VII 1285
Anschütz-Kaempfe, H., VII 842
Anson, George, IV 579
— Lord, IV 269
Antigonus II, II 566
Antipater of Thessalonica, II 593
Antoninus Pius, II 30
Antonius, Marcus, II 572, 573
Anville, J. B. Bourguignon d', IV 597
Apollodorus, II 512, 636
Appert, Francois, V 38–9, 42
— N., VII 1399, 1421
— Raymond Chevalier, V 39
Appian, II 516
Appius Claudius, II 28, 500, 512, 670
Applegath, Augustus, V 696, 698
Appold, John George, V 525, 545, 560
Arago, Dominique François Jean, scientist
    copper disc experiment, V 179
    daguerreotype process, V 720
    electromagnet, IV 653, 655; V 178
Arban, F., V 400
Arbuthnot, John, IV 2
Arcadius, II 601
Arcet, d', IV 253
Archer, Frederick Scott, V 725
Archimedes of Syracuse, mathematician
    density concept, III 22, 58, 66; V 394
    hydraulics, III 313
    mechanics exponent, II 604, 632–3, 634
    military machines, II 633, 698–9; IV 665, 667
    pulley inventions, II 633, 658
    screw principle, II 116, 631, 633, 676–7
    toothed-wheel invention, II 676
    water-lifting devices, II 676–7
Archytas of Tarentum, II 631–3, 677; IV 666
Arden, E., VII 1389, 1391
Argand, Pierre Ami, IV 263; V 104
Aristarchus of Samos, III 503
Aristobulus, II 199
Aristophanes, III 569
Aristotle, Greek philosopher, I 14
    alchemy, influence on, II 731
    chronology, table F
    Earth, spherical form, III 504
    Greek cities, report, II 528
    machinery and society, II 638
    motion, III 374
    plant-nutrition, IV 37
    post-mill figure, II 624
    pulley, origins, II 630, 658
    tin extraction, I 590
    town-planning, III 372
Arkwright, Sir Richard, inventor, V 786
    cotton industry, partnership, V 597
    'hydraulick machine', IV 154
    silk-spinning, IV 311, 314

spinning-machine, IV 277–8, 287, 295
steam power, IV 154, 283
textile-machinery, III 151, 280–2, 283–4, 285, 299
water-frame invention, V 840
Arlandes, Marquis d', V 395
Armsby, H. P., VI 309, 310
Armstrong (c.1839), IV 5
— E. H., VII 1091, 1097–8, 1237
— Neil A., VII 857
— Sir (Lord) William George, V 217, 534, 535
Armstrong Whitworth, V 121
Arnald de Villanova, II 138, 142
Arnemann, J., VII 1317, 1318
Arnold, Asa, IV 286
— J. O. (c.1890s), VI 492
— John (1736–99), IV 415
— M., VII 1470
— Thomas, V 821
Arphe, Juan, III 64–5
Arrol, Sir William, V 535
Arsenius, Walter, III 540, 586, 621
Artingstall, F. D., V 403 n. 1
Ascham, Roger, III 351
Ashby, A., V 566
— Lord, VI 146
Ashur-bani-pal, Assyrian king
    bed design, I 676, 677
    boats, coracle type, I 736–7
    chronology, I table F
    irrigation record, I 552
    water wells, importance, I 527
    wine list evidence, I 282
Ashur-nasir-pal I, I 282; II 495
Ashur-nasir-pal II, Assyrian king
    boats, coracle type, I 736–7
    chariots, I 718, 727
    chronology, I table F
    date-palm cultivation, I 552
    dyes, use of, I 247
    furniture, metal, II 235
    horse-bit design, I 723
    irrigation, gardens, I 553
    metal engraving, I 646, 649
    water wells, design, I 527
Ashworth, G. & E., V 573
Asoka, I 772
Aspdin, Joseph, IV 448, 497; V 483
— William, IV 448–9
Asphalt Mining & Kerosene Gas Co., V 105
Aspinall, Sir John, V 340
Astbury, John, IV 346–7, 349, 353
Aston, F. W., VI 230
— John, V 826
Atha, B., VI 497
Athenaeus of Attalia, II 674
— of Naucratis, I 266; II 133

Athénas, French chemist, IV 239
Atherton, J. B., VII 1074
Atholl, Duke of, IV 269
Atlantic Cable Company, V 225
Atwood, George, III 645
— Luther, V 105
Aubuisson, d', IV 159
Auer, J., VII 1334
Augustine, St, of Hippo, III 519
Augustus, Roman emperor, III 507
  aqueduct system, Rome, II 670
  building techniques, II 410–11
  chronology, II table II lvi
  Empire, survey, III 504
  height limit, buildings, II 530
  quarrying, II 28
  road works, II 508, 512–13
  Roman calendar, III 577, 578
  war-galleys, II 571–2
Augustus II, Elector of Saxony, IV 338
Aurelius, Marcus, II 530
Ausonius, II 31, 597, 600
Austen, John H. & George W., V 105
Austin, Herbert (Baron Longbridge), V 432
Aveling & Porter, of Rochester, IV 538
Avogardro, A., IV 225
Axon, IV 314
Ayrton, William Edward, V 418; VI 140; VII
  1257

Baatz, G., VII 810
Babbage, C., inventor and theorist
  analytical engine, VII 1175–9
  calculating machines, VII 1151, 1153, 1154,
    1159
  computers, VII 1151, 1183, 1184, 1192, 1194,
    1196
  difference engine, VII 1173–4, 1175
  management theories, VI 78–9, 80, 86
— H. P., VII 1176–7
Babbitt, I., VI 436
Babcock, G. H., V 138
— S. M., V 33, 34–5; VI 303
Babcock & Wilcox, V 138, plate 8A
Bachelier, Nicolas, III 459, 464
Backus, J., VII 1200
Bacon, Anthony, V 815
— Francis, II 604; III 679–80, 717–18
— John, IV 446
— Richard M., V 698; VI 615
Bacon, Roger, philosopher
  human flight, V 394
  lenses, III 230
  magnetic compass, III 524
  maps, III 521–2
  scientific progress, III 718–19
Bacqueville, Marquis de, V 394

Baddeley, John, of Shelton, IV 349
— Ralph, of Shelton, IV 349
Baden-Powell, B. F. S., V 398
Badeslade, T., III 316
Badische Anilin-und-Soda Fabrik, V 248, 276–8
Badius Ascensius, Jodocus, III 397–9
Baedeker, V 714
Baekeland, L. H., VI 554
Baeyer, Adolf von, 278, 788; VI 554
Bailey, R. W., VII 985
Bailey and Culley, IV 10
Baillie, Charles William, V 453, 463
Bain, Alexander, IV 654; V 224; VII 1213
— J. S., VI 50, 54–5, 57–9, 75
Baines, Edward, IV 249
Baird, J. L., VII 1259–60
Baker, D., VI 477
— Sir Benjamin, civil engineer
  Aswan dam, V 543
  bridges, IV 459; V 501, 506
  wind pressures, measurement, V 503
  tunnelling, Hudson River, V 519–20
— J. C., VII 1434
— M. N., VII 1377
— Matthew, III 479, 487–8
Bakewell, F. C., VII 1257
— Robert, III 15; IV 21, 40
Balard, A. J., V 319
Balchen, B., VII 806
Baldewin, Eberhardt, III 658–60
Baldwin, F. S., VII 1164
Ballaison, IV 538
Ballou, E. H., VI 665
Balme, de la, IV 633, 637
Balzer, S. M., VII 1009
Bancroft, E., IV 248
Banks, Sir Joseph, IV 28, 206, 262, 575; V 177
Barattieri, Giovanni Battista, III 315
Barba, Alvaro Alonzo, III 28–9, 63
Barbaro, Daniele, V 716
Barbarossa, see Frederick I
Barclay, R., VII 1269
Bardeen, J., VII 1118
Bardin, I. P., V 498
Barisino, Tommaso of Modena, III 230
Barker, S. W., VI 649
Barlow, A., V 579
— Edward, III 668–70
— Peter William, V 332, 347
— W. H., V 474, 492
Barman, Christian, III 114
Barnaby, Nathaniel, V 373
Barnack, Oskar, V 732; VI 1288
Barnes, D. E., VII 1332
Barnwell, F., VII 794
Barron, S. L., VII 1341
Barth, C., VI 84

Bartholomew the Englishman, III 676
Bartlett, IV 71
Bartolotti, III 314
Barton, Luke, V 598
Basch, S. S. K. R. von, VII 1344
'Basil Valentine', see 'Valentine, Basil'
Bass, George, IV 358
Bateman, J. F., IV 490, 497
Bates, of Kirklevington, IV 21
Bath, B. H. S. Van, see Van Bath
Batho, W. E., of Birmingham, IV 538
Batly, of Leeds, IV 323
Bauer, Andrew, V 695
— Georg, see Agricola
— W., VI 558
Baumann, E., VI 556
— K., VII 1030
Baumé, Antoine, III 22
Baur, Julius, V 65
Bauschinger, J., V 495
Bauwens, Liéven, IV 324–5, 560
Baxter, G., V 705
— J. P., VI 263, 265
Bayer, Karl Josef, V 92
— O., VI 561–2
Baylis, J. R., VII 1375
Bazalgette, Sir Joseph W., civil engineer
    drainage, London, IV 510–11; VII 1383, 1386
    sewers, materials, IV 449, 514–15
Bazin (c.1867), IV 631
— Henri Émile (1829–1917), V 540, 546
Beacham, T. E., VII 1061
Bealy, of Radcliffe, Manchester, IV 245
Beard, Richard, V 721, 722
Beart, Robert, VI 381
Beattie, Joseph, V 326–8
Beau de Rochas, Alphonse, V 158
Beaufort, Rear-Admiral Sir Francis, V 449, 452
Beaumont, Henry, V 633
— J. B. A. L. L. Élie de, IV 622
Beautemps-Beaupré, C. F., IV 618–19
Becher, Johann Joachim, III 682; IV 258
Béchereau, L., VII 794, 808
Beck, C., VII 1356
Bede, the Venerable, historian
    calendar, Christian, III 578
    cast-iron figures, II 7
    lead roofing, II 429
    stone-cutting skill, II 32
    time measurement, III 595, 596
    window-glazing skill, II 425
Bedford, Duke of (1765–1880), III 15; IV 21, 39
— Francis Russell, III 319
— L. H., VII 1105, 1261
Bedson, George, V 63, 621, 624, 625
— J. P., VI 495
Beecher, C., VII 1135

Béguin, Jean, II 741
Behaim, Martin, III 591, 614
Behrens, Peter, V 837
Beighton, Henry, IV 176–7
Beilby, George, V 250
Belevitch, V., VII 1112
Belgrand, Eugène, IV 498
Bélidor, Bernard Forest de, writer
    architectural theory, IV 443, 444, 445
    bridges, design, V 492
    canal flash-lock, III 440, 441
    dredging process, III 425
    machines, III 324, 333–4, 344
    pile-driving methods, III 422
    water-power, II 608
Belknap, III 117
Bell (c.1785), IV 165
— (c.1812), IV 246
— Alexander Graham, inventor
    aeroplane designs, V 409
    space frames pioneer, VII 952
    telephony, V 226–7; VII 1220–3, 1231 n.
— Henry, V 143
— Sir Isaac Lowthian, V 801–2, 804, 810
— Sir J., VI 465
— Patrick, IV 7–8; V 11
Bell Brothers, V 801, 812
Bella, IV 41
Bellairs, Henry, V 826
Bellin, J. N., IV 615
Belliss, G. E., V 136
Belliss & Morcom, V 136, plate 8B
Belpaire, Alfred, V 328
Ben-David, J., VI 151
Benedict XIV, Pope, IV 478
— Biscop, II 32, 326
Bennett, C. E., VII 1077
— Charles, V 728
— F., VII 806
Bensley, Thomas, V 694–6
Benson, Thomas, IV 347
Bentham, Sir Samuel, naval engineer
    machine tools, IV 418, 426, 437, 581–2; V 636
    ships, IV 579, 581
Bentley & Co., IV 422
Benton, Linn Boyd, V 688
Benz, Karl, V 167, 170, 427, 428
Berek, M., VII 1289
Berger, H., VII 1358
Berghaus, H. C. W., IV 614, 622, 626
Bergheim & MacGarvey, V 110
Bergier, Nicolas, IV 521, 527
Bergius, F., VI 541
Bergmann, T. O., IV 108, 374
Berliner, E. (fl 1887), VII 1262
— Emil (1851–1929), V 404 n.

Bernard, Claude, V 17
— St, III 198
— Sir Thomas, V 830
Bernini, Lorenzo, III 289
Bernoulli, Daniel, V 146, 537
— Jakob, V 491
— Johann, V 407
Berry, A., V 399
Bert, Paul, V 516
Berthelot, P. E. M., V 111; IV 541
Berthelsen, S., VI 657
Berthollet, Claude L., chemist
  chemical discoveries, IV 221-2, 253
  chlorine, bleaching property, IV 24-7
  dyestuffs, IV 249; V 257
  explosives, V 285
Berthoud, Ferdinand, IV 396, 410, 413-14; V
  458 n.
— Pierre-Louis, IV 414
Bertin, IV 31
Bertrand, J., V 387, 388
Bertsch, A., VII 1289
Berzelius, J. J., IV 225-7, 405; V 268, 269
Besnard, Philip, IV 369
Bessel, F. W., IV 601; V 442
Bessemer, Henry, industrialist
  casting method, steel, VI 497
  copper smelting, V 81
  glass-making, V 679
  industrial revolution, V 804, 809, 810, 818
  oil drilling bits, V 107-8
  oxygen enriched air patents, VI 489
  rail manufacture, V 331
  refractories, furnace, V 664
  steel converter, V 54-7, 58, 59, 60, 61, 70,
  606
  structural steel, V 477
  type-composing machine, V 685
Besserdich, W., VII 726
Besson, Jacques, engineer
  crane grab mechanism, IV 636
  screw-cutting lathe, III 334-5, 337; IV 425
  water-mill, III 328-9
  windmill, II 626-7
Betz, A., VII 793
Beukelszoon, William, II 124; III 22
Bevan, E. J., VI 554, 649
Bezinghen, Abot de, III 340
Biard, IV 302
Bickford, William, IV 80; V 284, 832
Biddle, Edward John, V 117
Bidoni, Giorgio, V 546
Bidwell, S., VII 1257
Bier, A., VII 1336
Biffen, R. H., VI 328
Bigelow, Erastus B., V 581-3
Biggs, J. B., VII 1346

Bilham, E. G., VII 1386
Binnie, Alexander, V 558
Biondo, Flavio, III 288
Birch, IV 319
Bird, John, IV 382, 390, 399, 603
Birdseye, C., VII 1423
Biringuccio, Vanoccio, metallurgist
  assaying, III 59, 62, 65
  author, *Pirotechnia*, III 27-8; IV 233
  boring-mills, III 367; IV 422
  cannon, III 364-5, 367, 368 n.
  copper alloys, III 36-42
  furnaces, metallurgical, III 50, 52, 53, 54-5
  gilding, III 48, 49
  glass-making, III 40, 210, 212, 214
  iron industry, III 30, 34
  metal-casting, II 475
  minting, III 47
  steel, III 34, 36
  tin alloys, III 43, 44
  type-founding, III 391, 392
Birkbeck, George, IV 445
Birkeland, C., VI 522, 523
Birkigt, M., VII 1012
Bishop, R. E., VII 828
Bismarck, Prince, VI 145, 146
Bissell, George H., V 109-10
Biswas, A. K., VI 196
— M. R., VI 196
Blache, III 165
Black, H. S., VII 1240
— Joseph, chemist, IV 444
  alkalis, discoveries, III 25; IV 217, 237
  'fixed' air discovery, IV 38
  hydrogen balloon proposal, V 395
  latent heat, discovery, IV 182, 673
  paper bleaching process, IV 255
  quantitative chemistry, II 744-5; IV 217-18
  scientific contribution, IV 673-4, 680; V 267,
  780
Blackett, P. M. S., VI 161
— W. C., IV 85; VI 361
Bladen, V. W., IV 351
Blaeuw, P. and J., III 402
— Willem Janszoon, III 400
Blagden, Sir Charles, IV 619
Blake, Eli Whitney, IV 538
— John, IV 499
— John, & Company, V 526
Blanc, M., VII 808
Blanchard, François, V 398, 400, 403 n., 411
— Thomas, IV 436, 440
Blancourt, H. de, III 678, 683, 698
Blériot, Louis, aircraft pioneer
  aeroplane designs, V 392, 409-10; VII 791,
  808
  channel crossing, VII 699, 1469

glider tests, V 407
motivation, VI 32
Bligh, William, V 23
Blith, Walter, III 134–6, 321–2
Bloch, J., VI 677
Blon, J. C. Le, III 408
Blondeau, Peter, III 341
Blondel, A., VII 1243, 1244
— François, III 256
Blood, W. Bindon, IV 486
Blumenthal, Cellier, V 112
Blumlein, A. D., VII 1263
Böckler, C. A., III 16–17
Bockway, F. J., VII 1321
Bodmer, J. G., IV 287
Bodyington, Stevyn, II 248
Boelcke, O., VII 794
Boëmus, Johannes, II 126
Boerhaave, Hermann, IV 233
Boëthius, III 502
Bogaerden, Adrien van, III 452
Bogardus, James, V 470
Boggs, M. S., VII 817
Bohn, René, V 279; VI 566
Bohr, N., VI 225
Boileau, Louis Auguste, V 471
Boizard, Jean, III 47–8, 59, 65
Bolas, Thomas, V 732
Bölckow, Vaughan, V 801, 812
Bollée, Amedée, V 423
— Leon (fl 1887), VII 1166, 1188, 1189
— Léon (fl 1895), V 435
Boller (c.1502), III 18
Bologna, Fioravante da, III 445
Bolsover, Thomas, IV 130; V 633
Bolton, J., VII 1390
— W. B., V 728
Bonde, Carl, III 455
Boneuil, Anthony Bourboulon de, IV 238
Bong, R., VII 813
Bonne, Rigobert, IV 625
Bontekoe, Cornelius, III 5
Bontemps, Georges, IV 360, 367, 375; V 678
Boole, G., VI 155; VII 1192
Boot, H. A., VII 1107
Booth, A. D., VII 1195, 1202
— Enoch, IV 348–51
— H. C., VI 294; VII 1129
— James, V 779
Borda, J. C., de, IV 397, 401, 601, 604; V 528
Borton, S., VI 682
Bosch, C., VII 1210
Boscovitch, Roger Joseph, IV 478
Bose, G. M., IV 648
Bottezat, G. de, VII 822
Böttger (c.1847), IV 255
— Johann Friedrich (1682–1719), IV 338–40

— Rudolf, V 86, 284
Böttiger, Paul, V 279
Bouchier, IV 318
Bouchon, Basile, III 166
Bougainville, Louis de, IV 598
Bouguer, Pierre, IV 478, 577; V 353, 386
Boulé, Auguste, V 543
Boulton, Matthew, manufacturer
    commercial talents, V 799, 800
    'Lunar Society' membership, IV 673
    machine-tools, IV 419, 422
    porcelain, IV 339
    'Sheffield plate', IV 130
    steam-engines, IV 162, 164, 183, 185, 352
— M. P. W., V 411
— & Watt, engineering firm, IV 183, 673
    coal-gas, IV 265–9
    cotton-mill, V 837
    engine improvements, IV 183–6, 187, 190, 191
    engines, navy use, IV 581, 582
    gas-lighting, IV 252, 262, 263
    glass industry, engines, IV 372
    machine-tools, IV 433; V 649
    paddle-steamers, machinery, V 143, 145
    pottery industry, engines, IV 352
    steam engines, IV 162, 164, 368
    steam-dredger, IV 641
    steam-driven factory, IV 283
    water-pumping engines, IV 77, 78
Bourbouze, J. P., V 737
Bourdon, Eugène, V 538
Bourgelat, C., IV 32
Bourn, Daniel, III 154–5; IV 277, 280, 288, 298
Bourne, Williams, V 142
Boussingault, J. B. J. D., IV 42; V 106
Boutheroue, François, III 463
— Guillaume, III 463
— Hector, III 465
Bouton, George B., V 169
Bouvier, IV 638
Bow, R. H., V 492–3, 501
— W. A., IV 486
Bowen, N. L., VI 632
Bowen Cooke, C. J., VII 762, 763
Bowles, R. C. M., VII 1347
Bowman, IV 303
Boxer, E. M., V 630
Boyce, Joseph, IV 7
Boyd, T. A., VII 989
Boyden, Uriah A., V 528, 529, 531, 537, 538
Boyle, C. P., VII 869
— H. E. G., VII 1334, 1335
— Richard, IV 470
— Robert, chemist, II 738; III 681–3, 718, 720
    blowpipe furnace, III 55
    Dalton's atomic theory, IV 223

Boyle, Robert, chemist (*cont.*)
  dyeing experiments, IV 248
  electricity, IV 648
  elements, definition, IV 215
  gases, IV 216
  hydrometer inventions, III 22, 68
  laboratory, II 749–50
  metallurgy, III 56, 57
  phlogiston problem, IV 218–19
  quenching experiments, III 37
  shellac lacquers, III 696, 697
  vacuum pump, III 637
  vision, discussion, V 735
— S. E., VI 58
Bozolus, IV 650
Brackelsberg, C. A., VI 475
Braconnot, H., V 284
Bradley, C. S., VI 522
— Humphrey, III 316–20, 459, 462
— James, III 554
Bragg, W. H., VI 623
Brahe, Tycho, Danish astronomer
  accuracy of instruments, III 621, 640; IV 380, 388
  clocks, astronomical, III 658, 659–60
  instruments, general, III 400, 582–3
  observatory instruments, III 588, 599
  planetary models, I 799
Braithwaite, J., VII 985
Bramah, Joseph, inventor, manufacturer
  machinery, lock-making, V 636, 800
  machine-tools, IV 418, 423–5, 434, 435
  press, hydraulic, V 535
  tubes, extrusion method, V 627
  water closets, IV 423, 507; VII 1127
  wood-planing machine, IV 423, 437
Bramante, III 288, 338
Brambilla, G. A., VII 1317
Bramer, Benjamin, III 626–7
Bramwell, W. C., V 571
Branca, Giovanni, architect
  chimney-wheel, design, IV 161
  rolling-mills, description, III 342
  steam, power source, IV 672; V 149, 150
  turbine designs, IV 161, 168; V 149, 150
Brancas, L. F. de, IV 339
Brand, Hennig, V 252
Brande, W. T., IV 256
Brandenburg, Margrave of, III 285
Brandt, Georg, IV 145, 374
Branley, E., VII 1232, 1233
Brattain, W. H., VII 1118
Braun, F., VII 1106
— W. von, VII 816, 848, 863
Brearley, H., VI 492–3
Brederode, Hendrik Van, III 302
Bredon, Simon, III 617

Bredsdorff, J. H., IV 614
Bregandyne, Robert, III 487
Bréguet, A. L., IV 414
— Louis François C., V 219
— brothers, VI 32; VII 792, 821, 822
Bremoutier, IV 33
Brendan, St, II 581
Brennan, L., 822
Bretonnière, L., V 279
Bretschneider, O., VI 562
Brett, Jacob, IV 660
— — and John Watkins, V 220, 224
— John W., IV 660
Breuil, Abbé, I 152
Brewster, Sir David, IV 376; V 727, 780, 781
Bridge, E., VII 904
Bridges, David, & Company, V 768
Bridgwater, Duke of, IV 348, 363–5, 367
Briggs, H., VII 1153
Bright, Sir Charles Tilston, V 225
Brindley, James, IV 348, 462, 571
— John, IV 348
Briot, Nicholas, III 341
British Aluminium Company, V 93
Brockhaus, F. A., V 683
Broglie, L. de, VII 1216
Brooke, Charles, V 186
— Homer, V 676
— J. M., IV 619
Brooks, J. C., VI 665
— John, IV 354
Brotherhood, Peter, V 134, 137
Broughton Copper Company, V 628
Browett Lindley, V 137
Brown, A. E., VI 472
— Andrew, IV 243
— C., VII 1071
— C. E. L., V 232
— H. P., V 201
— Sir John, V 612, 805
— John, & Sons, V 614
— Joseph R., V 653; VII 1041
— Robert, V 780
— Sir Samuel, IV 459–60, 585
— William, IV 178–9
Brown & Sharpe Company, V 653, 654, 656
Browne, Edward, II 22–4
— Henry, III 159
Brownrigg, William, IV 259
Bruce, David, V 683, 684
Brueghel, Pieter, the elder, artist, IV 639
  alchemical laboratory, II 749, plate 42A
  garments, leather, II 175
  medieval cart, II 546, 548
  scythe (painting), II 96
  shaduf figure, II 680, 681
Bruff, Peter, V 331

Bruges, Pierre Théodore de, IV 237
Brunel, Isambard Kingdom, engineer
  arches, structural theory, V 492
  biographical note, IV 427
  brickwork, iron-reinforced, V 488
  bridges, IV 461–2; VII 887, 888, 898
  Bristol Station roof, V 468
  caissons, IV 462; V 514
  Clifton suspension-bridge, V 507 and n.
  hydraulic appliances, V 534
  hydrodynamics, V 386
  Paddington station, V 332
  steam-boat pioneer, V 361–3, 364–5, 372, 375
  tunnel, Thames river, IV 463
— Sir Marc Isambard, engineer
  building materials, IV 449, 451
  hosiery machine, V 598
  machine-tools, IV 418, 426–7, 430, 435; V 636
  marine engine, V 146
  saw-mills, water-powered, IV 202
  tunnel, Thames river, IV 449, 463–4, plate 32;
    V 361 n.
  tunnel-shield patent, V 516
Brunellischi, Philippo, III 245–6, 248
Brunner (c.1850), IV 398
— (1842–1919), John, V 243
Brunner Mond & Company, V 803, 806, 833
Brunton, IV 123
Bruyère, Jean B. F. Giraux de la, IV 369
Brush, C. F., V 213, 215
Buache, Philippe, IV 611–13, 621; V 439 n. 2
Buchan, Alexander, IV 620
Buchanan, R., IV 425
Buck, J., V 656
Buckingham, Duke of, III 238; IV 362
Budd, James Palmer, IV 113
— William, V 561–2
Buddle, John, IV 84, 93, 97
Buffalo Bill, see Cody, S. F.
Buffon, Nadoud de, V 664
Bull, Edward, IV 190; V 129
Bullard, E. P., jun., VII 1036
Bullet, Pierre, III 293
Bullock, William, V 700
Bullough, IV 302–3
Bulmer, William, V 691
Bunch, C. C., VII 1349
Bunning, James Bunsten, V 471
Bunsen, R. W., chemist
  battery, electric, V 204
  electrochemical process, V 248
  gas-burner, IV 273, 325
  research, discussion, V 788, 821
Buontalenti, Bernardo Timante, III 309
Burdin, Claude, V 528
Burdon, Rowland, IV 457–8
Buresh, F. M., VI 670

Bureus, Andreas, III 455
Burgess, John, V 728
Burghley, Lord (William Cecil), III 317–18
Burgi, Jobst, III 658, 660–2; IV 382
Bürgin, Emil, V 190, 192
Burnham, D. H., VII 930
Burr, Aaron, IV 544
Burroughs, W. S., VII 1168
Burt, William Austin, V 689
Burton, E. F., VII 825
Bury, Edward, V 324
Buseman, A., VII 828–9
Bush, V., VI 121, 169, 271; VII 1154
Butler, Edward, V 429
— Pierce, III 385
Butterley Company, V 477
Buyten, Martin van, III 409
Byrd, R. E., VII 806
Byron, Lord, VII 1178

C. M. (name unknown), IV 647, 649–50
Cabral, Pedro Alvarez, V 264
Cadbury, G., VI 962
— R., VI 962
Cadburys, Messrs, V 830
Cadeby, John, II 248
Caesar, Caius Julius, Roman emperor
  cavalry, II 705
  Empire, survey, III 504
  leather, use by Britons, II 168
  metals, I 589; II 10, 52
  pottery, Britain, I 378, 410 (6)
  roads, I 712
  Roman calendar, III 576–7
  siege technique, II 716
  travel permits, II 514, 530
  water-supply, Rome, II 672
  wines, II 134
Cahours, A., V 106
Calabrais, Jean le, III 190
Caldecott, Randolph, V 705
Caligula, Roman emperor
  bread famine, Rome, II 600
  lighthouses, II 518, 522–3
  shipbuilding, II 569, 572–3
  vessel (ship), II 518
Callender, W., VII 1073
— W. O., VII 1073
Calley, IV 174
Callippus, III 571
Callon, Charles, V 531
Calvisius Sabinus, II 513
Cambridge Electric Lighting Company, V 139
Camelford, Lord, IV 340
Cameron, D., VII 1394
Camm, Sir S., VII 812, 834
Camman, G., VII 1347

Campano da Novara, III 522, 616
Campbell (Gibson and), IV 315, 321
— Colin, IV 470
— G. A., VII 1114
— H. H., VI 479
— Hector, IV 255
— J. R., V 34; VII 1401
Campbell Gas Engine Company, V 161
Campin, Robert, II 249, 653
Camus, Charles Étienne Louis, III 345
Canadian Niagara Power Company, V 523, 524
Candolle, A. P. de, I 355
Cannizzaro, S., IV 225
Cannon, W., VII 1328
Cantelupe, Walter de, III 201
Capitaine, Émile, V 162
Caracalla, II 418, 673
Caradosso, III 338
Cardalucius, J. H., III 41
Cardan, Jerome, III 17, 30, 658
Cardano, Girolamo, V 716
Carew, Richard, II 12
Carey (c.1840), IV 545
— G. (c.1880), VII 1257
Carl Theodore, IV 25
Carlberg, C. W., III 222
Carlisle, Sir Anthony, IV 227, 653
Carlsund, O., V 63
Carnay, IV 239
Carnegie, Andrew, V 68
Carnot, Lazare Nicolas Marguerite, IV 644
— Sadi, IV 163, 166; VII 1003
Caro, Heinrich, V 274, 276–7, 279, 281
— N., VI 523, 524
Carochez, IV 399
Carothers, W. H., VI 349, 560, 561, 650
Carrel, A., VII 1354
Carrett, Marshall & Company, V 421
Carson, J. A., VII 1115
Carter, H., I 695
Cartier, III 22
Cartwright, Edmund, IV 151, 295–7, 299–300, 314
Cartwright & Warner, V 598
Carvilius Pollio, II 227
Casablancas, F., VI 655
Casal, Gaspar, III 4
Case, T. W., VII 1312
Caslon, William, IV 272
Cassel Gold Extraction Company, V 250
Cassini, Jacques, V 596, 605, 615
— — Dominique, Comte de, IV 604
— Jean Dominique, III 544, 553–4; IV 596–7, 605
— de Thury, C. F. de, cartographer
  France, national survey, IV 605–6; V 440
  map, France, III 554; IV 606, 609, 611, 614
  survey instruments, IV 601, 603

Cassius, Andreas, III 225
Cassivellaunus, II 544
Castelli, Benedetto, III 313–15; IV 520–1
Castigliano, Alberto, V 495
Castner, Hamilton Young, V 248–51; VI 518, 519, 523
Castner-Kellner Alkali Company, V 251
Caswell, John, III 58, 544
Catchpole, III 101
Catherine de Medicis, III 10
Catley & Ayres, V 422
Catnach, James, V 702
Cato, Marcus Porcius, writer
  flour-mills, II 110
  copper-mines, Spain, II 52
  iron plough-shares, II 86
  oxen, feeding practices, II 95
  press-beam drawing, II 114
  viticulture, II 134–5
Caton-Thompson, G., I 511
Catullus, II 201, 561
Caus, Salomon de, III 45, 342; IV 168
Cautley, Sir Proby Thomas, IV 555; V 542
Cavalier, Jean, III 465
Cavendish, Henry, scientist
  balance, laboratory, IV 404
  gases, explosion experiment, IV 220–1
  hydrogen, IV 259, 674; V 395
  water, synthesis, IV 220–1; V 177
Caventou, J. B., V 318
Caver, W., VII 1114
Cayley, Sir George, aircraft designer
  aerodynamic theory, V 407, 408, 410–11
  airship designs, V 401 n. 2
  gliders, V 405, 413; VII 789
  helicopter designs, V 404 n.
Céard, Nicolas, IV 530
Cellini, Benvenuto, goldsmith, III 358, 712
  coinage process, III 341
  foils, gem, III 49
  furnaces, III 53, 54
  screw-press, description, III 338–9
  soldering method, III 48
Celsius, O., I 450–1
Celsus, Aulus Cornelius, II 121
Cennino Cennini, II 352–3
Censorinus, III 654, 575, 578
Cessart, Louis Alexandre de, IV 465, 538
Chadwick, Sir Edwin, IV 513; V 824, 828, 829
— J., VI 224, 269
— R., VII 814
Chaffee, E. M., V 863–5
— R., VII 857
Chailey, Joseph, IV 459
Chain, E. B., VII 1355
Chambers, Ephraim, IV 314, 674
Chambéry, Perret de, III 290

Chambrelent, IV 33
Chambrier, de, V 111
Champion, John, IV 130, 261
— Richard, IV 340–1, 352
— William, IV 130
Chan K'ien, II 131
Chance, Alexander, V 238–9, 830
— Sir James, IV 368
— Lucas, IV 360, 367
— Robert Lucas, V 677
Chance Brothers, IV 360; V 677, 679, 681
— & Hartley, glass-makers, IV 373
Chancil, IV 254
Chandler, A. D., VI 86
Chanute, Octave, V 406, 410
Chapelon, A., VII 766
Chapman, F., V 387
Chappe, Claude, IV 645–6, 651–2; V 329
Chaptal, J. A. C., IV 255
Charlemagne (768–814), emperor
    brewing practices, II 139
    Byzantine silk weave, II 761
    chronology, II table III lviii
    crown, II 459
    dyestuffs, textile, V 260
    metal-mining, II 65
    silk fabric, III 196, plates 11A, B
    stone-masonry, II 386–7
    water-mills, II 608–9
    weight standards, I 779
— (c.1793), III 18
Charles, J. A. C., IV 216, 255; V 395–6, 401
Charles I, English king, III 186, 320, 486, 494
Charles II, English king, III 297, 494, 497; IV
    667
— IV, emperor, III 205
— V, emperor, III 16, 290, 449
— VI (1711–40), emperor, IV 528
— VI (1380–1422), French king, II 691
— VII, French king, III 384
— VIII, French king, III 349
— XI, Swedish king, III 682
— of Anjou, II 489, 491
— the Bold, III 74
— Emanuel I, IV 529
— the Great, see Charlemagne
— the Hammer, II 720, 735
Charlevoix, IV 33
Charlton, E. E., VII 1332
Chaucer, Geoffrey, III 596, 609, 617; V 260
Chaulnes, Duc de, III 641; IV 391–4
Chaussier, François, IV 259
Cheape, William, IV 317
Cheetham, D., V 574
Cheops, I 356, 668, 714
Cherwell, Lord, VI 270
Cheshire, E., VII 1277

Chevalier, Charles, V 720
Chevel, A., VI 554
Cheverton, Benjamin, IV 436
Chevreul, M. E., IV 253
Chézy, Antoine de, III 464
Ch'iao Wei-Yo, III 439
Chiese, Philippe de, III 454
Childe, V. Gordon, historian
    battle-axes, I 513
    cultivation, effects of, I 80, 374
    dwellings at Kostenki, I 302
    fort at Finavon, I 322
    pot-throwing, I 378
Childeric, II 451, 453
Chodowiecki, D., III 156
Chosroes II, II 451
Christian VII, IV 560
Chubb & Sons, V plate 41 B
Church, A. H., IV 342, 345
— William, V 685
Churchill, C., VII 1050
— Sir Winston, English statesman
    atomic energy, development, VI 227
    MAUD Committee, VI 270–1
    metals, corrosion, VI 451
    post-war Britain, VI 281
    second world war, VI 19, 21
Churchward, G. J., VII 763, 764, 765, 774
Cicero, Marcus Tullius, II 514; III 501–2, 504,
    578
Cierva, J. de la, V 405; VII 822
Citroen, A., VII 718
Clanny, W. R., IV 96–7
Clapeyron, B. P. E., V 501
Clapham, Sir John, V 803, 812, 813, 833
Clarendon, Lord, V 784
Clark, D. K., IV 441; V 637, 639, 652, 654–5
— J. G. D., I 516
— Thomas, IV 503; V 566
— William (1821–80), IV 538
— William (fl 1857), V 678; VI 578
Clarke, A. C., VII 869
— A. R., V 442
— Chapman & Company, V 138, 139
— E. M., V 180, 181
— Roger, III 22
Claudius, Roman emperor
    chronology, II table II lvi
    drainage, Lake Fucinus, II 679
    port construction, Rome, II 518
    road transport control, II 530
    stone quarries, II 29
    water-supply, II 670, 671
— Clavus (Claudius Claussøn Swart), III 532–3
— Ptolemaeus of Alexandria, see Ptolemy
Claus, C. F., V 239
Clausius, R. J. E., IV 163, 680

Clayton (c.1738), IV 252
— John (1657–1725), IV 259
Clegg, Samuel, the elder, IV 267–70
— — junior, IV 272
Clement, Joseph, IV 418, 431, 434
— William, III 658, 664–5
Clement IV, Pope, III 522
— VII, Pope, III 311, 338–9
Cleostratus, III 570
Clephane, James O., V 687
Clerk Maxwell, James, physicist
   Cavendish laboratories, V 793
   electro-magnetic waves, V 227, 737
   mechanics, bridge trusses, V 492, 501
   'waves' prediction, V 227
   zoetrope, improvements, V 737
Clerville, L.-N. de, III 466
Clifton, R. B., VI 141
Clinton, de Witt, IV 555
Clive, Sir Robert (Baron Plassey), V 440
Clover, J. T., VII 1337
Clulow, IV 319
Clymer, George, V 692, 694
Coade and Sealy of Lambeth, IV 446–7
Coates, J. & P., V 806, 812
Cobden-Sanderson, T. J., V 694
Cochrane, Archibald, IV 238, 252, 260, 587
— Sir Thomas, tenth Earl of Dundonald
   asphalt-paving, IV 540
   bituminous products, V 105, 223
   caissons, compressed-air, IV 462
   marine boiler, V 153
   tunnelling, under-water, V 516
Cock, Christopher, III 234, 635
Cocker & Higgins, IV 285
Cockerill, John, IV 105
Cockrell, Sir C., VII 744
Cody, S. F., V 398, 409
Coehoorn, M., III 372
Coffey, Aeneas, V 303, 306–7
Cohn, A. S., VII 1341
Coignet, Edmond, V 489–90, 496, 502
— François, V 488–9
— et Cie, V 253
Coke, Thomas William, III 15; IV 20–1, 39
Coker, E. G., VII 992
Colaeus of Samos, II 568
Colbert, Jean Baptiste, French statesman
   building industry, IV 442
   canals, III 464–6
   cloth calender, III 178
   glass-making, Normandy, III 238
   roads, IV 525
   textile factory, III 152, 711
   tinplate manufacture, III 689
Colburn, Irving W., V 679
Colby, Thomas, IV 601, 608, 610; V 452

Cole, Humfray, III 586
Coleman, IV 4
Coles, Captain Cowper, IV 590
Collier (of Platt and), IV 297, 302, 305
— G., V 582, 583
Colling, Charles, IV 21
— Robert, IV 21
Collingwood, Cuthbert, IV 582
Collins, Greenvile, IV 616
— John, IV 209
Collison, John, IV 238
Collot, IV 405
Colmar, C. X. T. de, VII 1153, 1163, 1173
Colorado Smelting and Mining Company, V 79
Colt, Samuel, IV 438–9; V 638–9
Columbus, Christopher, III 3, 8, 477, 548–9; VII 869
Columella, II 111, 134
Colvin, F. H., VII 1049
Common, John, IV 7
Compagnie de l'Alliance, V 181–3
Compton, A., VI 236, 271
Comrie, L. J., VII 1154
Conant, J., VI 236
Condamine, Ch. M. de la, IV 253
Condie, John, IV 109–10, 116
Condy, Henry B., V 310, 315
Congreve, Sir William, V 412; VII 859
Conrad II, II 455
Conradis, H., IV 629, 636, 642
Considère, A. G., V 496
Constantine I the Great, Roman emperor
   chronology, II table II lvi
   cities, town planning, III 279
   flour-mills, II 599, 600–1, 639
   porphyry, use, II 29
   road-system, II 515
   Roman culture, II 761
— Monomachos, II 462
Constantinesco, G., VII 988, 1063
Contarini, G. M., III 535
Cook, H. (fl 1865), V 737
— H. W. (fl 1903), VII 1346
— Captain James (1728—79), explorer
   maps, IV 598
   navigational aids, V 455
   ships, design, IV 574–5, 579
   survey methods, III 557; IV 616, 617–18, 619; VII 1446
— Sir William (fl 1957), VI 282
Cooke, James (fl 1782–94), IV 5, 11
— M. L., VI 84
— Sir William F. (1806–79), IV 654, 656–9; V 218–20, 223
Cookson, of Jarrow, IV 238
Cookworthy, William, IV 340, 343
Coolidge, W. D., VII 1082, 1331

Cooper, C. M., VI 260
— Ronald, IV 345–6
— Thomas, IV 105
Cope, R. W., V 692–4
Copernicus, III 589
Copland, Patrick, IV 247
Corbulo, Gnaeus Domitius, II 680
Cord, E. L., VII 719
Cordner, E. J., V 398
Córdoba, Francisco Hernández de, V 263
Corliss, George H., V 131
Corneille, T., III 682
Cornelisz, Cornelis, II 625
Corneliszoon, Cornelis, of Uitgeest, III 106,
    307–8
Cornu, P., VII 821
Cornwallis, Sir William, IV 582–3
Corobus, III 572
Corsini, Ottavio, III 314–15
Cort, Henry, ironmaster
    coal, puddling process, III 81
    grooved rollers invention, IV 99, 106–7
    industrial organization, V 804, 810
    wire manufacture, V 620
    wrought iron process, IV 106–7, 114, 115
Cortes, Martin, III 548
Cortez, Hernando, III 6, 695
Cosimo de' Medici I, III 707
— — — II, IV 168–9
— — — III, IV 170
Cosmas Indicopleustes, II 55
Cosnier, Hugues, III 460–3
Coster, Laurens of Haarlem, III 377, 383
— Salomon da, III 663
Cotes, Roger, V 353
Cotton, Sir Arthur, V 542
— Williams, V 598–9
Cottrell, F. G., VI 471
Coulaine, de, IV 540
Coulomb, Charles Augustin de, scientist
    applied mechanics, IV 163, 479–80, 482–3
    friction, laws, V 118, 119
    structural theory, V 491–2, 496
Cournaud, A., VII 1357
Courtois, B., IV 255
Cowles, A. H., VI 489
— E. H., VI 489
Cowper, Edward (1790–1852), V 696
— Edward A. (fl 1857), V 58, 67
Cox, P., VII 1278
Coxe, IV 654
Coxhead, R. C., VII 1270
Cragg, John, 470
Cramer, J. A., III 60
Crampton, Thomas Russell, railway engineer
    locomotive designs, V 325, 328, 334, 341
    rotary cement kiln, V 484, 485

Cranage, the brothers, IV 106
Crane, George, IV 111
— Walter, V 705
Crapponne, Adam de, III 459
Craske, Charles, V 700
Crassus, II 706
Crates the Grammarian, III 518, 614
Crawford, J. W. C., VI 558
Cremona, Luigi, V 493, 501
Cresques, Abraham, III 544
Cresy, Edward, III 428 n. 2
Crewdson, E., VI 206
Crighton, IV 284
Cripps, Sir S., VI 169
Crisp, Sir Nicholas, III 680, 686
Crocco, G. A., VII 822
Croissant, E., V 279
Crome, A. F. W., IV 622
Crompton, R. E., VII 871
— R. E. B., electrical engineer
    arc-lamps, lighting system, V 211–12, 214
    current battle, A.C./D.C., V 200
    dynamo, development, V 191
    electricity, industrial use, VI 296
    heaters, electric, VII 1132, 1133
Crompton, Samuel, IV 279, 314; V 786
— William, V 579
Crompton & Company, V 192, 203
Cromquist, A. W., V 660
Cromwell, Oliver, III 46, 349
Cronstedt, A. F., IV 145, 374
Crookes, Sir William, V 214; VI 325; VII 1300,
    1327
Croone, William, III 59
Cropper & Company, V 710–11
Cros, C., VII 1262
Cross, C. F., VI 553, 649
— H., VII 945
— Viscount, VI 286
Crosskill, IV 3, 4
Crossley Brothers, V 160
Crowther, C., VI 311
Cruikshank, G., VII 963
— William, V 177
Crum, W., V 290
Cruquius, N. S., IV 612
Ctesias, I 246
Ctesibius of Alexandria, mechanician
    biographical note, II 708 n.
    catapult, improvements, II 712–13
    hydraulic organ, II 633, 634
    pump inventions, II 633
    water clock, III 601–2
Cubitt, Sir William, III 101; V 514
— — and Wright, IV 462
Culley (Bailey and), IV 10
Culmann, Carl, IV 486; V 493, 501

Culpeper, Edmund, III 635
Cumberland, Duke of, IV 342
Cummings, IV 507
Cunningham, Henry, IV 593
Curie, I., VI 224; VII 1333
Curtis, C. G., V 140; VII 1025, 1028
Curtiss, G. H., aircraft designer
  aeroplane designs, V 409; VII 791, 1010
  engines, aeroplane, VII 1012
  flying boats, VII 796
Curtuis, T., VI 548
Curutchet, P., VII 1318
Cuthbert (*fl* 758), abbot, II 326
— St (*d* 687), Bishop of Lindisfarne
  leathercraft, II 169
  metalwork, inlaid cross, II 451–2, 457
  tomb fabrics, II 756, 763; III 197
Cyrus the Great, II 495, table 1; III 567
— the Younger, II 496–7, 697
Czapski, S., VII 1354

Dadfords, the, IV 572
Daelen, R. M., VI 497
Daga, I 437–8
Dagan (Dagon), fishgod, I 253
Daguerre, L. J. M., V 720, 725, 736
Dahl, C. F., VI 346, 610
Daimler, Gottlieb, mechanical engineer
  motor cycle, V 427, 434, 436
  petrol-engine, V 164–5, 167, 169
Daimler Company, V 170, 171, 428
Daisenberger, J. M., IV 263
Dakin, H. D., VII 1354
Dalby, W. E., VII 762
Dalen, G., VII 1139
Dallin, John, IV 95
Dalrymple, Alexander, IV 617–18; V 449
Dalton, John, chemist
  atomic theory, IV 223–6; VI 223
  biographical notes, IV 267; V 780
  dyestuffs, discussion, V 267
  molecular structures, IV 225–6
Damian, John, II 746; V 394
Dampier, William, III 499
Dance, George, the younger, IV 449
Dancer, J. B., V 727
Dandolo, Count Vincenzio, IV 308
Danforth, Charles, IV 286–7, 290–1
Dangon, Claude, III 165, 189
Daniel, Ralph, IV 353
Daniell, J. F., V 204–5
Danisy, III 434
Danner of Nuremberg, III 337
Danti, Giovanni B., V 394
— Ignatio, V 716
Darby I, Abraham (1677–1717), ironmaster
  air supply, blast-furnace, IV 103

coke smelting-process, IV 99–100, 101, 165, 235; V 810
  zinc, IV 130
— II, Abraham (1711–63), IV 106
— III, Abraham (1750–91), IV 117, 455–6
— family, III 689; V 830
Darcy, Edmund, II 155
— H. P. G., V 539, 546, 563
Darius I the Great, Persian king
  brass, use of, II 54
  canals, III 439
  cubit measure, Persian, I 779
  cuneiform script, I 752, plates 28, 29A
  messenger-service, Persian, II 496
  qanaat construction, Egypt, II 666
Darlington, S., VII 1115
Darrius, G., VII 1033
Darwin, Charles, I 3; VI 329; VII 1340
— Erasmus, IV 673
— H., VII 1340
D'Ascanio, C., VII 822
Daubenton, L. J. M., IV 27
Davey, N., VI 217
Davis, A., VII 1280
— E. W., VI 475
— G., VII 1340
— John, III 522, 625
Davy, Edward, IV 654; V 223, 224
— Sir Humphry, scientist, IV 676
  Agricultural Chemistry lectures, IV 41
  alkalis, isolation, IV 227, 678–9
  computers, Babbage's letter, VII 1173
  electric arc, V 178
  electricity, IV 227, 655, 678–9
  metals, electrolytic process, V 72, 73; VI 518
  nitrous oxide, discovery, V 312
  photography, experiments, V 718–19
  safety-lamp, coal mines, IV 95–7; V 832; VI 373
  sodium, isolation, V 248; VI 518
  sugar-beet industry, IV 28
Dawson, Sir P., VII 771
Day & Son, V 706
Deacon, G. F., IV 501; V 536–7, 557, 560
— Henry, V 237, 242
Dean, L. W., VII 1349
Deane, Sir Anthony, III 482, 488
Dearborn, C. A., VII 863
Debaufe, Jacob, III 671
— Peter, III 671
Debray, Henry Jules, V 608
de Candolle, A. P., *see* Candolle, A. P. de
Dechen, H. von, IV 626
Deck, J. T., V 661
Deckel, F., VII 1285
Decoster, IV 294
De Dion-Bouton, V 435

Dee, John, III 231, 586
Deeley, Richard, M., V 339
Deere, John, V 4
Defoe, Daniel, V 829
Dehn, W. M., VI 548
Dehne, A. L. G., V 301
Deighton, William, V 69
Deinocrates, III 275
De la Beche, H. T., IV 622
Delambre, J. B. J., IV 402; V 441
Delaroche, of Amiens, IV 304
— Paul, V 716
Delcambre, A., V 685, 686
DeLeeuw, A. L., VII 1038, 1050
Delisle, Guillaume, IV 597, 599
Dellagana, James, V 700
Delocre, François-Xavier, P. E., V 522, 556
Deloney, Thomas, III 151
Delprat, G. O., V 75
Deluc, J. A., IV 605
Demachy, J. F., III 22
Demany, M. G., VII 1310, 1315
Demeny, Georges, V 742, 743
Demetrius I, II 567, 700, 702, 716
Demokritos, alchemist, II 732–3
— of Abdera, II 733
De Morgan, Augustus, V 781
Demosthenes, II 4–5, 58
Denison, E. F., VI 116
Densmore, James, V 687, 690
Dent & Priestman, V 161
Dereham, Richard, III 162
Deriaz, P., VI 206
Derry, T. K., VI 48, 57
Desaguliers, J. T., III 687; IV 158, 173, 176, 444
Des Barres, J. F. W., IV 616
Descartes, René, III 5, 232, 234; IV 671
Désormes, C. B., V 239
— Nicolas Clément, V 239
Deutsche Gold-und-Silber-Scheide Anstalt, V 250
Deveral, IV 116
Deverill, Hooton, V 603
Dewar, Sir James, V 295
Diades, II 698
Dibdin, W. J., VII 1388
Dicaearchus, III 505
Dicas, John, III 22
Dick, Alexander, V 628
Dickinson, H. W., IV 207
— J., VI 618, 650
— William, IV 302
Dickson, J. T., VI 561
Didelot, IV 324–5
Diderot, Denis, encyclopaedist, IV 674–5
  cannons, boring, III 367
  guns, rifling explanation, III 360
  leather processing, II 152, 153
  machine-tools, IV 417, plate 28
  road-roller, IV 537–8
  tools, home-made, III 113
  trades, description, III 116
  windmills, description, III 89
Didot, Firmin, V 691
Diesbach (c.1704), IV 249
— H. de (fl 1927), VI 568
Diesel, Rudolf, engineer
  engine, invention, V 120, 163; VII 1003–6
  marine engines, VII 759, 1006
  railway locomotives, VII 1008, 1009
Digges, Leonard, mathematician
  optical instruments, III 231, 586
  surveying instruments, III 541, 542; IV 402
  tele-lens combination, V 717
— Thomas, III 541, 542
Diggle, Squire of Bury, IV 303
Dijkstra, E. W., VII 1200
Diller, James, IV 260
Dio Cassius, III 565
Diocletian, II 55, 639, table II lvi
Diodorus Siculus, Greek historian
  beer, Egyptian, I 280
  gold extraction, Egypt, I 580–1
  houses of reeds, Egypt, I 473
  mummification, I 266, 267
  screw, water-lifting, II 677
  vegetable oils, I 288–9
Diogenes Laertius, II 612
Dion, Comte de, V 169, 423, 435
Dionysius I, II 699
— Exiguus, III 578
Diophantus, I 803
Dioscorides Pedanius, Greek physician
  antimony, chemistry, I 588
  indigo, origins, I 247
  medicines, preparation, II 370, 371
  vermilion, II 5–6
Dircxz, William, III 167
Diringer, D., I 761, 762, 763, 769
Disney, W., VII 1308
Divini, Eustachio, III 234
Dixon, George, IV 259
Dizé, IV 239
Dobbs, Edgar, IV 497
Döbereiner, J. W., V 247; VI 524
Dobler, A., VI 206
Dobo, IV 295
Dobrovolsky, G. T., VII 857
Dodds, Sir C., VII 1406
— Ralph, V 512; VII 879
Dodson (c.1750), III 555
— William (fl 1665), III 321
Dolivo-Dobrowolsky, M. von, V 232
Dollond, G., IV 361
— John, III 233, 555; IV 358, 603; V 717

Dollonds, the opticians, IV 390
Domagk, G., VII 1326, 1355
Dombasle, Mathieu, IV 38, 41
Domenico da Viterbo, III 446, 451 n.
Domitian, II 134
Donald, David, III 157
Donati, G., V 662
Donato d'Agnolo, *see* Bramante
Dondi, Giovanni de, III 653–5, 658
Donis, *see* Nicholaus Germanus, Dominus
Donisthorpe, G. E., IV 296–8
— Wordsworth, V 743
Donkin, Bryan, engineer
    computers, mechanical, VII 1174, 1175
    paper-making machine, III 415–16; VI 614–15, 618
    printing-press, rotary, V 690, 698
Donkin and Hall, canners, V 39
Doolittle, J. H., VII 817
Dorman Long & Company, V 477, 499, 806, 812
Dorn, Gerard, IV 242
— Johann F., V 305
Dornberger, W. R., VII 865
Dornier, C., VII 810
Dorothea Slate Quarry Company, V 129
Dorr, Samuel, IV 305
Dossie, R., III 226, 229
Doubleday, Thomas, IV 238
— & Easterby, IV 238
Douglas, James, IV 458, 484
Douhet, G., VII 811
Doulton, V 480
Dowai, Walter de, IV 209
Dowen, Jr, Samuel, V 105
Downs, J. C., VI 519
Dowsing, H. J., VII 1133
Dowty, G., VII 1064
Doyne, W. T., IV 486
Drais von Sauerbronn, Freiherr Karl, V 414
Drake, Sir Francis, III 477, 494, 679
— E. L., V 105, 107–8, 110, 116
Draper Company, V 586
Drebbel, Cornelius, II 366; III 679–80, 695; IV 242
Dreyfus, C., VI 554, 650
— H., VI 554, 650
Driffield, V. C., VII 1283
Dritzehn, Andreas, III 384
Driver, G. R., I 768
Drummond, Dugald, V 323, 325, 345
— R. J., V 34; VII 1401
— Thomas, IV 608
Drusus (38 B.C.–9 B.C.), I 778
— Marcus Livius (*fl* 91 B.C.), II 46
— Nero Claudius (38 B.C.–A.D. 19), II 680
Duarte (Edward), III 533
Dubie, Maître, III 454

Dubrunfaut, A. P. and Leplay, V 304
Duchosal, P., VII 1342
Duckwall, E., V 43
Duddell, W., VII 1235
Dudley, Dud, IV 99
— Sir Robert, III 640
Duesbury, William, IV 342
Dufay, L. D., VII 1297
Duff, IV 319
Dufrénoy, O. P. A. P., IV 622
Duhamel du Monceau, Henri Louis, IV 38, 577
Dukas, Paul, IV 633
Dumas, clockmaker, IV 414
— J. P. A., chemist, V 106, 237, 672
Dummer, Edmund, III 484
Dumont, F. M. A., V 223–4
— Thomas H., V 737
Dumoulin, IV 409
Duncan, O. E., VI 254
— & Wilson, V 701
Dundas, Henry (Viscount Melville), V 143
— Lord, IV 238
Dundonald, Earls of, *see* Cochrane
Dunlop (eighteenth century), IV 21
— J. B. (1840–1921), IV 523; V 772, 822
— Rubber Company, V 773
Dunmore, F. W., VII 818
Dunne, J. W., V 409
Dunning, J. H., VI 61
Dunwoody, H. H. C., VII 1236
Dupain-Triel, J. L., IV 613
duPont de Nemours, E. I., V 821, 833
Durand, W. F., VII 1021
Durant, F. C., VII 863, 869
Dürer, Albrecht, III 370, 407
Durr, L., VII 799
Durrer, R., VI 489
Dwight, John, IV 331, 347
Dyar, H. G., V 241–2
Dye, W. D., VII 1213
Dyer, J. (*fl* 1833), V 587
— J. G. (*fl* 1829), IV 286–7

Eads, James B., V 510
Eannatum, I 749
Earle, G. & T., V 486
Earnshaw, Thomas, IV 415
East London Waterworks Company, V 127, 129, 524
Easterby (Doubleday and), IV 238
Eastman, G. P., VI 677–8; VII 1328
— George, V 730–1, 741, 747
Eastman Dry Plate & Film Co., V 729, 731
Easton, Alexander, V 526
— and Amos, IV 195
Eberth, K. J., V 561

Eccles, W. H., VII 1188
— S. & J., V 579
Eck, B., VII 1134
Eckener, H., VII 799
Eckert, J. P., VII 1188, 1190, 1200
Edgeworth, D. R., V 445–6
— Richard L., IV 535
Edison, Thomas Alva, inventor
    arc-lamps, V 209
    cinematography, V 742, 743–5, 746–7
    direct current systems, V 200–1, 207, 229
    electric power, IV 166; V 133–4
    electronic emission, VII 1092
    incandescent lamps, V 214–15, 216, 217–18,
        220; VI 285
    microphone invention, V 227
    phonograph, VII 1262, 1263, 1309
    research centre, VI 146–7
    sound, cinema, VII 1309–10, 1312
    steel tube, comments, V 64
    telephones, development, VII 1222–3
— & Swan United Electric Light Company, V
    216
Edward, the Black Prince, II 177–8
— I, II 532; III 283–4
— III, II 525, 726
— VI, II 141; III 46
— VII, VII 1129
— the Confessor, II 426; III 198
Edwards, G., VII 828
— Humphrey, IV 191
— William, III 435–6
Eele, Martin, III 688
Egmond, Count Van, III 302
Egmont, Count Lamoral, II 684
Ehrlich, Paul, V 282; VII 1326
Ehrmann, F. L., IV 260
Eichengrün, A., VI 554
Eiffel, Gustave, V 62, 476
Eijkman, C., VI 301
Einhorn, Alfred, V 315; VII 1336
Einstein, A., VI 118, 164, 268
— S., VII 1038
Einthoven, W., VII 1340
Eisenhower, D. D., VI 124; VII 917, 1201
Ekmann, C. D., VI 610
Elder, John, & Company, V 149
Eleanor of Aquitaine, II 624
— of Castile, II 478
Electric Telegraph Company, V 218, 221–4
Electrical Company of Sweden, V 191
Elers, David, IV 345, 347
— John Philip, IV 345, 347
Eliot, Edward James, V 830
Elizabeth I, English queen
    coinage, III 46
    cross-bows, hunting, II 723

furniture, painted style, II 248–9
    London, planning, III 295
    mining, Cumberland copper, II 13
    state coach, IV 522
    woad-mills, V 263
— II, reigning queen, VII 905, 917
Elkington, George Richard, and Henry, V 633
— James Balleny, V 84–6
Ellet, Charles, IV 459
Ellicott, John, IV 350
Elliot, George, V 240
— Obadiah, IV 523
Elliotts Metal Company, V 85
Ellis, E. B., VII 1356
Ellman, John, of Glynde, IV 21, 40
Ellyott, William, III 79
Elmen, G. W., VII 1123
Elmer, P., VII 1207
Eloi, St, II 76, 77
Elsberg, J., VII 1334
Elster, J., VII 1092, 1105
Emanueli, L., VII 1076
Emerson, H., VI 84
— William, IV 444–5
Emme, E. M., VII 865
Emmert, H. D., VII 1033
Empedocles, II 678
Emperger, Fritz von, V 495
Enderby Brothers, V 219
Engelbach, R., I 570
Engels, Friedrich, V 828
— Jakob, V 287
England, William, V 732; VII 1285
Engler, C., V 111, 115
English and Irish Magnetic Telegraph Company,
    V 218, 225
Engues, M., V 707
Ennion, II 322, 325, 337
Enos, J. L., economist
    financing innovation, VI 55, 56, 70–1
    industrial innovation, VI 32, 51, 52
Entemena, I 652, 653
Enten, Pieter Pietersz, IV 632
Entrecolles, Père d', IV 339
Épernon, Duc d', III 319
Équem, IV 539
Eratosthenes, III 505–7, 509 n. 1, 522, 547
Ercker, Lazarus, metallurgist, III 28
    assaying, III 59, 60, 62–3
    balances, assay, III 61, 67–8
    brass-founding, III 37
    furnaces, III 50, 53
    speculum-metal, III 41
    steel production, III 34
Ericsson, John, IV 589; V 147, 461–2, 819
Erlach, J. E. T. von, of Vienna, IV 178
Erlinger, Georg, III 535

Erxleben, Christian Polykarp Friedrich, IV 261
Esarhaddon, I 554, 713; II 495
Esnault-Pelterie, Robert, V 392, 409, 410, 411
Essen, L., VII 1213
Estienne, Charles, III 13
— Robert, IV 523–5
Ethelbert II (d 762), king of Kent, IV 199
— of Kent (560–616), II 610
Etzlaub, Erhard, of Nuremberg, III 534–6
Eucratides, II 487
Eudoxus, III 504; IV 666
Euler, Leonhard, Swiss mathematician
  crushing factor, materials, IV 485–6; V 491
  elasticity, materials, IV 479, 483
  hydrodynamic theory, V 407
  ship-design, IV 577
Eupalinus of Megara, II 668
Eupatrids, family of Attica, II 129
Euripides, II 630
Eurysaces, II 118
Eutropius, III 575
Evans, Sir Arthur, I 758
— Edmund, V 705
— Sir Frederick, V 453
— Mortimer, V 743
— Oliver, American engineer
  aeolipile experiment, V 149–50
  boiler invention, V 137
  milling, III 16, 19
  steam-dredger, IV 641
  steam-engines, IV 165–6, 188–9, 195–6; V 126
— T. M., IV 294
Evely, R., VI 53, 54
Evelyn, John, diarist, III 681; IV 671
  London replanning proposals, III 296, 297, 298
  London smoke, III 77, 686
  Richelieu (Indre-et-Loire), description, III 291
  roads, Dutch cities, IV 529
  shipbuilding, III 491–2
  varnish, 'English', III 696
Everest, Sir George, V 442
Everett of Heytesbury, IV 304
Ewing, Sir Alfred, V 139
Exmouth, Viscount, see Pellew, Edward
Eyde, S., VI 522, 523
Eyrinis, Eyrinis d', IV 539
Eytelwein, Johann A., V 545

Faber du Faur, A. C. W. F., IV 113
Fabre, H., VII 816
Fabrizio, Geronimo, III 231
Fabry, IV 93
Faccio de Duillier, Nicholas, III 670–1
Fageol, F., VII 728, 729
— W., VII 728, 729
Fairbairn (c.1846), IV 293, 323
— P., & Company, V 651

— Sir William (1789–1874), ship-engineer
  boilers, ships, V 137, 366
  bridge tube structures, IV 460; V 504
  iron beams, strength, IV 485–6
  iron ships, V 351, 355, 362
  machine tools, V 636, 650
  riveting-machines, V 366, 504
  structure, ships, V 360, 364
  weight distribution, ships, V 353–4
  wrought iron, V 473
— William, & Company, V 133
Fairlie, Robert F., V 345
Falcon, III 166; IV 318
Falda, Giovanni Battista, III 312–13
Fantoni, III 314
Faraday, Michael, scientist
  benzene, discovery, V 269; VI 536
  biographical notes, IV 674, 676; V 780, 795
  'coal oil', V 104
  electric motor, V 231
  electricity discoveries, VI 117, 444; VII 991
  electrolytic phenomena, IV 227
  electromagnetism, IV 655, 679; V 178–9
  electroplating, factory tour, V 633
  galvanising effect, V 624
  gas-light hazards, IV 272–3
  generators, electricity, V 182, 184, 187
  glass-making experiments, V 673
  insulation, electrical, V 221
  optical glass, IV 360–1
  relay action, electrical, V 223
  research, discussion, VI 65
  steels and alloys, V 64–5, 66
  transformers, electrical, V 198 and n.
Fardoil, IV 384
Farey, John (1760–1826), IV 206
— — (1790–1851), IV 165, 206, 423
Farkashagy-Fisher, Moritz, V 662
Farman, H., VI 32; VII 792
Farmer, Moses G., V 187
Farr, W., VII 1174
Farrer, W., VII 327
Faure, C. A., VI 489
— Camille, V 206, 344, 418
Fawcett, Benjamin, V 705
— E. W., VI 558
Fay, J. A., IV 437
Fayol, Amédée, IV 275
— H., VI 77
Fefel, H. H., VI 682
Fell, John, III 402
Fellows, E. R., V 656
Felt, D. E., VII 1168
Fenizi, Ascania, III 311
Fenner, C. N., VI 623
Ferber, F., VII 792
Ferdinand, Grand Duke of Tuscany, III 314

Ferguson, H., VII 1065
Fermi, E., VI 224, 236, 239, 240
Féron of Paris, III 115
Ferranti, Sebastian Ziani de, electrical engineer
  alternators, V 193–4, 195, 196, 200
  current battle, A.C./D.C., V 200, 230
  Deptford power station, V 199–200, 201
  transformers, design, V 198, 199
Ferrier, Claude M., V 725
Fessenden, R. A., VII 1236, 1237, 1239
Festus, II 212
Fibonacci of Pisa, see Leonardo da Pisa
Fick, A., VII 1321
Field, Cyrus W., IV 661
— Joshua, IV 424, 429; V 145
Filastre, Cardinal Guillaume, III 532–3
Filenes, T., VI 84
Finiguerra, Maso, III 406
Fink, A., VI 79, 82
Finlay, C. L., VII 1297
Finley, Judge James, IV 458
Finson, N., VII 1329, 1330
Fischer, E., VI 301
— F., VI 540
— G., VII 1054
— Johann Conrad, IV 108; V 66, 608
— Otto Philip, V 279
— R., VII 1297
Fisher, Edna, I 5
— J. W., VII 1346
— R. A., VI 330, 331
Fison, J. P., IV 9
Fitch, John, V 142, 147, 152
Fittig, R., VI 558
Fitz, Henry, V 721
Fitzherbert, John, IV 201
Flabianus, Petrus, III 209
Flaminius, Gaius, II 500, 512
Flamsteed, John, III 552, 667
Fleming, A., VII 1855
— J. A., VI 448; VII 1093
Fletcher, H., VII 1349
— Thomas, V 632
Flexner, A., VI 161
Flinders, Matthew, IV 598; V 449, 455
Flitcroft, Henry, IV 470
Florey, H. W., VII 1355
Floris V, II 619
Focke, H. K. J., VII 823
Fokker, A. H. G., VI 32; VII 808, 847
Folland, H. P., VII 794
Fonck, R., VII 795
Fontaine, Jean, III 460
Fontana, Carlo, III 248
— Domenico, III 289
— Giovanni, IV 642
Fonvielle, Louis C. H., V 563

Ford, Henry, manufacturer
  cars (automobiles), early designs, V 433
  machine-tools, VII 1042
  mass-production methods, V 640, 818; VII
    714–15, 986, 1056
  time-study, application, VII 1057
— Motor Company, V 433
Fordyce, Alexander, IV 237
— George, IV 237
Forest, L. de, VII 1091, 1095
Forrest, R., VI 424
— W., VI 424
Forrester, J. W., VII 1195, 1197
Forshaw, J. H., VII 968
Forssman, W., VII 1357
Fortin, Jean, French instrument-maker
  balances, IV 404, 405, 407
  comparators, IV 397
  dividing-engine, IV 394
  screw-cutting method, IV 388
  triangulation instruments, IV 402
Fossier, A. E., VII 1319
Foster, Sir Michael, V 796
Foucault, J. B. L., V 210, 211
— Léon, IV 399, 408–9; V 737
Foulis, Andrew, V 699
Foullon, Abel, III 540
Fourché, IV 405
Fourcroy, A. F. de, IV 222
Fourdrinier, Henry, III 416; V 700; VI 614–15,
  618
— S., VI 614, 615, 618
Fournel, IV 539
Fourneyron, Benoît, engineer
  friction principles, rejection, V 119
  Niagara Falls turbines, V 533
  water turbine, V 528 and n. 2, 529, 531
Fowle, IV 71
Fowler, E. P., VII 1349
— G., VII 1389
— Sir John, V 503, 519
— Sir Ralph, VI 161
— Robert, IV 40
Fox, Sir Charles, V 610
— Henderson & Company, IV 477
— James, of Derby, IV 433
— W., VII 1312, 1313
Fox & Company, V 624
— Talbot, W. H., photographer
  paper process, V 702, 722–5, 736
  photographic plate, V 836
  photogravure process, V 708
Foyn, Svend, IV 60–1
Francesco I de' Medici, III 217, 707
Franchet, L., I 409
Francini, III 462
Francis, James B., V 529–30, 534; VI 203–4

Franciscus Paduanius of Forli, III 621
Francke, Meister, II 652
François I, III 290, 449, 459
Frank, A., VI 523, 524
— B., VI 681
Fränkl, M., VI 489
Frankland, E., VII 1388
— Percy F., V 562
Franklin, Benjamin, I 1; IV 619, 678; VII 1127
— C. S., VII 1244
— Sir John, V 41
Franz II, IV 553
Frasch, H., VI 516
Fratta et Montalbano, Marchése della, III 50
Fraunhofer, Joseph von, IV 360–1, 399; V 673
Frazer, Sir James, II 25
Frederick, archbishop of Bremen, II 683
— C., VII 1135
— I Barbarossa (1152–90), II 75
— II, the Great (1740–86), III 2; IV 23, 25–7
— William (1640–88), III 454, 682
— — III (1797–1840), IV 28
Freeman, C., economist
    innovation, discussions, VI 52–3, 66–7, 72, 114
    petrochemical industry, VI 55, 56
    research, discussion, VI 62–3, 64
Freemantle, William, IV 195; V 125
Frescobaldi, II 65
Fresnel, A. J., V 241
Freudenberg, H., VI 523
Freyssinet, E., VII 879, 892, 936
Friedman, B., VII 1142
Friese-Greene, William, V 743
Frisch, O. R., physicist
    atomic bomb, VI 226–7, 269
    atomic fission, VI 225
    uranium isotopes, VI 232
Frisi, Paolo, III 315
Fritz, John, V 612
Fromanteel family, III 663
Froment, Gustave, IV 409
Frontinus (fl A.D. 300), II 338
— Sextus Julius (c. A.D. 30–104), II 405, 419, 671–4
Froude, William, V 386–90, 546–7, 549
Fry, Elizabeth, V 824
— Peter W., V 725
Frye, Thomas, IV 341–2
Fuchs, K., VI 243
— Leonhart, III 4
— W. C., VII 1331
Fuggers of Augsburg, IV 665
Fulcher, G. S., VI 575, 576
Fuller, G. and J. C., V 205
Fulton, Robert, inventor
    canals, IV 570–1
    paddle-steamer, V 143

steam-boat, V 152
steam-power, IV 165–6, 548
Furnival & Company, V 708
Fust, Johann, III 377, 383–6

Gabriel (Voisin and), V 392, 406, 409
— Jacques (IV, V and VI), French architects, III 425 and n.
Gaddi, Taddeo, III 428
Gaffky, G. T. A., V 562
Gagarin, Y. A., VII 832, 857, 865
Gahn, J. G., IV 147; V 252
Gaillait, John, IV 508
Galantier, III 165
Galaup, J. F. de, see La Pérouse
Galbraith, J. K., VI 40, 50
Gale, L. D., IV 660
Galen, II 121, 674; III 24
Galileo, Galilei, astronomer, engineer
    atmospheric pressure, IV 169
    balances, assay, III 67
    catenary curve, III 258
    hydraulic engineering, III 303, 313–15
    lens-grinding, III 234
    mechanics, III 252; IV 149
    microscopes and telescopes, III 231–2, 552, 631, 634
    motion, laws, III 374
    pendulum-escapement, III 662, 663
    sector invention, III 627–8
    strength, material, III 58–9, 252, 432; IV 471, 478
    structural theory, III 432
    vacuum, analysis, III 332, 344
— Vincenzio, III 662
Gall, H., VI 521
Gallatin, Albert, IV 537
Gallonde, IV 404
Galton, Samuel, IV 673
Galvani, Luigi, IV 649, 652–3
Gama, Vasco da, V 259
Gambey, Henry, instrument-maker
    astronomical instruments, IV 399
    balances, IV 405
    comparators, IV 397
    dividing-engine, IV 390, 394–5
    lathe, improvement, IV 385
    triangulation instruments, IV 402
Gamble, J. H. (c.1851), V 39, 40
— John (c.1795), IV 646
— — (fl 1801), III 416; VI 614
— Josiah C., IV 239, 241; V 802
— W., VII 1280
Gannery, IV 414
Gans, Robert, V 566; VII 1380
Gantt, H. L., VI 84
Garbett, Samuel, IV 243–4

Gardiner, Sir Alan H., I 747, 754–5, 762
Garnerin, A.-J., V 397–8, plate 23 A
Garnier, Jules, V 86
Garrett (c.1850), IV 6, 9
— William (1844–1903), V 621
Garros, R., VII 794
Gascoigne, William, III 640; V 445
Gauchet, P., VII 808
Gaudry, Charles, IV 275
Gaulard, Lucien, V 198, 199, 200
Gaumont, L., VII 1310
Gauthey, Emiland Marie, IV 480–1, 556
Gauthier, A., VII 1285
Gautier, Henri, III 429, 436; IV 443, 521, 527
Gayley, J., VI 477
Gay-Lussac, Joseph-Louis, IV 163, 217, 221; V 245
Gear, Andrew, IV 437
Geber, see Jābir ibn Ḥayyān
Ged, William, V 699
Geddes, P., VII 962
Geitel, H., VII 1092, 1105
Geissenheimer, W., IV 110–11
Gellerat et Cie of Paris, IV 538
Gemini, Thomas, III 586, 621
Geminus, III 570
Gemma, Reiner Frisius, mathematician
    astrolabes, III 586
    camera obscura, V 717
    Gerard Mercator, pupil, III 550
    Jacob's staff (dioptra), III 591
    longitude, measurement method, III 503
    triangulation principle, III 539
Gemmell, R. C., VI 413
General Oceanic Telegraph Company, V 224–5
Gennes, M. de, III 168–9
Gensanne, de, IV 260
George III, III 645; IV 28
— V, VII 905, 922
Gerard, James, IV 238
— John, III 1, 3
— of Cremona, III 521, 585
Gerber, H., V 502
— N., V 33
Gerbert (Sylvester II), III 594–5, 608, 609
Gerritszoon, Jan, III 318
Gerson, Levi ben, III 528, 546, 591
Gesner, Abraham, V 104–5, 115, 116
Gibbons, John, IV 112–13
— M., VI 52
Gibbs, James E. A., V 589; VI 682
— John Dixon, V 198, 199, 200
— Josiah Willard, V 660
Gibson, R. O., VI 558
Gibson and Campbell, IV 315, 321
Giddy (or Gilbert), Davies, IV 189 and n.
Giedion, S., VII 1128

Giesen, W., VI 493
Giffard, Henri, V 325–6, 401, 402
Gigli, L., VII 1358
Gilbert, Davies, see Giddy
— Sir Joseph Henry, IV 42; V 10, 17
— William, IV 648
Gilbreth, F., VI 84; VII 1056
— L., VI 84
Gilchrist, Percy Carlyle, metallurgist
    industrial change, discussions, V 804, 810, 820
    steel-making process, V 60, 802, 809; VI 633
Gill, Sir David, V 441
Gillespie, W. M., IV 537
Gilman, A. D., V 481
Gillmore, Q. A., IV 537
Gilpin, George, III 22
Gimingham, C. H., V 218, 222
Ginns, D. W., VI 241
Giordano of Pisa, III 230
Giovanni da Fontana, II 692
Girard, Philippe de, IV 292–4
— Pierre Simon, V 531–2
Giroux, Alphonse, V 720
Glaisher, James, IV 490; V 400, 558
Glaser, Christoph, III 682
Glass Sellers' Company, V 672
Glasser, O., VII 1321
Glassford, John, V 815
Glauber, Johann Rudolph, iatrochemist, III 720
    acids, IV 231, 233, 243
    assay methods, III 68
    copper alloys, III 37, 40
    density, metals, III 58
    furnaces, II 743–4; III 52, 55, 685; IV 231
    medicinal research, II 745–6; III 682; IV 231, 243
    metallurgy, III 56
    tin alloys, III 43, 44
Glaukos of Chios, II 58
Glazebrook, R. T., VII 1214
Glehn, A. G. de, V 340; VII 762, 763, 766
Glover, John, V 245
Gluck, Jean, III 695
Glynn, Joseph, IV 205
Gmelin, J. F., IV 375
Godard, of Amiens, IV 297
Goddard, E. C., VII 862
— Robert H., physicist
    rockets, liquid fuel, VII 857, 861–3, 865
    space-flight, principles, VII 859
Godfrey, Ambrose, II 750
Godowsky, L., VII 1297
Goerz, Carl P., V 732; VII 1284, 1285
Gohory, Jacques, III 8
Goldenberg et Cie, III 117
Goldner, Stephen, V 39, 41–2 and n.

Goldschmidt, B. L., VI 260, 262
— Hans, V 97
Goldstine, H. H., VII 1190
Gölsdorf, K., VII 762
Goltz, T. von der, IV 19
Gonelin, IV 256
Gonzaga, Charles of, III 290
Gooch, Sir Daniel, V 324, 325
— John, V 324
Goodwin, Hannibal, V 729, 747
Goodyear, Charles, V 766–8
Göransson, Goran Fredrik, V 55, 70
Gordon, Andrew, IV 648
— Cuthbert, IV 249
— George Phineas, V 710
— J. E. H., V 194, 197, 200, 202–3
— L. D. B., V 781
— Lewis, IV 486
Gorrie, John, V 46
Goss, N., VI 446
Gossage, William, IV 241; V 236, 238–9, 242
Gosset, W. S., VI 330
Goulding, John, IV 298–9; V 571
Goupy, A., VII 792
Gourief, IV 545
Gouyn, Charles, IV 342
Goward, F. K., VII 1332
Gower, John, II 126
— Richard, IV 579
Gracchus, Gaius, II 505, 507
Graebe, Karl, V 261, 276
Graseff, Michel I, Auguste de, V 556
Graham, George, instrument engineer
    dilatometer, IV 396
    escapements, III 665, 671, 673–4
    horology, improvements, III 557; IV 410–11
    observatory clock, III 553
    precision mechanics, IV 382, 389–90
    quadrant, III 641
    telescope, III 553–4
— Thomas, V 660, 796; VI 557
Gram, Hans Christian Joachim, V 282
Gramme, Zénobe Théophile, electrical engineer
    alternators, V 193, 195, 212
    dynamos, V 188–9, 190–2, 196, 210
    electric motors, D.C., V 231
    machine-tools, electrification, V 641
    rotary transformer, V 230
— & Cie, V 189
Grant, G. B., V 656
— John, V 486
Gratiot, General, IV 537
Gravatt, William, V 444
Gray, E., VII 1220–1
— Stephen, IV 648–9; V 177
Greatbach, John, IV 350
Greathead, J. H., V 347, 520

Green, Charles (1785–1870), IV 260; V 400
— — (fl 1838), V 625–9
— Guy, IV 354
— James, IV 570, 572
— Mansfield, tinsmith, IV 286
— W. (fl 1778), V 445
— William (1725–1811), IV 602
— William (fl 1929), VI 103
Green & Pellatt, IV 361
— & Sons, IV 538
— & Wigram of Blackwall, IV 590
Greenaway, Kate, V 705
Greenewalt, C. H., VI 239
Greenhough, H. S., VII 1354
Greenway, IV 268
Greenwood, Frans, III 229
Greenwood, of Leeds, IV 322–5
— & Batley, IV 323; V 131, 647
Greg, R. H., V 829
Gregory, David, IV 358
— Hutton, V 329
— James, III 552 n., 667
Gregory XIII, Pope, III 577–9
— of Flax Bourton, IV 210
— of Tours, II 608
Greig, J. W., VI 632
Grenet, Eugène, V 203, 204
Gresley, Sir N., VII 766
Grew, Nehemiah, IV 37
Griess, Peter, V 273 and n. 1, 281
Griessman, B., VII 1349
Griffin, A. E., VII 1378
— J., VII 1204
Griffith, A. A., VII 1018, 1019
Griffiths, D. W., VII 1302
— Hannah, of Norwich, III 115
Grignon, P. C., IV 112
Grissom, V., VII 857
Gritzner, Max, V 589
Grondahl, L. O., VII 1116
Grönwall, E. A. A., VI 490
Gropius, Walter, V 837
Grose, Captain Samuel, V 129
Grossart, C., V 755, 756
Grosseteste, Robert, III 230
Grouse, IV 249
Grout, Jonathan, IV 647
Grove, Sir William, V 208
Grover, W. O., V 589; VI 682
Groves, General, VI 227, 238, 239, 252
Grubenmann, Hans Ulrich, III 427–8
Grundy, IV 640
Gruschke, G., VII 1349
Guarini, Guarino, III 254
Gudea, king of Lagash
    bitumen, I 252
    chronology, I table F

cubit, Sumerian measure, I 778
door-socket, stone, I 194
silver and lead, I 583
statue, inscribed, I 780
viticulture, I 282
Guénolé, St, of Brittany, II 139
Guericke, Otto von, III 637; IV 169–70, 648
Guest, J. J., VII 1042
Guettard, Jean-Etienne, IV 340, 621
Guggenheim, H. F., VII 863
Guglielminetti, IV 542
Guglielmini, Domenico, III 315
Guibal, IV 93
Guido da Vigevano, II 620, 725–6
Guillaume, Ch.-E., IV 415
Guillemin, E. A., VII 1115
Guillotte of Spitalfields, IV 318
Guimet, IV 256
Guinand, Henri, IV 360
— Pierre Louis, IV 359–61
Guise, Duc de, III 353
Guisan, IV 529
Gullstrand, A., VII 1354
Gunter, Edmund, III 542
Gusmão, Bartholomeu Lourenço de, V 395, 401 n. 1, 405
Gustavus Adolphus, III 349
— Vasa, III 31, 455
Gutenberg, John, III 44, 377, 381, 383–5
Guthrie, F., VII 1092
Gutta-percha Company of London (later Telegraph Construction and Maintenance Company), V 221
Gutzkow, IV 142
Guyldforde, Sir Richard, III 473
Guyon, Jacques, III 463
Guyton de Morveau, Louis Bernard, IV 108, 222–3, 238–9
Gwathney, J. T., VII 1334
Gwynne, James Stuart, V 525
— John, V 525, 560
Gyllenborg, Gustavus Adolphus, IV 39

Haber, L. F., chemist
innovation, discussion, VI 61
instruments, VII 1210
nitrogen fixation, VI 522–3, 524–7, 528
Habermel, Erasmus, of Prague, III 621
Habrecht, Isaac, III 656
Hackworth, Timothy, V 322
Hadfield, Sir R., metallurgist
manganese steels, V 65; VI 428, 455
silicon steel, VI 492; VII 1081, 1123
Hadley (c.1864), IV 325
— John (1682–1744), mathematician
engineering works, rivers, III 458
octant, III 556; IV 400; V 457

quadrant, III 555–6; IV 400, 604; V 457; VII 837
Hadrian, Roman emperor, II 598
aqueducts, II 670
hot baths, mines, II 9
road-construction, II 498, 507
ship model, bronze, II 575
Haeff, A. V., VII 1195
Haen, E., VI 518
Hagedoorn, A. L., VI 304
Hahn, O., VI 118, 224, 268
Ḥaidar, Abu Manṣūr, II 762
Haigh, B. P., VII 985
Haines, Richard, III 162
Hajek, Václav (Wenceslaus Hagecius), II 611
Hakluyt, Richard, III 477; V 266
Halban, H. von, VI 226, 241
Haldane, J. B. S., VII 1337
Hales, Stephen (1677–1761), IV 37–8, 216–17, 258
— (fl 1733) VII 1344
Haley of Houghton Tower, IV 279
Halfpenny, William, see Hoare, Michael
Hall (c.1876), IV 173
— Charles Martin, V 91–2, 249; VI 518
— Chester Moor, III 233, 633; IV 358
— Horace, IV 292
— John, & Sons, V 284, 292
— Joseph (fl 1839), IV 114; V 472
— — (fl 1870), V 328, 341
— Samuel, V 152–3
Halley, Edmond, Astronomer Royal, III 499, 544
charts, III 555; IV 612, 620–1
magnetic measurements, sea, VII 1462
quadrant, IV 616–17
Halsey, F. A., VI 80
Hamburg, D., VI 35
Hamilton, Sir William, V 780
Hammond, J. L. and B. (quoted), IV 349–50
— James, VII 1270
— John, VI 305
Hammurabi, Assyrian king
chronology, I tables E, F
coinage, I 585
irrigation, Tigris valley, I 548–9
stones, ring settings, I 659
sumerian script, I 751
travel speed, II 495
Hanbury, John, III 689
Hancock, John, V 765
— Joseph, IV 130
— Robert, IV 354
— Thomas, rubber pioneer
insulation, gutta-percha, V 221–2
rubber industry, V 757–62, 764–5, 770–1
rubber tyres, carts, IV 523
vulcanization, V 767–70

Handley-Page, F., VII 808
Handson, Ralph, III 551
Hannibal, II 133
Hanno the Carthaginian, II 568
Hannyngton, J., VII 1153
Hansen, E. C., V 303
Hanson, Timothy, V 5 n.
Hansson, N., VI 310
Hansteen, C., IV 621
Harcourt, William Vernon, IV 361; V 673
Harding, Joseph, V 34
Hardwick, Thomas, IV 472
Hardy, Sir W., VII 1403
Hargrave, Laurence, aeronautics pioneer
   aeroplanes, flying models, V 408
   bi-plane structure, V 409–10
   box-kite invention, V 398, 399, 406; VII 789
Hargreaves, James, IV 278–9, 281; V 577
Harington, Sir John, IV 507 n.
Harmer, J., of Sheffield, IV 304
Harmisch, F. C., VII 1331
Harness, H. D., IV 623–4, 626
Harper, J., VI 618
Harries, C. D., VI 557
Harrington, Sir J., VII 1127
Harriot, Thomas, III 9, 551
Harris, John, III 682
— Walter, III 24
Harrison, J. (fl 1859), V 137
— James (?1697–?1767), III 670, 672
— — (fl 1856), V 45–8
— John, clockmaker
   balances, IV 404
   chronometers, III 503, 557, 672–3; IV 410–14,
      598; V 458 and n.
   grasshopper escapement, III 670, 672–3; IV
      412, 413
   gridiron pendulum, IV 395, 411
   timepiece, invention, III 672; V 458; VII
      838
— Thomas, IV 82
— William, II 244; III 43
Hart, C. A., VII 1385
Hartley (Chance &), IV 373
— R. V. L., VII 1248
Hartlib, Samuel, III 677, 678, 680–1
Hartmann, E., VII 812
— Georg, III 615
Hartness, James, V 646
Hartree, D., VII 1191, 1194
Hārūn al-Rashīd, II 735, table III lix
Harvey, John, IV 319
— & Company, V 128–9
Hassall, Arthur, V 51, 834
— John, VII 1131, 1132
— Joseph, IV 209
Hassler, Ferdinand, V 440

Hathaway, H. K., VI 84
Hatshepsut, Egyptian queen
   basketry collection, I 422
   myrrh-trees, I 544
   ships, I 55, 735–6
   wood craftsmanship, I 701
Hattersley, R. L., V 579
Hatton, Sir Christopher, III 494
Hattušiliš III, I 595
Hauksbee, Francis, III 637–8
Haupt, Herman, IV 486
Hausbrand, E., V 113
Haussmann, Baron, VII 966
— Jean-Michel, V 257, 296
Hautefeuille, P. G., V 660
Hawke, Lord, III 499
Hawkshaw, Sir John, V 332
Hawksley, Thomas, V 523, 557, 558
Hawthorn, R. & W., V 340
Hawthorne, W. R., VII 992
Hayford, J. F., V 443, 444
Haynes, E., VII 1037
Haywood, William, IV 515; VII 1386
Hazeldine, William, V 63
Hazeltine, L. A., VII 1101
Heald, J. H., VII 1042, 1044
Heath, Thomas of Lane Delph, IV 347
Heathcoat, John, IV 310; V 600–3, plate 34B
Heaviside, Oliver, physicist
   circuits, design theory, VII 1113
   electromagnetic theory, VII 1247
   pendulum observations, V 444
   telephony, theory, VII 1221, 1242
Hedge, Lemuel, IV 437
Heeren, Friedrich, V 242
Hefner-Alteneck, F. von, V 189, 191, 192
Heidegger, M., VII 163
Heilmann, Ducommun, & Steinlein, V 210
— Josué, IV 296–8
Heim, L. R., VII 1042, 1044
Heister, L., VII 1317
Helena, II 29
Hellbrügge, F., VI 489
Hellot, Jean, IV 340
Helmholtz, H. L. F. von, IV 163; V 226, 788
— Richard von, V 335
Helmont, J. B. Van, see Van Helmont
Helvetia Milk Condensing Company, V 36
Hemaka, I 696, 701
Hemming, John, V 241–2
Henckel, Johann Friedrich, III 30
Henderson, L. J., IV 165
— V. E., VII 1338
— Y., VII 1337
Henley, W. T., V 223
Hennebique, François, V 490, 512
Hennell, T., III 111

Henri II (1547–59), French king, III 341
— IV, French king
  canals, III 459–61, 464
  indigo dye, V 261
  land reclamation, III 318–19
  machine-knitting industry, III 185
  roads, IV 525
  town-planning, III 293
Henry, Joseph, IV 660
— William, IV 265
Henry I, English king, II 611
— II (1002–24), emperor, II 468; III 198
— II (1154–89), English king, V 260
— III, English king, II 37, 478; IV 449
— V, English king, II 588; III 203
— VI, emperor, III 198
— VII, English king, II 306; III 497
— VIII, English king
  archery, Artillery Company, III 351
  coinage, III 46
  furniture style, II 248
  gloves, II 178
  hops, brewing ban, II 141
  ships, III 471, 478, 480
  sporting-guns, III 358
Henry the Lion, II 491
— the Navigator, III 477, 544
— de Yevele, II 385
Henshaw, D. E., VI 658
Henson, William S., V 408, 410, 411
Henut-tawi, I 670
Henz, IV 641
Heraclius (610–41), emperor of the East
  craftsman's handbook, II 351
  glass-making, II 315, 329, 332
  lead-glazes, II 300
— (? tenth century), III 207
Hercules, mythical Greek hero, III 572
Hergt, F. D., VII 808
Herissant, L. A. P., V 755
Hermannus Contractus ('the Lame'), III 609
Hero of Alexandria, mechanician
  aeolipile, heat-engine, IV 168, 193, 672; V 149–50
  'Antikythera' machine, III 618
  automata, II 635, 638
  biographical notes, II 634–6
  cam-shaft principle, II 635, 643
  cranes, II 660–2
  dioptra invention, III 609–12, 641
  gravity, mechanics, II 634
  hydraulic turbine, II 648
  mechanical advantage, theory, II 634–5
  military machines, II 636, 708–11
  press designs, II 114, 115–17
  screws and screw-cutting, II 631–2, 635; III 334
  siphons, explanation, II 668, 669
  surveying instruments, III 512
  wind-organ, II 614–15, 618
Herodotus, Greek historian
  alum, Egypt, I 262
  aqueduct of Samos, II 667–8
  Babylon city, III 270–1
  boat-bridge, construction, I 715
  brass-manufacture, II 55
  building materials, I 468; III 270–1
  capacity measure, I 783
  cargo from Tartessus, II 568
  cotton from India, I 449
  dyke, Memphis region, I 529
  Euphrates, canal network, I 555
  fertility, Babylonia, I 550
  food preservation, I 263, 264
  Greek learning, I 105
  hemp for garments, I 373; II 197
  houses, Crete, I 484 n.
  mummification, I 266, 267–8
  pyramid construction, I 482
  roads, Egypt, I 714
  salt deposits, I 257
  Tyrian purple (dye), I 247
  vegetable oils, I 288
  water purity, II 674
  water supply, Persian army, I 527
  weaving loom, II 212
Herold, J. G., IV 339
Heron of Alexandria, I 803
Héroult, P. (fl 1880–1900), VI 490
— P. L. T. (1863–1914), V 91–3, 249, 608; VI 518
Herrad of Landsperg, II 89, 115
Herschel, Clemens, IV 501; V 537, 560
— Sir John, IV 361; V 723; VII 1151
— Sir William, astronomer
  optical glasses, IV 361
  reflecting telescope, III 233, 237; IV 398–9
  silver salts, photosensitivity, V 718
Hertz, Heinrich R., physicist
  electromagnetic radiation, V 227–8; VII 838, 1113, 1231, 1233
  electronic effects, VII 1092, 1247
  innovation, discussion, VI 65
Hesiod, I 116; II 85, 549–50
Hesi-Re, I 697–8, plate 26B
Hesse, Esban, III 342
Hetep-heres, Egyptian queen
  joinery techniques, I 692, 695, 699
  mummification, I 268
  soldering, example, I 653
  tomb furniture, I 653, 698–9, 700
Heumann, K., V 278
Hevelius, Johann, III 236, 625, 640
Hevesey, G., VII 1333

Hewett, Samuel, of Barking, IV 53
Hewlett, R. G., VI 254
Hexham, John, III 317
Heycock, Charles T., V 607
Heyden, Jan, Van de, *see* Van de Heyden
Heylyn, Edward, IV 341
Heywood, Sir Benjamin, V 777
— R., VI 681
Hezekiah, I 530–1, 764, 765; III 510
Hibbard, H. L., VII 813, 825
Hick, Hargreaves, V 131
Hickman, Henry Hill, V 312
Hiero II, of Syracuse, II 27; IV 666
Higgins (Cocker &), IV 285
— Bryan, IV 238
Higgs, J. W. Y., III 149
Hildegard of Bingen, II 141
Hilderbrand & Wolfmüller, V 434
Hildernisse, W., IV 636
Hill (Roberts, & Company), IV 300
— A. V., VII 1473
— Abraham, III 157
— David Octavius, V 723
— G. H., IV 497
— J., V 579
— Sir Leonard, V 516
— Sir Rowland (1795–1879), V 816, 700
— Rowland (*fl* 1931), VI 558, 561
— William, of Liverpool, IV 238
Hillier, J., VII 1352
Hilpert, S., VII 1123
Himilco, II 568
Hindle, C. F., VII 1341
Hindley, Henry, IV 391
Hine & Mundella, V 598
Hinton, C., VI 160, 250, 263, 265
Hipp, M., VII 1213
Hipparchus, astronomer, I 264, 802
    astrolabe, III 603
    calendar studies, III 559, 564, 571
    celestial globe, III 612
    distance and direction, III 517
    Eratosthenes' globe, III 506–7
    instruments used, III 583, 586–7, 591
Hippocrates, III 569
Hippodamus of Miletos, II 528; III 272–3
Hiram, I 633, 664
Hirn, Gustave-Adolphe, V 114, 115, 118–19
Hitler, A., VI 16–23, 164; VII 863
Hjärne, Urban, III 682
Hjorth, Sóren, V 183–4
Hoare, Michael (William Halfpenny), IV 470
Hobson, William, IV 538–9
Hochstadter, M., VII 1076
Hochstetter, Daniel, III 63
Hock, J., V 160
Hodgkin, D., VII 1217

Hodgkinson, Eaton, IV 460, 485–6; V 491, 504
Hodgson, John, IV 95
Hoe, Robert, V 697
—— & Company, V 701
Hoffman, A. J., VI 684
— Friedrich, V 667
Hofmann, A. W., chemist
    benzene, tar extract, VI 536
    biographical notes, V 269–70, 782 n., 821
    coal-tar ingredients, V 270
    dyes, V 270, 272, 273, 281
— J. J., III 682
Hogarth, William, III 11, 120
Hohenheim, T. B. von, *see* Paracelsus
Hoke, Major, VII 1041
Hokusai, II 107
Holbein, Hans, III 586
Holden, H. O. L., V 436
— Richard, III 176
Holdsworth, S. D., VII 1414
Hole, I., VI 490
— James, V 778
Hole-Shaw, H. S., VII 1061
Holker, Jean Louis, IV 245
Holland, Samuel, IV 616
Hollerith, H., VII 1179–81, 1182, 1200
Holley, A. L., VI 480
Hollway, J., V 80
Holman Brothers, V 129
Holmes, Frederick Hale, V 181–3, 191–3, 210
— Randle, II 254
Holmgren, L., VII 1353
Holst, G., VII 1102
Holt (Ramsbottom and), IV 302
Holtz, F., VII 1038
Holtzapffel, IV 431
Holwerda, K., VII 1378
Homberg, Wilhelm, III 683
Home, Francis, III 176; IV 39, 244, 246
Homer, Greek poet
    brass manufacture, II 55
    iron, a valued metal, II 58
    ivories, decoration, I 677
    leather, II 148–9, 166–7
    metal-work, Hellenic, I 660
    road haulage, II 545
    shipbuilding, II 563–7
    wool, processing, II 193, 194
Homolka, B., VII 1282
Honnecourt, Villard de, III 648–9
Honorius, II 601
Hood, Thomas, V 831
Hooke, Robert, scientist
    architecture, III 252, 256, 257, 297–8
    aeronautics, V 394, 405
    balance-wheel system, III 669; IV 409
    bearings, III 41–2

construction theory, IV 482–3
dyeing experiments, IV 248
furnaces, III 55
instrument-makers, III 631, 635
lead-shot, manufacture, III 45
metallurgy, III 56, 57
optical instruments, III 635, 640–1
quadrant principle, III 555
research and Royal Society, III 681
sound, transmission, V 226
steel-making, III 35
structural theory, IV 472; V 494, 496
telescope, IV 645
tools, IV 382, 420
Hooker, Sir Joseph Dalton, V 774
Hooper, Stephen, III 100–1
Hoover, W., VI 295
Hopetoun, Lord, IV 674
Hopkins, F. G., VI 301
Hopkinson, B., VII 989
— John, V 200, 202, 203, 229
Hopper, G., VII 1157, 1199
Hor, I 686, 699
Hordry, E. J., VI 55
Hore, John, III 457
Hornblower, Jonathan Carter, IV 166; V 133
Hornby and Kenworthy, IV 300
Horner, William G., V 736
Hornsby, IV 6
— Company, V 131
Horrocks, William, of Stockport, IV 300
Horsford, E. N., V 35
Houghton, John, III 43, 682
Houldsworth, Henry, of Glasgow, IV 285–6
Houlière, M. de la, IV 101
Houston, Sir Alexander, V 567; VII 1375
Howard, Charles, of Effingham, III 481
— Ebenezer, VII 962, 964, 971
— Edward, V 285, 296
— J. & F., V 137
Howe, Elias, V 588–9, 819
— Frederick Webster, IV 435, 439; V 646, 653
Howells, G. R., VI 263
Hoxie, C. A., VII 1312
Huddart, Joseph, V 462
Hudig, D., & Company, of Rotterdam, IV 639
Hudson, George, V 804
Hufnagel, C., VII 1358
Hughes (c.1820), IV 319
— (c.1849), IV 409
— Howard, VII 722
— Thomas (c.1838), IV 536
Hugill, H. R., VI 631
Hulett, G. H., VI 464
Hull, A. W., VII 1106
Hulls, Jonathan, V 141
Hulot, IV 436

Hulsbos, Symon, of Leiden, III 305
Humboldt, Baron Alexander von, cartographer, V 821
    altitude measurements, IV 612
    barometric observations, S. America, IV 605
    isodynamic zones, S. America, IV 621
    isotherms, use, IV 620
    plant maps, IV 614, 620
Hummel, Hans, of Basel, III 168
Hunt, T. Sterry, V 111, 112
— William, & Son, III 116
Huntsman, B., scientific instrument-maker
    carbon steels, V 606, 608
    steel, crucible process, III 35; IV 99, 107–8
Hurd, Thomas, V 449
Hurry, E. H., V 485, 486
Hurter, Ferdinand, V 821; VII 1283
Husband, H. C., VII 990
Hüssener, A., VI 537
Hussey, IV 7
Hutchinson (c.1810), of Felling, IV 238
— J. (c.1829), IV 291
— John (1825–65), V 238
Hutin, M., VII 1239
Hutt, F., VII 808
Hutton, Charles, IV 445, 613
— James, IV 491
Huxley, A., VI 163
— T. H., V 783, 790
Huygens, Christiaan, scientist, IV 382, 671
    ballistics, calculations, III 374–5
    clocks and pendulums, III 344–5, 551, 662–6, 668–9; IV 409
    Earth shape, deductions, V 442
    engine, pressure principle, V 159
    gunpowder-engine, IV 171, 672; V 119, 157
    lenses, preparation, III 232, 234, 236, 637
    level design, III 543
Hyatt, I., VI 552
— John Wesley, V 729, 737, 747; VI 552
— Thaddeus, V 490
Hyde, J. F., VI 563
Hylacomylus, see Waldseemüller

Ibn al-Baiṭār, II 370, 372
Ibn al-Haitham (Alhazen), III 230; V 716, 735
Ibn al-Khaṭīb, II 139
Ibn al-Nabdī, III 613
Ibn al-Zarqālī, III 521
Ibn Khurdādhbih, II 524
Iddesleigh, Lord, VII 1388
Ideler, C. L., III 564
Illingworth, J., VI 497
Im-Dugud, I 640, 653
Imhof, K., VII 1394
Imhotep, I 537
Inbau, F. E., 1343

Ingenhousz, Jan, IV 38
Inman, James, IV 584
Innocent XI, Pope, III 312
— XII, Pope, III 313
Irwin, J., VII 868
Isherwood, Sir J., VII 741
Isidore, St, III 519, 522
Isimkheb, II 164
Isocrates, I 408
Ivanov, I. J., VI 306
Ives, F. E., VII 1295, 1296
— H., VII 1259
Iveson, IV 315

Jābir ibn Ḥayyān, II 356, 736, 738–9, 742–3
Jablochkoff, Paul, V 196, 212–13, 215
Jackling, D. C., VI 413
Jackson, C., VII 1334
— H. B., VII 1232, 1233, 1252
Jacob (c.1820), IV 414
— Emil (fl 1870), V 248
Jacobs, C. B., VII 1042
Jacobson, A. C., VII 1358
Jacobssen, Jacob, IV 639
Jacquard, J. M., IV 317–19
Jacques Gabriel, see Gabriel, Jacques
Jacquier, IV 478
Jäderin, Edward, V 447
Jaffé, Max and Moritz, V 704
James, C. C., VII 1394
— Duke of York, III 297
— J., IV 295
James I, English king, III 9, 295
— II, Scottish king, II 727
— IV, Scottish king, II 746
Janssen (or Jansen), Johannes, III 231
— P. J. C., V 737–9, 741
Jansz, Joost, IV 639
Jantzen, P., of Elbing, IV 539
Jars, Gabriel, III 86; IV 101, 178–9
Jaspers, K., VI 163
Jecker, IV 394
Jedlik, Anyos, V 187
Jefferson, Thomas, II 90; IV 437; V 4, 440
Jeffrey of Paisley, IV 300
Jenatzy, Camille, V 420
Jenkin, H. C. Fleeming, V 536
Jenkins, C. F., V 742; VII 1259
Jenner, Edward, II 254
Jenney, William Le Baron, V 478
Jennings, W., IV 319
Jenson, Nicolas, III 384
Jerome, St, II 97, 106, table II lvii
Jessop, William, IV 467–8, 484, 641
Jessops, the, IV 572
Jevons, F. R., VI 69
— W. S., III 87

Jewell, W. M., VII 1376–7
Jewkes, J., VI 34, 66
Joachim II, III 454
Johann Thölde, see 'Valentine, Basil'
Johansen, K. W., VII 946
Johansson, C. E., VII 1041
John, Prince of Portugal, III 547
John V Palaeologus, II 585
— the Calabrian, II 656
— Campanus of Novara, see Campano da Novara
— of Garland, II 641, 653
— of Linières, III 585, 616
— of Northampton, III 614
Johnson, C. E., VI 682
— I. C., V 484
— J., VII 813
— J. B., VII 1248
— L. C., IV 448
— Matthey & Company, IV 141
— R. D., VI 52
— Samuel (1709–84), V 776
— — Waite (1831–1912), V 339
— Thomas, IV 299–300
Johnson & Brothers, V 621
Joliot, F., VI 224; VII 1333
Jones (c.1825), IV 319
— F. Wood, I 10, 12
— H., VII 1336
— Inigo, architect
　　Renaissance, architecture, III 256–7
　　roofs, low-pitched, III 256, 265; IV 470, 472
　　town-planning, III 295–6
　　windows, mullioned style, III 267
— P. M., VII 1329
Jones & Lamson, V 646
Jongh, de, of Alsace, IV 322
Jordan, F. W., VII 1188
— Johann, V 524
— T. B., IV 436
Joseph II, IV 528–9
Josephus, II 715
Josse, St, II 762; III 191
Jouffroy d'Abbans, Marquis Claude de, V 142
Joule, James Prescott, IV 163, 603
Jousse, Mathurin, metallurgist
　　ferrous metals, III 33, 36, 57
　　windmill specifications, III 89, 92, 106
Joy, David, V 337–8
Jules, Pierre, V 162
Julian the Apostate, II 140, 515, table II lvii
Julius II, Pope, III 288, 338
Julius Africanus, II 376
Julliot, H., VII 796
Junghans, S., VI 497
Junkers, H., VI 32; VII 808
Jürgen, Master, III 160
Justin II, II 452, 765

Justinian I, Emperor of the East
  aqueducts, II 670
  chronology, II table III lviii
  silk production, II 197, 763
  water storage cisterns, II 664, 666

Kachkaroff, P., VI 516
Kahlbaum, G. W. A., VI 558
Kalthoff, Caspar, III 680
Kaltoff, Caspar, IV 672
Kammerer, Jacob F., V 252
Kapitza, P., VI 243
Kaplin, V., VI 205
Kapp, Gisbert, V 203
Kartveli, A., VII 814
Kater, Czn, S. T., IV 642
Kawit, I 698
Kay, James, IV 293
— John, III 153, 161, 168–70; IV 277
Keeler, L., VII 1343
Keir, James, IV 237, 673; V 607
Keith, P. C., VII 233
Kekulé, Friedrich August, V 274–5, 283, 788, 821
Keller, C., VI 490
— Helen, I 18
Kellerman, K. F., VII 1377
Kellner, Carl, V 251; VI 519
— O., VI 309, 310
Kellogg, W. K., VII 1333
Kells. C. E., VII 1325
Kelly, E., VI 135
— William (c.1790), IV 279, 287–8
— — (1811–88), V 53–4, 57, 70
Kelso, J. L., I 409
Keltridge, William, III 490
Kelvin, Lord, see Thomson, William
Kempelen, Wolfgang von, V 138
Kendall, Larcum, III 557; IV 411
Kendon, A., VII 865
Kendrew, J. and Porthouse, A., IV 291–2
Kennedy, Sir A. B. W., V 200, 796
— G., VII 1377
— John, IV 280, 288
— Julian, VI 476
— T., VII 1377
— Thomas, IV 501–2
Kent, William, IV 470
Kenworthy (Hornby and), IV 300
— and Bullough, IV 302
Kenyon, Kathleen, I 372
Kepler, Johann, astronomer
  accuracy, discussions, III 640, 658
  camera obscura, use, V 717
  compound microscopes, III 232
  planetary models, discussion, I 799
Kerr, W. J., VII 1348
Kerst, D. W., VII 1332

Kettering, C. F., VII 719, 1001
Keynes, J. M., VI 116
Kha, I 440
Khalaf ibn 'Abbās al-Zahrāwī, II 371
Khnumet, I 657
Kidinnu (or Kidenas), III 568, 574
Kier, Samuel M., V 109, 113
Kierkegaard, S., VI 163
Kilburn, T., VII 1193, 1196
Killeen, M., VI 478
Kingsbury, A., VII 988
Kingsford-Smith, C., VII 806
Kipping, F. S., VI 563
Kircher, Athanasius, V 717, 736
Kirchhoff, G. R., VII 1113
Kirk, A. C., V 149
Kirkham, C. B., VII 1012
Kirkwood, James P., V 564
Kirtley, Matthew, V 327, 345
Kitsons, V 801
Kjellin, F. A., VI 489
Klaproth, Martin H., IV 221, 375; V 257
Klatte, F., VI 556
Klemm, H., VII 804
Klie, K., VII 1271
Klingenstierna, Samuel, of Uppsala, IV 358
Knibb, Joseph, III 667
Knietch, R., VI 517
Knight, Charles, V 702
— Gowin, III 555; V 454
— John H., V 422, 430
Knocker, G. M., II 294 n.
Knowles, J., V 579
Knox (Macintosh), & Company, IV 251, 256
Koch, C., VII 1293
— Robert, V 281, 561–2
Koenen, Mathias, V 489, 495, 502
Koenig, E., VII 1282, 1297
— Frederick, V 694–6, 715
Köhler, W., I 14
Kohts, Mme, I 14
Kolbe, Hermann, V 316, 788, 821
Kolff, W., VII 1359
Koller, Carl, V 315
Köller, Franz, V 65
Komarov, V., VII 857
Konsortium für electrochemische Industrie, V 251
Koops, Matthias, III 415; VI 613
Kopp, H. F. M., V 286
Koppy, Freiherr von, IV 28
Korotkoff, N. S., VII 1345
Kotter (or Kutter), III 358
Kowarski, L., VI 226, 241
Kozhedub, I., VII 813
Kratzer, Nicolas, III 586
Kress, W., V 404 n. 1, 409
Krische, W., VI 553

Kritz, N., VII 1171
Krupp, Alfred, IV 108; V 57, 615, 822
— Friedrich, V 764
Ktesibios of Alexandria, I 113
Kublai Khan, III 379–80
Kuffler family, III 695
Kühn, H., VII 1293
Kuichling, E., VII 1385
Kunckel, Johann (Baron Löwenstjern)
    biographical comment, III 682–3
    blast-lamp description, III 55
    cobalt, glass-colouring, III 218
    glass-making, III 222, 224–5, 229, 698
    japans, lacquering, III 696, 697
    mercury fulminate, V 296 n. 3
Kutter, see Kotter
Kyeser, Konrad, II 691–2
Kynoch, George, V 630–1

Labelye, Charles Dangeau de, III 425; IV 467
Labouchere, Henry, see Taunton, Lord
La Condamine, Charles de, V 754–5
Laddon, I. M., VII 828
Laennec, R., VII 1346
La Hire, Philippe de
    applied mechanics, IV 163, 444–5, 478
    arches, mechanics, III 433–4; V 492
    astronomical fixes, IV 605
    map of France, IV 606
    survey-instruments, III 543
Laing, J., V 737
Laird, John and William, V 351
Laithwaite, E. R., VII 1088
Lambert, J. H., IV 621, 624–5
Lamberville, C., III 86
Lambot, J. L., V 488
Lambton, William, IV 610; V 440, 442
Lamé, G., V 494
Lamoral, Count of Egmont, see Egmont
Lampadius, Wilhelm August, IV 261
Lampitt, L. H., VII 1403
Lana-Terzi, Francesco de, V 394
Lanchester, F. W., engineer
    aerodynamic theory, V 407; VII 793
    engine, 1897 design, V 175–6, 431
    hobbing machine, V 656
    road-vehicles, V 430–2, plate 19 A
    tension roof patents, VII 952, 953
    vibration dampers, VII 987
Lanchester Engine Company, V 432
Land, E. H., VII 1294
Landore Siemens Steel Company, V 373, 629
Landsteiner, K., VII 1356
Lane-Fox, St George, V 215, 218, 219
Langenheim, F. and W., 725, 737
Langer, Carl, V 88
Langland, William, II 126 n., 214, 541

Langley, Batty, IV 470
— S. P., V 409, 411; VII 789, 1009–10
Langmuir, I., VII 1083, 1093, 1097, 1331
Lankester, Sir R., VI 154
Lanoix, J. B., of Lyons, IV 261
La Pérouse, (J. F. de Galaup), Comte de, IV 598
Laplace, Pierre Simon, Marquis de, IV 396–7
Lapostolle, of Amiens, IV 260
Lardner, D., VII 1159
Larsen, C. A., IV 60; VI 357
— J. A., VII 1343
— R., VI 348
Lash, H. W., VI 497
Laurentius Vitrearius, II 326
Lauritsen, C. C., VII 1332
Laussedat, Aimé, V 448–9
Lauste, E., VII 1311, 1312
Laval, Carl Gustav de, V 33, 140, 150
Lavoisier, Antoine Laurent, chemist
    balances, IV 404, 405, 406–8
    biographical notes, III 25; IV 674; V 821
    carbonic acid, IV 38
    caustic alkalis, IV 227
    combustion experiments, II 736; IV 218–21
    dilatometer, use, IV 396, 397
    dyestuffs, discussion, V 267, 268
    elements, nomenclature, II 745; IV 115, 220–2
    measuring rules, IV 397
    metals, postulate, V 72
Law, R. R., VII 1119
Lawes, Sir John Bennet, chemist
    animal nutrition, V 17
    Rothamsted Experimental Station, IV 42
    superphosphate fertilizer, IV 20, 254; V 10,
        254; VI 529
Lawrence, B., VI 657
— Richard S., IV 439–40
Lawson, Harry John, V 416, 417
— and Westly, IV 293, 315
Leach, B., I 409
— G., V 570
Leared, A., VII 1346
Leatham, A., VII 1348
Leavers, John, V 602–4
Lebaudy brothers, VII 796
Le Bel, J. A., V 111, 113, 118
Le Blanc (c.1785), IV 437–8
Leblanc, M. (fl 1886), VII 1236, 1238–9
— Nicolas (1742–1806), IV 238–9; V 257, 671, 802
Lebon, Philippe, IV 258, 261–5, 275
Le Châtelier, Henry Lewis, V 660
Leclanché, G., V 204
Ledwinka, J., VII 717
Lee, Edmund, III 98
— George, IV 265, 266, 268
— John Bridges, V 448
— William, III 185–6; V 596–7

Leeghwater, Jan Adriaanszoon, engineer
 drainage, Lesparre marshes, III 319
 drainage-mills, III 303, 305-7
 scoop-wheels, II 688-9; IV 157
Leeuwenhoek, A. van, III 232-4, 631, 633
Lefêbvre, L. J. P. B. (*fl* 1859), VI 522
Lefebvre-Gineau, IV 408
Lefèvre (*c*.1600), chemist, IV 242
Legendre, A.-M., VII 1150
Le Gray, Gustave, V 724
Lehmann, J. G., IV 611-12
Leibniz, Gottfried Wilhelm, mathematician
 binary scale, VII 1192
 calculator, VII 1151, 1160, 1161-3, 1171
 scientific discoveries, IV 671
Leicester, 1st Earl of (1532-88), III 304
— Earl of (*c*.1631), III 296
Leigh, E. (*fl* 1858), V 573
— Evan (1811-76), IV 290
— John, V 269
Leipmann, W., VII 1358
Leland, H. M., VII 1039
Lelièvre, IV 253
Lely, Cornelius, V 545; V 907
Lemercier, Jacques, III 291
Lémery, N., III 682, 696; IV 242
Lemoine, IV 538
Lenel, IV 397
Lenigre, J., VII 1342
Lenin, K., VI 10, 12, 14
Lenk, Baron Wilhelm von, V 284, 290, 294
Lenoir, Étienne (1744-1832), V 157, 158
— — (1822-1900), IV 394, 397, 401-2, 604
Lenormand, L. S., V 398
Leo I, Pope, II 599
— VI, Pope, II 376, 516
— X, Pope, III 288, 311
Leonardo (Fibonacci) da Pisa, II 766-7; III 599, 608
— da Vinci, III 250-1
 alchemical furnace, II 744, 745
 arches, III 432-3
 biographical notes, II 651-2; IV 442, 667, 669
 buildings, III 250-2
 canals, III 286, 447-8, 450, 459
 clocks, III 657
 coinage machinery, III 47, 340-1
 dredger designs, IV 637, 638
 dynamics, analysis of problems, III 344
 file-maker, automatic, II 395
 flying machine, V 394, 398, 403 n., 404 n.
 furnaces, III 54
 gear-teeth studies, II 648, 654
 gig-mills, III 172
 hodometer, III 512, 513
 hot-air turbine, IV 160-1
 hydraulic engineering, III 286, 311

 lock mitre-gates, III 447-8, 450
 machines, illustrations, II 654-7
 metallurgy, III 39, 40, 41
 military engineering, III 285, 370
 mirror-finishing machines, III 256
 paddle-wheel propulsion, V 141-2
 roasting (cooking) devices, III 20
 roller-bearings, III 41, 327
 rolling-mill, metals, V 611
 screws, III 335, 657; IV 425
 shearing-machine, designs, III 174, 175
 spinning-machinery, II 205; III 161-2
 strength, materials, III 250-2, 432-3
 town planning, III 285-6
 turret windmill, III 330
 vision, V 735
 war-machines, II 699, 726
 weaving-loom, III 167
Lepaute, Jean André, III 674
Leplay (Dubrunfaut, A. P. and), V 304
Lerebours, J. N., IV 360, 399
Leroy, Julian, III 557
— Pierre, clock-maker
 balance wheel escapement, IV 410
 chronometers, IV 410, 598; V 458 n.
 duplex-escapement, III 674
 horology, improvements, III 557; IV 413
Le Sage, IV 650
Lescallier, Daniel, IV 580
Leschot, Rudolf, V 108; VI 381
Lescot, P., III 253
Le Seur (*c*.1742), IV 478
Lesseps, Ferdinand de, IV 561-2
LeSueur, Ernest (*fl* 1892), VI 519
Lethaby, W. R., III 240
Leupold, Jacob, IV 188, 537
Levavasseur, L., V 409; VII 808
Lever, William Hesketh (Lord Leverhulme), V 806
— Brothers, V 806
Levinstein, Ivan, V 274
Levy, Maurice, V 493
— Max, V 704
Lewis, J., of Brinscombe, IV 305
— John L., VI 108
— Sir T., VII 1340
Ley, W., VII 863
Li Yueh, III 440
Libavius, Andreas, II 748-9; III 30; IV 233
Liebermann, Karl, V 261, 276
Liebig, Justus von, chemist, V 821
 bleaching-powder, comments, IV 248
 education, contributions, V 785, 788
 farming, continental, V 14, 17
 organic compounds, analysis, V 268
 plant nutrition, IV 39
 soda, comments, IV 237

Liebig, Justus von, chemist (*cont.*)
  superphosphates proposal, IV 254; V 254
  tar, constituents, V 106, 115
Liernur, C., VII 1384
Lightfoot, John, V 273
Lilienthal, Otto, V 405–7, 410, 411; VII 789
Liljendahl, J., VII 1384
Lillienfeld, J., VII 1118
Lindbeck, A., VI 60
Lindbergh, C. A., VII 806, 862
Lindblad, A. R., VI 490
Linde, C. von, VI 489
Lindelof, Andrew, IV 538
Lindemann, F. A., VI 230
Lindsay, Sir Coutts, V 198
Linnaeus (Carl von Linné), III 6, 159
Linotype Printing Company, V 688
Linperch, P., III 89
Linschoten, Van, *see* Van Linschoten
Linstead, R. P., VI 568, 591
Lippisch, A. VI 32; VII 829
Lippmann, E. O. von, III 44
— G., VII 1340
Lister, Joseph (Lord Lister), V 300, 315; VI 540;
  VII 1355
— Martin, III 703; IV 544
— S. C. (Lord Masham), IV 296–8, 321–3, 325;
  V 574–5
Liston, Robert, V 312
Litterer, J., VI 81
Little, I. M. D., VI 53, 54
Littledale, H. A. P., I 658 n.
Littlepage, J. D., VI 416
Littler, William, IV 342–3, 349
Liverpool Dry Plate & Photographic Company, V
  728
Livesey, John, V 603
Livy, II 27
Ljunström, F., VII 1030
Lloyd, R., VI 474
Lloyd & Summerfield, V 681
Lloyd-Davies, D. E., VII 1385
Lloyd-George, D., VI 285
Lloyd's, Edward, Coffee House, V 805, 815
Lloyds, Fosters & Company, IV 113
Lobsinger, Hans, of Nuremburg, III 32
Locke, Joseph, V 323
Lockett, W. T., VII 1389, 1391
Lockyer, Sir N., VI 147, 148
Locomobile Company of America, V 425
Locquenghien, Jean de, III 452
Lodge, Henry, IV 268
— Sir O., physicist
  electrostatic cleaning system, VI 471
  radio pioneer, V 227; VII 1232, 1233
  syntony, VII 1113, 1234–5
Löhneiss, Georg Engelhard von, III 28, 51

Lohrke, J. L., VI 659
Lombe, John, IV 311
— Sir Thomas, IV 311, 313
Lomond, M., IV 650
Lomonossoff, Professor, VII 767
Lonce, de, IV 641
London Electric Cab Company, V 419
London Electricity Supply Corporation, V 199
Long, Crawford W., V 312
Longbridge, Baron, *see* Austin, Herbert
Loomis, Elias, IV 621
L'Orange, P., VII 1008
Lorini (*b.* 1540), IV 635–8
— Buonaiuto (*c.*1592), III 286
Losh, William, IV 238
Lou Shou, II 220
Louis VII, II 608
— IX (St Louis), French king
  bastides, fortress towns, III 283
  crusade-ships, II 585
  farm-lands, France, II 681
  metal craftsmanship, II 453, 457
  pyrotechnics, military, II 377
— XIII, III 293, 463; IV 525
— XIV, III 290, 293, 465–6; IV 442, 525
— XV, III 553; IV 525
— XVI, IV 27, 645
— Philippe, V 260
Lovejoy, R., VI 522
Lovelace, A. A., Countess, VII 1178, 1194, 1199,
  1201
Lovell, Sir B., VII 990
Low, IV 72
Lowenstein, F., VII 1097
Löwenstjern, Baron, *see* Kunchel
Lowitz, Theodor, V 116
Lowther, James, IV 258
Lu Ch'ih, III 440
Lubbers, John H., V 678
Lübeck, Petter von, III 455
Lucas, A., I 373, 409
— J. H. W., VII 1338
Lucian of Samosata, II 630; VII 858
Lucretius, I 216; II 9, 596
Lud, Walter, III 538
Ludlum, A., VI 497
Luft, A., VI 555
Luke, St, II 188
Lull, Raymón, II 142, 356; III 526
Lumière, A. L., VII 1296
— A. M. L. N. and L. J., V 282, 742, 745–8, 750
Lundström Brothers, V 253
Lunge, George, V 241, 247
Lupicini, Antonio, III 310–11
Lürmann, Fritz W., V 68
Luscombe, D. A., VII 805
Luthi, R., VII 1342

Lutwyche, Thomas, IV 238; V 236
Lycurgus, II 3

McAdam, John L., road-engineer
    books, IV 535
    camber, roads, IV 531
    macadam system, IV 534–6, 537
    road, cross-section, IV 527, 528
    roads, construction principles, IV 532–5; VII
        873
    stones, road-making, IV 534, 538, 544
McCallum, D. C., VI 79, 82, 89
MacArthur, John S., V 95; VI 424
Macarthy, J. F., VII 1359
Macary, Antoine, IV 635
McConnell, James E., V 327, 328, 334, 341
McCormick, Cyrus Hall, IV 7–8; V 4, 11, 817
McCowan, J., V 546
MacCulloch, John (1773–1835), IV 622
McCulloch, John R. (1789–1864), V 674
McCullum, N. E., VI 488
McDowall, John, IV 437
Macdundonald, Earls of, see Cochrane
Macfarlane Gray, J., V 384–5
McGee, J. D., VII 1260
McGowan, H., VI 59, 60, 288
Mach, E., VII 853
Machiavelli, Niccolò, III 370
Macintosh, Charles, chemical manufacturer
    alum manufacture, IV 250–1
    rubber, shoe material, V 765
    rubber-proofing processes, IV 253; V 761–2,
        769
— George, IV 249; V 824 n.
— Knox & Company, IV 251, 256
Macintosh & Company, V 762, 766–70
McIntyre, John, V 371, 375
Macirone, Colonel, IV 544
MacKenzie, J., VII 1344
Mackenzie, Murdoch, IV 616–18; V 450
McKesson, E. I., VII 1334
Mackie, Alexander, V 685
— W., VII 1330
Maclaurin, W. R., VI 38
Maclure, William, IV 622
McMillan, E. M., VII 1332
Macmillan, Kirkpatrick, V 414
McNaught, John, IV 186–7, 194–5
MacNaught, William, V 133
MacNeill, IV 539
McPhee, H. C., VI 305
Macquer, P. J., chemist
    dyes, IV 248–9
    porcelain, IV 340, 342
    rubber, solvents, V 755
    stoneware, comments, IV 346
Macrobius, III 575

Maddox, Richard Leach, V 728
Maertense, Willem, III 453
Magellan (c.1780), III 12
— Ferdinand (?1480–1521), III 4, 477
Magill, I., VII 1334
Magnus, H. G., V 247
Mahan, A. T., IV 582
Maiben, John, IV 274
Maillart, R., VII 934
Mainz, John of, III 393
Maire, IV 553
Malam, John, IV 270
Malaspina, IV 598
Malherbe, IV 238–9
Malhère, Eugene, V 604
Mallet, Anatole, V 345
Malo, IV 540
Malpighi, Marcello, IV 37–8
Malthus, T., VI 26
Mandeville, Bernard, III 706
Manhès, Pierre, V 82
Manly, C., VII 1010
Mannes, L., VII 1297
Mannesmann, Max, V 629
— Reinhard, V 63–4, 629
Mannock, E., VII 795
Mansart (or Mansard), François, III 253, 255
— J. Hardouin-, III 254–5, 425
Mansell, Sir Robert, IV 362
Mansfield, Charles, V 270
— E., economist
    change, dynamics, VI 73
    expenditure, R and D, VI 62, 63–4
    industrial research, VI 36
    innovations, output, VI 41, 51–2
Månsson, Peder, Swedish metallurgist
    glass-making, III 209–10, 212, 214, 220
    publications, III 28, 209
    speculum-metal, III 40
Manston, T., VII 1272
Mantoux, Paul, III 85
Manzoni, W. T., IV 461
Mao Tse-Tung, VI 16, 25; VII 873
Marais, Eugene, I 5
Maraldi, G. D., IV 605
Marcaire, IV 631
Marcellus, II 714; IV 665–6
March, IV 433
Marco Polo, II 156; III 73, 379, 524; V 102
Marconi, Guglielmo, inventor
    aerials, radio, VII 1243–4
    finance, research, VI 45
    innovation, discussion, VI 65
    marine radio, VII 1252
    radio pioneer, V 227–8; VII 838, 1231–5,
        1248–50
Marcus, Siegfried, V 426

Marcus Graecus, II 351, 379
Marey, E. J., V 737, 739–43
Marggraf, Andreas Sigismund, III 30; IV 28
Maria Theresa, Empress, IV 27, 528
Mariano, Jacopo, II 610
Marindin, F. A., VI 285
Marino, Ieronomo, III 290
Marinus of Tyre, III 507–9
Marion & Company, V 730
Mariotte, Edmé, III 464; IV 37, 163
Marius, Gaius, II 703, 705
Markham, Sir Clements, V 773
— Gervase, III 134–5
Marlborough, John Churchill, III 347, 350
Marsaut, IV 98
Marseille, H. J., VII 813
Marsh, A. L., VII 1133
Marshall, Alfred, V 807
— C. (possibly 'C.M.'), IV 649–50
— John (1663–1725), III 633–4
— — (c.1790), IV 292–3
— William, III 15; IV 21, 40
Marston, A., VII 1386–7
— W. M., VII 1343
Martial, Roman poet, I 284
Martin, A. J. P., VII 1210
— C., of Verviers, V 572
— C. H., VII 723
— Émile and Pierre, V 59, 70, 499
— G., VII 829
— H. E., VII 1356
— John, IV 272
Martineau (Taylor and), IV 197; V 130
Martini, Francesco di Giorgio, III 286
Marton, L. L., VII 1351, 1352, 1353
Marum, Van, see Van Marum
Mary I, IV 522, 524
— the Jewess, II 733–4
Māshāllāh of Basra, see Messahala
Masham, Lord, see Lister, S. C.
Maskelyne, Nevil, III 557; VII 1150–1
Mason, A., VI 552
— Joseph, III 165; IV 317
— Sir Josiah, V 799
Mason & Elkington, V 85
Maspero, H., I 597 n.
Massanez, J., VI 479
Massey, Edward, V 459–61
Matguzaki, K., VII 1158
Mather & Platt, V 203
Mathesius, Johannes, III 28
Mathieson Alkali Company, V 252
Matignon, Mareschal de, III 318
Matthes, A., VI 497
Matthiesen, A., V 86
Mauchly, J. W., VII 1188, 1200

Maudslay, Henry, engineer, IV 288
    block-making machinery, IV 427–8
    engineers employed, IV 428, 429–30, 431, 434;
        V 800–1
    lathes, improvement, V 644, 800; VII 1041
    machine-tools, IV 162, 418, 423–6, plate 26; V
        636
    micrometer screw, VII 1050
    screw-cutting lathe, IV 388, 424, plate 25A
    table engine, IV 196, 426; V 125
— Joseph, V 145–6
— Sons & Field, IV 424; V 133, 146, 147
Maughan, Benjamin Waddy, IV 273; VII 1142
Maupertuis, P. L. Moreau de, III 553
Maurer, E., VI 492
Maurice, P., VII 1342
— Prince of Nassau, III 304, 306, 349
Mauro, Fra, III 530
Maurolico, Francesco, III 231
Maury, Lieutenant M. F., IV 619–20; VII 1446
Maxim, Sir Hiram, V 409, 410; VII 1129–30
Maximilian I, Elector, III 358; VII 1127
— I, Emperor, II 389, 491; III 352–3
Maximus, II 136
Maxwell, J. C., VII 1112, 1113, 1231, 1247
May (Ransome and), IV 11
Maybach, Wilhelm, V 169, 171; VII 999–1000
Maydwell and Windle, III 225
Mayer, J. T., IV 401; V 458
— Julius Robert von, IV 163
— K., VI 668
— Tobias, IV 604
Mayhew, Henry, IV 516
Mayo, E., VI 85
Mayow, John, IV 216
Mazzoleni, Marcantonio, III 627
Méchain, P. F. A., IV 402
Meda, Giuseppe, III 449
Medici, Giuliano de', III 311
Medina, Bartolomeo, IV 138
Medina Sidonia, Duke of, III 480
Megasthenes, II 666
Mège-Mouriés, Hippolyte, V 13 n.; VII 1400
Mégnié, IV 394, 404, 407
Meijer, Cornelis Janszoon, III 312–14
Meikle, Andrew, III 98, 100–1; IV 10
Meinesz, F. A. V., VII 1463
Meisenbach, Georg, V 704
Meister Lucius & Brüning, V 247
Meitner, L., VI 224
Melckenbeke, Dominicus van, III 305
Meldolla, Andrea, III 407
Melen, E., V 570
Méliès, Georges, V 749, 750
Mellen, G., VI 497
Mellet, IV 548
Mellor, J. W., V 660; VI 623, 624

Meltzer, S. J., VII 1334
Melville, Viscount, *see* Dundas, Henry
Mendel, G., VI 304, 327–8
Mendeleev, D. I., V 111
Menes, I 267, 529, 536
Menzies, Michael, IV 9
Mercator, Gerard, III 536, 550, 586
Mercer, John, V 587
Mercier, Le, III 462
Mergenthaler, Ottmar, V 687
Mérian, IV 540
Meritens, Baron, A. de, V 192–4, 206, 210
Merklein, IV 384
Merret (or Merrett), Christopher, metallurgist
    copper alloys, IV 37, 41
    glass-making, 218–19, 220–1, 682, 698
    lead flux, glass-making, V 672
    research proposals, III 681
Merrill, Joshua, V 115
Merrow, J., VI 682
Mersenne, Marin, III 49 n. 1, 58
Mersey, Mattias, III 79
Merz, C., VI 285; VII 1068, 1086
Meryet-Amen, I 422, 690, 701
Mesha, I 764, 765
Mes-kalam-dug, I 644, 645
Mesrell, III 341
Messahala (Mesahallah, Māshāllāh) of Basra, III
    529, 609
Messel, Rudolf, V 247
Messerschmitt, W., VII 810
Mestrell, Eleye, III 341
Metagenes, II 660
Metcalf, A. W., VI 657
— John, IV 531, 535
— H., VI 80
Métherie, de la, IV 239
Meton, I 795; III 568, 570
Metrodorus, II 600
Metropolitan Water Board, V 129
Meuseler, IV 97
Meusnier, J. B. M., IV 406–7
Meuvret, J., IV 31
Meyenberg, J. B., V 35
Meyer, W., IV 642 n.
Miall, Stephen, IV 242
Michel, A. G. M., VII 988
— de Nostradame, II 741
Michelangelo, II 28; III 246, 248, 289
Michelin, André, V 773
Michels, A. M. J. F., VI 558
Michelson, A. A., VI 120
Midgly, T., VII 989
Miethe, A., VII 1282
Mihály, D. von, VII 1259
Miles, G. W., VI 554, 650
— Thomas, IV 345

Mill, Henry, V 689
Miller, Patrick, V 142–3
— Robert, IV 300
— Samuel, V 147
Miller & Richard, V 683
Mills, E. W., VI 252
— F. W., VII 1285
— J., VII 1336
— Peter, III 297
Milne, Thomas, IV 615
Milton, John, V 776
Miner, W. M., VII 1240
Minkelers, J. P., IV 252, 260
Miseroni, Dionisio, III 223
Missaglia, Benedetto da, III 448–50
Mitchell, R. J., VII 812
— W., VII 811
Mithridates VI the Great, II 593
Moberly, George, V 785
Modena, Filippo da, III 445
Moede, W., VII 1058
Mohr, Otto, V 493, 495, 501
Moissant, Charles, V 745
Moisson, H., VI 523, 541
Molineaux, IV 324
Moll, Tilleman de, III 455
Möller, Anton, III 167
Mollweide, K. B., IV 625
Moncke, IV 656
Monconys, III 680
Mond, A., VI 59
— Ludwig, chemical industrialist
    carbonic acid, production, V 239
    industrial skill, V 239, 802, 807
    nickel refining process, V 88–9
    Solvay process, improvements, V 243
    sulphur, reclamation process, V 238
Monge, Gaspard, IV 482; V 493
Monier, Joseph, V 488, 489, 502
Monnartz, P., VI 493
Monotype Corporation, V 712
Monte Corvino, Archbishop John of, III 380
Monteith, Henry, IV 249
Montgolfier, Jacques Étienne, V 395
— Joseph Michael, IV 499; V 395, 398, 526
Montgomerie, William, V 221
Montlaur, A. de, VI 521
Moore, Charles T., V 687
— G. T., VII 1377
— Sir John, V 824 *and* n., 831
Morant, R. L., VI 143–4
Morat, F., VI 668
Mordey, W. M., V 194–5, 198, 200
Moreton, Charles, III 157
Morgan of Battersea, IV 143
Morgenthaler, O., VII 1272
Moritz, Carl Philipp, IV 530

Morland, Sir Samuel, IV 672; VII 1161
Morley, Thomas, III 409
Morphy, D. W., VI 295
Morrice (or Morris), Peter, III 317; IV 207
Morris, William, V 694; VII 1274
Morrison, Charles (possibly 'C.M.'), IV 649
Morse, Samuel F. B., IV 659; V 218, 220
Mort, Thomas Sutcliffe, V 47–8
Mortimer, John, III 134
Morton, G. A., VII 1105
— James, VI 566, 567
— John, IV 5
— W. T. G., V 312
Morveau, Guyton, de, see Guyton de Morveau
Moseley, Henry, V 492
Moser, E., V 587
Moss, S. H., VII 1020–1
Motel, IV 414
Mouchel, Louis Gustav, V 490
Moulton, Lord, VII 1179
— Stephen, V 764, 766–7, 769
Mountaine, III 555
Mouras, L., VII 1394
Moxon, Joseph, metallurgist and writer
    carpentry methods, II 242, 250–3
    printing presses, III 401
    steel quality, III 35
    strength of materials, IV 471–2, 473
    tool-kit, carpenter's, III 116
    type-casting, III 44
    type-founding, III 381, 391–6
Mudge, Thomas, III 673–4; IV 415
— William, IV 607
Mueller, M., VII 819
Muhammad, Prophet of Islam, III 580
Muhammad Ali (1805–49), V 542–3
Muhammad ibn Qala'un, III 203
Muir, Malcolm, IV 437
Mulgrave, Baron, see Phipps, C. J.
Mullen, B. F., VI 471
Müller, Johann, see Regiomontanus
— John, IV 444
— P., VI 324
— W., VII 717
Mulot, H. M. J., V 555
Mumfort & Moodie, V 670
Munro, J. M. H., VII 1391
Munsing, G., VI 682
Münster, Sebastian, III 540
Munsterberg, H., VI 85
Munters, C., VII 1146
Muntz, A., VII 1387
— G. F., IV 130; V 607
Murdock, William, engineer
    coal-gas, IV 165, 262–3, 265, 266–8
    crankshaft invention, IV 185, 186
    gas-lighting, IV 246, 252, 261, 262–3, 265–6

    oscillating engine, V 146
    rust cement, invention, V 476
    slide-valve invention, IV 186
Murray, James, V 254
— Lord George, IV 646
— Matthew, IV 186, 292, 433; V 322
Mushet, Robert Forester, V 55, 65, 70, 640
Mushin, W. W., VII 1348
Muspratt, James (1793–1886), manufacturing
        chemist, IV 239
    ammonia-soda process, V 241–2
    Leblanc soda processes, IV 239, 241; V 236,
        238, 239, 802
    sulphuric acid process, V 245
— James Sheridan (1821–71), V 241–2, 245
Musschenbroek, Pieter van, III 59; IV 479; V
        736
Mussolini, B., VI 12, 17, 20
Muybridge, E., V 737, 739–40
Muys, Cornelis Dircksz, IV 156, 638–40
Mycerinus, I 668
Myddleton, Sir Hugh, IV 492
Mylius, J. D., II 740
Mylne, Robert, IV 453–5

Nabonassar, III 564–6
Nabonidus, I 282
Nabopalassar, I 256
Napier, Sir Charles, IV 587
— David, V 145, 696–7
— James, IV 128
— John, Baron of Merchiston, III 58; VII
        1151–3
— Robert & Sons, IV 587
Napir-aṣu, I 631
Napoleon Bonaparte, Emperor, VI 1
    afforestation, IV 32
    canals and waterways, IV 555, 556
    electric battery, IV 226–7
    farming, IV 27–8
    roads, IV 526, 529, 536
    textile machinery, IV 292
    visual telegraphy, IV 647
— III, V 91, 248, 809; VII 1400
Narmer, I 754–5, table F liv
Nash, Isaac, III 116
— John, IV 447, 474; V 837
Nāṣir al-Dīn al-Ṭusī, III 584
Nasmyth, James, engineer
    biographical notes, IV 418; V 801, 838
    industrial change, discussion, V 801, 804
    inverted steam engine, V 133, 134
    machine tools, IV 428, 429–31, 435
    plane surfaces, importance, IV 424–5
    safety machinery, V 832
    slot-drilling machine, V 643
    sluice-valve invention, IV 501

small arms industry, IV 440; V 639
steam hammer, IV 116, 430–1, plates 24 A, B;
  V 610, 636
valve design, hydraulic, V 536
Nassau, Prince of, IV 260
Natta, G., VI 559, 560
Naugle, H. M., VI 496
Navier, C. L. M., IV 444, 480; V 491, 496,
  502 n.
Nearchos, I 371
Neb-Hepet-Re, I 435
Nebuchadrezzar, Assyrian king
  Babylon, sketch-plan, III 270
  bridge construction, I 715
  cast-bronze fittings, I 634
  chronology, I table F
  defences, Babylon, I 555
  'hanging gardens', I 551
  mastic, building, I 256
  script discovery, I 766
  wine imports, I 282
Neckam, Alexander, II 124, 532; III 523
Neddermayer, S. H., VI 274
Needham, Joseph, historian
  China, historical dates, II 768–71
  iron ore, smelting process, III 73–4
  rockets, gunpowder, VII 859
  stirrups, use in China, II 556 n. 2
Nef, J. U., IV 233
Nehou, Lucas de, III 238
Neilson, James Beaumont, engineer, V 804
  gas jets, lighting system, IV 272
  Glasgow Gas Workers Institution, V 777
  iron, hot-blast furnace, IV 109–10, 253; V 67
Nelson, Horatio, IV 575, 582–3
— J., VI 649
Neri, A., writer, III 682; IV 234
  flux, lead constituent, V 672
  glass-making, III 217–19, 221, 229, 681
  speculum-metal, III 41
Nernst, W., VI 522, 526
Nervi, P. L., VII 935, 951
Nesfield, V., VII 1377
Neugebauer, O., III 564, 603
Neuman, J. von, electrical engineer
  computers, VII 1190, 1191–3, 1194
  'parallel' computer design, VII 1196–7, 1198
  programming techniques, VII 1199
Neumann, Caspar, III 692
— Paul, V 495, 502
Neuville, IV 461
Neville, Francis H., V 607
New York Kerosene Oil Company, V 105
Newall, J. W., VII 1053
Newberry, William, IV 437
Newcastle & District Electric Lighting Company,
  V 139

Newcomen, Thomas, engineer
  atmospheric engine, IV 79; V 141
  chronology, steam-power, IV 165, 672–3;
    VI 52
  cotton-mill engines, IV 283
  engines, Cornish mines, IV 78–9; V 124
  pumping engine, IV 171, 173–6, 178–82, 186;
    V 124
  pumps, beam use, VI 398
  water-wheel v steam-power, IV 199
Newell, R. E., VI 241
Newnes, V 835
Newton (c.1856), IV 323
— Sir Isaac (1642–1727), mathematician
  atomic concept, physics, IV 223
  ballistics, III 375
  dynamics, III 344
  Earth, shape, III 553; V 441
  frictional machine, IV 648
  'fusible metal', IV 120
  hydrodynamical theory, V 407
  instrument mechanics, III 642, 644, 645
  lenses, III 232, 233, 234; IV 358
  mechanics, third law, VII 859
  photographic arts, V 735, 738
  quadrant principle, III 555
  ships, V 353, 387
  speculum-metal, III 41, 634
  telescopes, III 41, 551 n., 632, 633–4
  vision, persistence, V 735
Niagara Falls Paper Company, V 533
Niagara Power Company, V 533
Nicholas, I., IV 656
Nicholas of Cusa, III 231, 533, 581
Nicholaus Germanus (Donis), Dominus, III
  531–2
Nicolle, Eugene D., V 47
Nicholson, Edward C., V 272
— Peter, IV 472
— William, IV 227, 653; V 695, 698
Nicot, Jean, III 8, 10
Niépce, Joseph, Nicéphore, V 702, 719–20, 725,
  plates 42 A, 43
Nightingale, Florence, V 824
Nipkow, P., VII 1258
Niquet, Antoine, III 467, 468 n.
Nishizawa, H., VII 813
Nixon, IV 93
Nobel, Alfred, chemist, V 821
  explosives, V 285, 292–3, 295
  petroleum, refining stills, V 114
  pipe-line pioneer, V 122
— brothers, VI 408
Nobile, U., VII 797
Noble, IV 297
Noël, Dom, IV 398
Nolf, A. L., V 251

Nollet, F., V 181, 183
— Jean Antoine, IV 649; V 207
Norfolk Estuary Company, V 545
Normand, Benjamin, V 149
Norris, Richard Hill, V 728
North, Simeon, IV 438
North American Kerosene Gas Light Company
  (*later* New York Kerosene Oil Company), V
  105
Northrop, E., VI 489
— J. H., V 584–7; VI 661
— J. K., VI 32; VII 810
Norton, C. H., VII 1035, 1041–4, 1048, 1050
— Thomas, II 741, 743–5, 747
— W. P., VII 1036
Norway, N. S., VII 1064
Norwood, Richard, III 544, 550
Nostradamus, *see* Michel de Nostradame
Nouaille, IV 312
Novate, Bertola da, III 445–6
Nowotry, W., VII 813
Numa, III 575
Nuñez, Pedro, III 548
Nūr al-Dīn, III 596
Nyquist, H., VII 1248

Oakes, T., IV 112
Oberg, J. E., VII 865
Oberth, H., VII 859, 860
Octavian, *see* Augustus
Odier, IV 650
Odling, W., VI 141
Oeglin, Erhard, of Augsburg, III 406
Oemichen, E., VII 822
Oersted, Hans Christian, physicist
 aluminium, isolation, V 72, 90, 248
 electrolytic preparations, V 73
 electromagnetism, discovery, IV 653, 654; V
  178
Ogilby, John, IV 525
Ohain, H. von, VII 819, 1019
Ohdner, W. T., VII 1164
Olaf Geirstad-Alv, II 580
Olaf Trygvisson, II 583
Olaus Magnus, IV 56–7, 154
Oldenbarnevelt, Johann Van, III 302
Olds, R. C., VII 714
Olsen, O. N., IV 614
Olympiodorus, II 733
Omar I, II 615
Onions, Peter, IV 106
Oordt, G. von, VI 524
Opel, F. von, VII 821, 859
Opnaim, Mayer, III 229
Oppenheimer, J. R., VI 271–2, 277; VII 858
Oppert, J., I 779
Orford Copper Company, V 88

Orleans, Duke of, IV 239
Orme, Philibert de l', III 253–4, 266; V 469
Ormsby, M. T. M., VII 1385
Orosius, III 518
Ortell, Joachim, III 317
Osorkon II, I 537
Ostromislensky, I., VI 556
Ostwald, W., VI 527
Otis, Elisha Graves, V 478, 498
Otis Company, V 664
Otto, N. A., V 120, 158–9, 160
Otto the Great, II 455
— & Langen, V 159
Otzen, R., VII 872
Oughtred, William, III 631
Ountash-gal, I 631
Outram, IV 572
Ovid, III 575
Oviedo y Valdes, G. F. de, V 102
Owen, Robert, V 830
Owens, M. J., V 675–6; VI 580, 582
Oxford Electrical Company, V 230
Oxland, IV 124

Paabo, M., VI 665
Pace, Girolamo di, III 310
Pacinotti, Antonio, V 188
Pacioli, Luca, III 250
Packe, Christopher, IV 611–12, 614, 621
Paeonius, III 603
Paine, Thomas, IV 457
Paixhans, H. J., IV 588
Palissy, Bernard, potter and writer
 ceramic industry, France, III 678–9; IV 329
 plant-nutrition, IV 37
 experimental method, IV 234
 town-planning, III 290
Palladio, Andrea, architect
 bridge structures, III 417, 426, 427, 429
 building style, IV 469–71
 road-plan, IV 520–1
Palladius, II 97, 601
Palmer, William, III 458
Panciroli, Guido, III 677
Panhard & Levassor, V 429–30
Pantheus, J. A., III 27, 29, 40–1, 47–8
Papanicolaou, G. N., VII 1356
Papin, Denis, French physicist
 digesters ('pressure-cookers'), V 39 n.
 gunpowder engine, experiments, V 157
 paddle-steamer proposal, V 141
 steam inventions, III 687; IV 165, 171, 173
 steam-power, IV 171, 173, 672; V 124
 vacuum pump, II 634; III 637
Pappus, III 586, 590–2
Pappus of Alexandria, II 603, 636, 645

Paracelsus (T. B. von Hohenheim), IV 665
  alcohol, I 293
  drugs, medical, V 318 n. 1
  metals, III 29, 30 *and* n. 1, 690
  mineral remedies, II 372
  purity, substances, II 744–5; V 318 n. 1
  sulphuric acid, IV 242
Paré, Ambroise, III 374
Parent, IV 405
Paris (*c*.1750), IV 398
— J. A. (*fl* 1826), V 735, 751
Parker, C. W., V 647
— James, IV 448
— John, VII 1038
— R. A., III 561–5
— W., VI 49
Parkes, Alexander, inventor
  celluloid, V 729, 747 n. 1
  furnaces, V 80–1
  nitrocellulose lacquer, VI 600
  'Parkesine' plastic, VI 552
  rubber, V 768
Parma, Duke of, III 315
Parmelee, J. M., VII 1168
Parnell, Henry, IV 543
Parry, George, IV 113; VI 475
— Sir William Edward, V 39, plate 4 B
Parsons, Sir Charles A., engineer
  gear pump, VII 987
  industrial change, discussion, V 822
  marine turbines, V 150–2; VII 1025, 1027,
    1033
  steam-turbines, V 138–9
  turbines, general, VII 1025, 1028, 1030, 1069
  Turbinia, demonstration, V 150–2, 156; VII
    1033
Parsons Marine Steam Turbine Company, V 152
Pascal, Blaise, philosopher, mathematician
  barometric pressure experiments, III 543: IV
    169
  calculating-machine, III 627; VII 1151, 1159–
    60, 1161
  hydrostatic law, V 535
  vacuum, analysis, III 332
Pasley, Sir Charles, IV 448
Passemant, IV 398
Pasteur, Louis, chemist and bacteriologist
  agriculture, V 17
  bacteriology, V 315, 561, 825
  canning process, V 42
  fermentation process, V 302, 303
  food preservation, VII 1401
  sewage treatment, VII 1387
  silkworms, diseases, IV 309, 319–20
  sterilizers, discussion, VII 1355
Pastor, C., VI 664
Paterson, W., VII 1375

Patron, J. G., VI 558
Patsayev, V. I., VII 857
Patterson, A., V 421
— J. H., VI 84
Pattinson, H. L., IV 137–8
Pattison, M., VI 162
Pattle, St J., VII 813
Paul, Lewis, III 153–4, 162–3; IV 151, 277, 280
— R. W., V 746
Paul III, Pope, III 288
— IV, Pope, III 288
— V, Pope, III 289
Paulus Alemannus, III 657
Pausanias, II 6, 25, 30, 498
Paxton, Sir Joseph, IV 476; V 837
Payen, of Paris, IV 143
Peacock, James, IV 501
Peal, Samuel, V 756
Pearsall, H. D., V 527
Pearson, K., VII 1368
Peckston, T. S., IV 270
Peierls, R. E., VI 226–7, 232, 235, 269
Pekelharing, C. A., VII 301
Péligot, E. M., VI 257
Pellatt (Green &), IV 361
Pelletier (*c*.1797), IV 253
— P. J. (1788–1842), V 318
Pellew, Edward (Viscount Exmouth), IV 582, 585
Pelling, III 5
Pelouze, T. J., IV 376; V 106, 114, 284
Peltier, Martin, IV 638, 641
Pelton, Lester Allen, V 532–3; VI 206–7
Pénaud, Alphonse, V 397, 404 n., 408, 409, 410
Peneau, A., VII 808
Penn, J., & Son of Greenwich, IV 589
— John, V 146, 148, 154
— William, V 829
Pennant, Thomas, IV 580
Penney, W. G., VI 281–2
Pennsylvania Rock Oil Company, V 109
Pepi I, Egyptian king
  chronology, I table F
  copper statues, I 639–41, 642
  sheet metal-work, I 637
Pepys, Samuel, diarist
  beverages, III 5, 6
  guns, III 358
  leather, gilding-process, II 174
  London, Great Fire, III 297
  multiplication, difficulties, VII 1151
  ships, hull design, III 488, 490
Perabsen, I 283
Percy, Lord E., VI 152, 167
— Henry, II 749
— John, IV 135; V 617 n. 2, 633
Peregrinus, Peter, of Maricourt, III 524
Péreire, Isaac, V 820

Pericles, III 272
Perier, J. C., V 142
Perin, IV 437
Perkin, A. G., VI 673
— Sir William Henry, chemist
    alizarin process, V 276–7
    aniline dyes, V 235, 270–2, 274
    biographical notes, V 790
    coumarin, synthesis, V 317
    dyestuffs industry, V 281, 300, 795
    mauveine, discovery, VI 117
— Thomas Dix, V 271
Perkins, Jacob, V 45–6, 133
Perrault, Claude, III 58, 256; IV 474
Perret, J. J., VII 1317
Perrin, M. W., VI 558
Perritt, R., VII 1353
Perronet, Jean Rodolphe, engineer
    arch-ring principle, III 436
    bridges, IV 451, 452, 453–4
    bridges, pier thickness, III 428–9
    training-school, IV 442, 444, 526
Perrot, Bernard, of Orleans, III 225
— Humphrey, of Bristol, III 226
Perry (c.1791), IV 580
— J. (fl 1880), VII 1257
— Captain John (c.1715), III 422
Perthes, Justus, IV 622
Peruzzi, B., III 288
Pescara, R. de Pateras, VII 822, 991
Peter, abbot of Gloucester, II 479–80
— B. H., VII 779
Peter I, the Great, Czar of Russia, III 453;
    IV 553; VII 1322
— IV, king of Aragon, III 544
Petermann, August, IV 624
Petrie, W., V 209
— Sir W. M. Flinders, Egyptologist
    Egyptian hieroglyphic, I 752–3
    glass-making, evidence, II 332, 336, 341
    houses, ancient Egypt, I 305
    iron, ancient Egypt, I 597 n.
    measures, I 779, 783
    pottery kilns, I 411
    stools, wooden, I 687
    tools, Assyrian, II 228–30
Petrucci, of Venice, III 406
Pett, Peter, III 485
— Phineas, III 485
Pettus, Sir John, III 677
Petty, Sir William, III 681, 692–5; IV 248
Pettyt, Thomas, II 536
Petzval, J. M., V 721, 722
Peugeot, Armand, V 429
— Frères, V 171–2
Pfanter, Herman, V 656
Pfister, Albrecht, of Bamberg, III 405

Pheidon, II 485
Phidias, I 661, 681; II 470
Philip II, Macedonian king, II 697, 700
— II Augustus, French king, II 691
Philips, Edouard, IV 415
— Reuben, IV 270
Phillips, Peregrine, V 246–7
Phillips & Lee, IV 265–6, 268; V 832
Philo of Byzantium, mechanician
    machinery inventions, II 633, 634, 708 n.
    war-machines, II 636, 708 n., 709–10,
        712–13
    water-mills, II 614
Philoponus, III 603, 605–6, 608
Phipps, Captain C. J. (Baron Mulgrave), IV 619;
    V 459
Physick, H. V., V 223
Picard, Jean, cartographer
    arc of the meridian, III 544, 553, 554; IV 596;
        V 441
    astronomical fixes, IV 605, 615
    astronomical tables, III 557
    levelling treatise, III 543
    map, France, IV 605
    quadrant, improvement, IV 603
— L., I 633 n.
Piccard, A., VII 1456
Piccolpasso, Cipriano
    kilns, II 297–300
    potter's kick-wheel, II 289, 291
    potter's terms, medieval, II 303 n.
    pottery, N. Italy, II 285
    screw-stoppers, pottery, II 293, 294
Pickel, Johann Georg, IV 261
Pickering, William, V 713
Pielsticker, Carl M., V 120–1
Pieters, Pieter, III 306
— William, III 306
Pietro dei Crescenzi, II 138–9
Piggott, S., I 492, 494
Pignatelli, IV 554
Pilcher, P. S., V 406–7, 411
Pilkington, William, V 242
Pinas, J., II 749
'Pincher Jack', III 113–14
Pinckney, Eliza, V 261
Pinson, K. P., VII 1337
Pischof, Alfred de, V 392, 409; VII 792
Pitcairn, J., VI 574
Pitentino, Alberto, III 442
Pitot, Henry, V 539; VII 841
Pitt-Rivers, A., I 313
Pittsburgh Reduction Company, V 93
Pius IV, Pope, III 288
Pixii, Hippolyte, V 179–80, 191
Planck, M., VII 1092
Planté, R. L. G., V 206

Plantin, Christophe, III 43, 391–2, 394–6
— and Moretus, Antwerp, III 402, 404
Plassey, Baron, *see* Clive, Sir Robert
Plateau, J. A. F., V 735, 736, 739
Platière, Roland de la, IV 304
Plato, Greek philosopher
    Athens, quarrying effect, II 27
    Earth, spherical form, III 504
    instruments, criticism, II 632; IV 666
    lathe, wood-turning, II 233
Platt, Sir Hugh, III 80, 135, 679
Platt Brothers, V 610, 801
— and Collier, IV 296–7
Plattes, Gabriel, III 63, 681
Plattner, C. F., IV 142
Platz, R., VII 794
Plautus, II 134
Playfair, Lyon, V 782, 785, 789
Pliny the Elder, Roman writer
    agricultural implements, II 86, 88, 96, 97
    alum, Egyptian, I 262
    antimony, chemistry, I 588
    bitumen, softening, I 254
    brewing, II 140–1
    chemistry, II 350, 355; III 676
    cloth, preparation, II 215–16, 218
    cotton from India, I 449; II 199
    fibres, II 192–5, 197, 199
    flour and bread, II 104, 105
    furniture, wooden, II 226
    glass-making, II 323, 328
    leather, preparation, II 151 *and* n.
    Low Countries, drainage, II 681
    maps, III 504
    metallurgy, II 42, 46; III 40
    military engines, II 700
    mining, II 3, 8, 19
    naphtha (crude oil), I 251
    natron production, I 260
    Nile valley, silt deposits, I 539
    parchment, II 189
    pigments, I 238, 240, 247
    presses, types, II 113–14, 116–18
    quarrying, II 27, 29
    radish cultivation, I 356
    Rome, drainage system, II 531
    salt production, I 257
    stones, colouring art, I 243
    threshing, wheat-grains, II 106
    timbers, II 226, 233
    tin ore, I 589
    vegetable oils, I 288–9
    viticulture, II 134–5
    wind-rose (direction system), III 517
Plot, Robert, II 35; III 34
Plumier, C., III 336
Plunkett, R. J., VI 562

Plutarch, philosopher and biographer
    date, foundation of Rome, III 578
    machines of Archimedes, II 604, 633; IV 665–6
    Roman roads, II 505, 507
    Syracuse, siege, II 714
Pochabradsky, B., VII 1033
Pocock, G., V 398
Pococke, R., IV 374
Poisson, S. D., V 495
Poitiers, Diane de, IV 522
Polhelm, Christopher, inventor, III 344 n.
    canal improvement, III 455
    metallurgy and mining, III 31
    metal-working, methods, III 342–4; IV 106–7;
        V 63
    water-driven factory, III 19
Pollard, S., VI 77
Pollux, II 199
Polo, Marco, *see* Marco Polo
Polybius, II 5
Polzanius, F. E., VI 523
Poncelet, J. V., IV 203; V 493, 529, 546
Ponte, Antonio da, III 424
Poole, B. W., VI 677, 688
Popkin, M. E., VI 677, 688
Popov, A. S., VII 1233
Porro, Ignazio, IV 602
— J., V 446
Porsche, F., VII 719
Porta, Giambattista della, writer
    assay method, III 68
    foils, colour, III 49
    furnace pump, III 53
    iron and steel, III 33, 34, 35–6
    lenses reference, III 231
    microscope, III 232
    speculum-metal, III 40
— Giovanni Battista della, IV 168; V 716
Portal, Lord, VI 77
Porte, J. C., VII 796
Porter (Aveling &), IV 538
— C. T., V 131
Portevin, A. M., VI 493
Porthouse, A. (Kendrew, J. and), IV 291–2
Potter, C. V., V 75
— John, IV 177
Poulsen, A., V 666
— V., VII 1235, 1264, 1315
Power, Henry, III 57
Powers, J., VII 1181–2, 1200
Prandtl, L., VII 793, 987, 1019
Pratt, J. H., V 443–4
Preller, IV 297
Prescott, Sir W., VII 1372
Priestley (*c.*1889), IV 323
— Joseph (1733–1804), chemist
    dietetic science, III 25

Priestley, Joseph (1733–1804), chemist (*cont.*)
  dye industry, discussion, V 267 *and* n.
  electrochemistry, research, V 177
  gases, isolation, IV 216, 228, 247
  laboratory, II 750, plate 43 B
  'Lunar Society' membership, IV 673
  oxygen, discovery, IV 38, 218, 220, 221
  steel, carbon, content, IV 108
Priscian (*fl* 350–80), II 355
— (sixth century), III 67
Pritchard, T. F., IV 455
Proclus, III 586, 588, 606
Procopius, Byzantine historian
  cataphracts (mailed cavalry), II 706–7
  fire-pots, military, II 375
  paving, Appian Way, II 28
  shipbuilding, II 569
  water-mills, II 601, 607–8
Profatius, III 599
Progin, Xavier, V 689
Prony, Gaspard François, IV 444, 480
Pryce, W., IV 78
Ptolemy (Claudius Ptolamaeus of Alexandria)
  anaphoric clock, III 604–5, 616
  armillary sphere, III 591–2, 612–13
  astrolabe, III 603, 605–6
  astronomy, Babylonian, I 802
  calendar, III 564, 565, 566, 567
  cartography, principles, III 521, 522, 530–9
  engravings, maps, III 409
  geography, III 502–3
  instruments, III 583, 584, 586–92, 599
  optics, III 229–30
  planetary theory, III 615–16
  world map, III 508–10, 511; IV 596
— Euergetes, III 564
Puckle, O. S., VII 1261
Pullars of Perth, V 271–2
Pullein, S., IV 310
Pullman, George M., V 342
Pulsometer Engineering Company, V 564
Pupin, M. I., VII 1114, 1236, 1242
Purbach, Georg, III 608
Pyrrhus, II 670
Pythagoras, III 504, 536, 626
Pytheas, II 569

Qaisar ibn Abu'l-Qāsim, II 614
Quare, Daniel, III 669–70
Quartley, IV 21
Quiglay, W., V 583
Quinson, IV 323
Quintus Candidius Benignus, II 599

Race, J., VII 1378
Radcliffe, William, IV 299–300, 302
Radegund, II 139, 452, 765

Rajehman, J., VII 1196
Raky, Anton, V 111
Raleigh, Sir Walter, explorer
  Ark Royal, description, III 481
  asphalt lake, visit, V 102
  experiments, IV 667
  navigation, discussion, III 551
  potatoes, discussion, III 2
  tobacco, III 9 n. 2
Ramage, Adam, V 691
Ramelli, Agostino, mechanician
  coffee-dam, III 421, 422
  cross-bows (*ballistae*), II 723–4
  milling machines, III 16, 18
  pumping machines, III 331, 332; IV 638
  screws, III 337
  ship, windmill proposal, V 142
  tower-mills, II 626
  windmills, general, III 89–91, 96, 329–30
Rameses II, Egyptian king
  chronology, I table F
  pectoral, inlaid-gold, I 659
  roads, I 714
  water-supply, Wadi Alaki, I 526
  wines, I 284
— III, I 286, 441, 543
Ramond, L., IV 605
Ramsay, Sir W., VI 147
Ramsbottom, John, V 325, 335–6
— and Holt, of Todmorden, IV 302
Ramsden, Jesse, instrument-maker
  balances, IV 404–5
  chain measure, V 447
  dilatometer, IV 397
  dividing-engine, III 555, 641–3; IV 390, 392–5,
    plate 1
  screw-cutting lathe, IV 387–8
  sextant, improvements, V 457
  telescope parts, IV 399
  theodolites, IV 402, 403; V 444
Randall, J. T., VII 1107
Randolph, Charles, V 422
Ranger, William, V 487
Ranke, L. von, V 788
Rankine, W. J. M., engineer and physicist
  arches, design theory, IV 486; V 502
  buoyancy, ships, V 354–5
  dams, design principles, V 556–7; VII 1368
  publications, IV 445; V 491, 492, 781
  sewers, strength, VII 1387
  stress and strain, V 494, 501
  thermodynamics, confirmation, IV 163
  waves, fluids, V 546
  work ratings, prime movers, IV 159
Ransome, E. L., V 490–1
— Frederick, V 484–5
— Robert, IV 3

Ransome and May, IV 11
— & Sims, V 490
Raper, H. R., VII 1328
Rappaport, M. B., VII 1348
Ratdolt, Erhard, III 405
Rateau, C. E. A., V 140; VII 1025, 1028
Rathbone, Aaron, III 541–3
Rathenau, Emil, V 837
Rattier & Guibal, V 765
Raud the Strong, II 583
Ravenscroft, George, III 219, 221; V 672
Rawcliffe, G. H., VII 1087
Rayleigh, Lord, VII 1239
Raymond, A. E., VII 825
Read, J., II 747
Reade, J. B., V 723
Reader, W. J., VI 59–61, 601
Readman, J. B., V 253–4
Reason, R. E., VII 1050
Réaumur, R. A. Ferchault de, chemist
    digestion process, stomach, III 25
    furnaces, IV 101
    gold-leaf, III 49 n.
    metals, structure, III 57
    porcelaine, experiments, IV 339
    steel-making process, III 28, 29, 33
    tinplate, France, III 689
Recklinghausen, H. von, VII 1345
Reddaway, W. B., VI 72
Reden, von, IV 235
Redpath Brown, V 477
Reece, J., VI 683
Reece Corporation, V 591
Reech, Frédéric, V 387, 388
Rees, Abraham, IV 427
Règemortes, Jean Baptiste de, III 464
— Noël de, III 464
Regiomontanus (Johann Müller), astronomer
    Alphonsine tables, III 535, 547, 617
    cartography, III 535
    instruments, scientific, III 585, 591, 598
    maps, III 532
Regnault, H. V., IV 163; VI 556
Reichart, Christian, IV 40
Reichenbach, Georg, IV 394, 399, 402
— H. M., V 729, 730
Reifenberg, A., I 633 n.
Reinecker, J. E. R., VII 1039
Reis, J. Philipp, V 226
Reisch, Gregor, III 538
Rekhmire, I 453, 544, 627, 690 n.
Rembrandt, III 407
Remington Company, V 689
Remsen, Ira, V 318
Renard, Charles (1847–1905), V 402, 404; VII 822
— Colonel (fl 1880), VII 1052

Renault, Louis, V 433–4
René II, Duke, III 538
Rennell, James, IV 610, 619; V 440
Rennequin, of Liége, IV 155
Rennie, George, IV 484; V 147–8
— John, civil engineer, IV 444
    bridges, IV 455, 456, 458
    canals, IV 571–2
    cast-iron, strength, IV 484
    harbour works, IV 466–8; V 539
    iron-mill, design, IV 105
    marine engines, V 147–8
    saw-mills, design, IV 202
    steam-dredger, IV 641
    watermills, design, IV 209
Repsold, J. G., IV 390, 399
Reuleaux, Franz, III 324 and n.; IV 150
Revenet, Simon-François, IV 354
Revière, J. A., VII 1358
Reynaud, F., V 743, 745
Reynolds, Osborne, V 118, 526, 547–50
Rhazes, see Abū Bakr Muḥammad ibn Zakariyyā al-Rāzī
Rhee, Samuel, IV 388
Rheita, Schyrlaeus de, III 633
Rhijnsburch, Jan van, III 442
Rhoissy, Marquis de, III 461–2
Rhodes of Bridgnorth, IV 244
Ricardo, H. R., VII 989, 1008, 1014
Riccobaldi of Ferrara, III 382
Rice, R., VII 814
Richard I, English king, II 70
— II, English king, II 727
— of Wallingford, III 585, 593, 617
Richards (c.1821), IV 319
— C. F. (fl 1930), VI 295
— Dickinson (fl 1937), VII 1357
Richardson, Clifford, IV 540
— O. W., VII 1093
Richelieu, Armand du Plessis de, III 291
Richer, IV 394
Riches, W., VI 681
Richet, Charles, V 404, 409; VII 821
Richmond, Duke of, IV 607
Richter, Gisela M. A., I 409
Richthofen, M. von, VII 795
Rickett, Thomas, V 420–1
Rickman, Thomas, V 470
Rickover, H. G., VI 244, 252–4
Ridley, H. N., V 774
— Mark, III 543
Riedel, K., VII 863
Riefler, S., III 671
Riggenbach, N., V 346
Riley, James, V 88, 373
Rinaldi, Georges, III 453
Ringman, III 538

Rippe, W., VI 541
Riquet, Jean-Mathias, III 466
— Pierre-Paul, III 464–8
Ritter, J. W., IV 227; V 205, 718
— Wilhelm, V 493, 501
Riva-Rocci, S., VII 1344
Rivaz, de, IV 410
Rivers, Lieutenant, IV 577
Robais, Van, see Van Robais
Robbia, Luca della, IV 334
Robbins & Lawrence Company, IV 435–6,
  439–40; V 645–6, 653
Robert (c.1783), IV 255
— Nicholas-Louis (1761–1825), III 415–16; VI
  614
Robert of Chester (12th cent.), II 738
Roberts, A., VII 1057
— Dale & Company, V 274
— Hill & Company, IV 300
— Richard, machine-tool maker, IV 418
  loom-making partnership, IV 300–2, 303
  machine-tools, IV 423, 428–9, 433–4, 435
  mule, steam driven, IV 288–90
  planing machine, V 650
Roberts-Austen, Sir William Chandley, V 607
Robertson, John, V 458
— J. Drummond, III 633 n. 2
Robins, Benjamin, III 360
Robinson (c.1874), of Papplewick, IV 154
— Thomas (c.1732), V 240
— — (1814–1901), III 458
Robinson & Russell, V 22
Robison, John (1739–1805), IV 444–5, 484
— Sir John (1778–1843), V 350, 351
Rockefeller, John D., V 122; VI 189
Rocque, John, IV 614–15
Roderick, Bishop, III 43
Rodney, George Brydges, IV 580
Roebling, John August, IV 459; V 508–9
Roebuck, John, industrialist
  sulphuric acid, manufacture, IV 231, 237,
    243–4; V 815
  Watt's steam-engine, IV 182–3
Roemer (Römer), Ole, III 345, 543
Roger, Bishop of Sarum, II 428
Rogers, Samuel, IV 114
— Woodes, III 499
Rohm, O., VI 558
Rohrbach, A., VII 810
Rolfe, John, III 9
Rollins, W. H., VII 1328
Römer, Ole, see Roemer
Romer, P. H., VII 1330
Romilly, Sir Samuel, V 831
Romulus, III 575
Ronalds, Sir Francis, IV 651–2, 658
Rondelet, Jean, IV 480, 482

Röntgen, W. K. von, VI 223; VII 1092, 1299–
  300, 1327
Roosevelt, F. D., American president
  atomic energy, VI 252
  New Deal policies, VI 15
  nuclear weapons, VI 227, 271, 279, 281
  Quebec Agreement, VI 227
  research, American, VI 121
— T., VI 4, 6
Root, Elisha K., IV 439; V 639, 646
Roozeboom, H. W. B., VI 625
Roscoe, Sir Henry Enfield, V 796
Rosen, I. K., VI 668
Rosenthal, E., I 409
Rosetti, G. V., III 679; IV 233
Rosing, B., VII 1258
Ross, Sir James Clark, IV 619; VII 1446–7
Rosse, Lord, IV 399
Rossi, I., V 497
Rossman, R. H., VI 664
Roth, G., IV 232
— L., VII 1288
Rothery, W. H., VI 460
Rouelle, IV 404
Round, H. J., VII 1097, 1101, 1244
Rovenzon, John, III 79, 80, 83
Rowbotham, E. S., VII 1334
Rowntree, Joseph, V 830
Rowson, Benjamin, of Bradford, IV 244
Roxbury Rubber Company, V 763
Roy, P. Le, see Leroy
— William, cartographer
  measuring instruments, IV 397, 601, 604; V
    447
  triangulation, S. E. England, III 554; IV 607–8
Rozier, J. F. Pilâtre de, V 395, 397, 398
Rubel, I., VII 1269
Ruchonnet, E., VII 808
Rudborne, Thomas, II 428
Rudenberg, G. R., VII 1350
Rudge, John A. R., V 743
Rudolph, Paul, V 722
— II, Emperor, III 621
Rue, Thomas de la, V 816
— Warren de la, V 40, 204, 208, 209
Ruff, O., VI 562
Ruggles, Stephen, V 709–10
Rühmer, E., VII 1310
Rumford, Count (Benjamin Thompson), IV
  674–6; V 830–1; VII 1126–7
Rumsey, James, V 142, 152
Rupert, Prince, III 45, 367, 407; V 142
Rupert-Elsner, W., V 660
Rushton, J. L., VI 655
Russell, John Scott, engineer
  coal consumption, ships, V 377
  Institution of Naval Architects, V 372

ship, armoured, IV 588
steam-engine, improvement, IV 196
structures, iron ships, V 357, 359–65
waves, experiments, V 546
— Richard, II 739
Rutherford, Ernest (Lord), V 227; VI 222; VII 1235, 1469
Ruysch, J., III 535
Rziha, Franz von, V 502

Sablonnière, de la, IV 539
Saby, J. S., VII 1119
Sachs, Edwin O., V 482
— Hans, III 40, 43
Sacrobosco, III 547, 613
Sadler (late 18th century), IV 581
— John (1720–89), IV 354
Sahure, I 482, table F
St John Hornby, C. H., V 694
St Louis, see Louis IX
St Luke, see Luke, St
Sainte-Claire Deville, Henri Etienne, chemist
aluminium, refining process, V 90–2, 248–9
aluminium-bronzes, V 608
metals, aluminothermic reaction, V 97
petroleum oil, research, V 106
sodium, refining process, V 248, 249
St-Fart, IV 475–6
Saint-Fond, Faujas de, IV 260–1
Saint-Rémy, P. Surirey de, III 34, 38, 361, 369
Saint-Victor, Abel Niepce de, V 725
Salman, P., VII 1052
Salmon, Robert, IV 5, 8
Salt, R., VII 1335
Salter, W. E. G., VI 67–74
Salvá, Don Francisco, of Barcelona, IV 651, 653–4
Samuelson, Bernhard, V 789, 794
Sanctorius, III 24
Sandrart, III 222, 224
Sandys, Carne & Vivian, V 129–30
— William, III 458
Sang, E., VII 1153
Sangallo, III 288
Sanson, IV 606
Santos-Dumont, Alberto, aviator
aeroplane flight, Europe, V 392–3, 441; VII 791
airships, V 402
Sanudo, Marino, II 725
Sardanapalus, see Ashur-bani-pal
Sargon I the Great, Assyrian king
bitumen production, I 253
bronze head, I 627
chronology, I table F
economic imperialism, I 56–7
silver and lead, I 583

timber resources, I 371
— II, Assyrian king
building materials, I 469
chronology, I table F
brass ('white bronze'), II 54
glass vessel-making, II 321, 336
horse-riding practice, II 495
metal spoils, I 564, 597
parks and gardens, I 554
weight standards, I 780
wines, I 282
yellow pigments, I 239
Sarti, Vittorio, V 404 n.
Sassenay, de, IV 539
Sauerbronn, Drais von, see Drais . . .
Saulnier, Jules, V 474–5, 837
Saunders, C. E., VI 327
Saussure, H. B., de, IV 605
— Nicolas Théodore de, IV 38; V 106
Savage, W., V 705
Savalle, Pierre A. D., V 114, 307
Savery, Roelandt, IV 639
— Thomas, engineer
chronology, steam power, IV 165, 166
'fire' engine, IV 79, 170
Newcomen-Savery engine, IV 174–5, 177, 178, 672–3
steam-pump, V 124, 141; IV 171–5
water-supply engine, III 687; IV 493
Savigny, J. H., VI 1317
Savorgnan, Giulio, III 286
Savot, L., III 37, 57
Sawers, D., VI 34
Sawitch, Alex, V 443
Saxby, John, V 330
Saxe, Maréchal de, V 142
Saxony, Elector of (d.1763), IV 27
—— (c.1799), IV 261
Sayce, B. J., V 728
Sayen, H. L., VII 1330
Scaliger, Julius Caesar, III 581
Scamozzi, Vincenzo, III 286; IV 520
Scanlon, Jessie, III 186
Scarlett, IV 398
Scaurus, Marcus Aemilius, II 679
Schaefer, K. L., VII 1349
Schaeffer, C., I 762–3
Schank, Captain John, IV 579
Scheele, Carl Wilhelm, chemist
chlorine discovery, IV 247
fats and oils, IV 253
green pigment, IV 256
manganese compounds, IV 147
organic chemicals, discoveries, V 268
'phosphoric taper' invention, IV 254
phosporus, production, V 252
silver salts, and light, V 718

Schiele, Christian, V 656
Scheiner, Christopher, III 632
— J., VII 1283
Scherer, F. M., VI 38
Scheutz, E., VII 1174
— G., VII 1174
Schilling, Baron, IV 654–6, 658
Schindler, C. C., III 63
Schlak, P., VI 650
Schlesinger, G., VII 1051
Schlick, O., VII 986
Schlickeysen, Carl, V 665, 669
Schliemann, H., I 660
Schlösing, T., VII 1387
Schlumberger, C., VI 394
Schlüter, III 28
Schmerber Frères, V 670
Schmidt, O. L., VII 1331
— Wilhelm, V 340–1; VII 762
Schmidts, Michael, III 454
Schmoockler, J., VI 49, 72
Schmued, E., VII 814
Schneider, V 66
Schnitzer, Abraham, V 102
Schoch, C., III 563
Schoeffer, Peter, III 377, 383–4, 386, 393
Schönbein, C. F., V 284, 290, 292; VI 551–2
Schönher, O., VI 522
Schoonbeke, Gilbert van, III 453
Schopper, Hartman, III 336, 391
Schorlemmer, C., V 106
Schott, Caspar, III 637; V 394, 717
— Otto, V 673, 682
— & Sons, V 673
Schottky, W., VII 1101
Schreittmann, C., III 60–1
Schreyer, H., VII 1193
Schroeder, Johann, III 29
Schroetter, Anton, V 253
Schrötter, IV 255
Schubert, Johann Christian, IV 40
Schuckers, V 687
Schule, IV 325
Schulze, Johann Heinrich, V 717–18
Schumpeter, J., VI 49, 65
Schürer, Christoph, III 678
Schutter, J., VII 799
Schutzenberger, P., VI 553
Schwab, Anton von, III 30
— Max, III 341
Schwartz, C. V., VI 489
— Johann Nepomick, IV 19
'Schwarz, Berthold', III 366
— David, V 402; VII 797
Schweigger, J. S. C., IV 655–6
Schwerz, von, IV 41
Scipio Africanus, II 121

'Scorpion King', I 51, 537, table F
Scott, Basil, V 477
— L., VII 1262
— R. W., V 600
— W. D., VI 85
Scultetus, J., VII 1317
Seaborg, G. T., VI 260–2
Sealy (Coade and, of Lambeth), IV 446–7
Seaman, H. J., V 485, 486
Seaward & Capel, V 145
Sécretan, IV 539
Seebeck, T. J., V 178
Seerig, A. W. H., VII 1317
Seger, H., V 660–1
Segner, Johan Andreas von, V 528
Ségolène, Ste, II 139
Seguin, Marc, engineer
    boilers, locomotives, V 137
    helicopter design, V 404 n.
    inventions, discussion, IV 676–7
    ornithopter proposal, V 403 n.
    railways, France, V 322
Segum, L., VII 792, 1011
Seleucus II, II 701
Selfridge, T. E., V 409
Sellers, Coleman, V 737
Semon, W. L., VI 556
Semple, George, IV 449
Senebier, IV 38
Seneca (d. A.D. 65), III 504
— Lucius (55 B.C.–A.D. 39), II 27, 211
Senefelder, C., IV 626
— J. A., V 706
Seneferu, I 283, 482, table F
Senlacq, M., VII 1257
Senmut, I 122, 124, 441, plate 13 B
Sennacherib, Assyrian king
    aqueducts and conduits, I 531, 533–4
    building materials, I 469
    chariots, I 718
    chronology, I table F
    moulds, metal-casting, I 632–3
    park with 'wool-bearing trees', I 554
    reed plantation, I 552
    rope-making, I 452–3
    viticulture, I 282
    water-supply, I 524, 527, 529
Senot, IV 388, 425
Senusret I, I 699
— III, I 49
Seppings, Sir Robert, IV 583–4
Serlio, Sebastiano, III 257, 429
Serpollet, Léon, V 424–5; VII 721
Serres, Olivier de, III 13
Serullas, Georges S., V 315
Servius, II 98
Seti I, I 286, 526, 529

Seusenhofer, Conrad, III 352–3
Severance & Holt, V 108
Severus, Lucius Septimius, II 137, 519, table II lvi
— Sēbōkht, III 521, 606, 608
Sforza, Ludovico, III 286
Shalmaneser III, Assyrian king
  bronze craftsmanship, I 646, 649
  chronology, I table F; II table I lv
  stirrups, horse-riding, II 556 n. 1
Shanks, James, V 239–40
Shannon, C. E., VII 1187, 1248
— Richard, IV 238
Shaphatbaal, I 763
Shapleigh, John, IV 530
Sharp (nineteenth century), IV 273
— James, IV 6, 10
Sharp Brothers, IV 288
Sharpe, IV 654
Shaw, P., VII 872
— Ralph, of Burslem, IV 348–9
— William T., V 737
Shearer, J. S., VII 1330
Sherbourne, Robert, IV 371
Sherman, W. T., V 341
Shiemann, M., VII 732
Shilkhak-in-shushinak, I 632
Shipway, F. E., VII 1337
Shirley, J. W., II 749
— Thomas, IV 258
Shish, III 491
Shockley, W., VII 1119
Sholes, Christopher Latham, V 687, 689
Sholl, IV 319
Shore, I., VII 1384
Short, James, III 233, 237; IV 398
Short brothers, VII 810
Shortt (c.1920), III 671
Shotbolte, John, IV 537
Shub-ad, queen of Ur
  animals, domestication, I 340
  metal craft, I 626, 644
  sledge-chariot vehicle, I 709, 710
Sickels, Frederick E., V 131
Sickle, S. Van, see Van Sickle
Siegen, Ludwig von, III 407
Siemens, Charles William, steel-maker
  arc furnace, V 66
  generator, demonstration, V 185
  open-hearth furnace, V 58–9, 499; VI 462
  pyrometry, V 660
  regenerative furnace, V 680, 681
  steel production, V 70, 108
— E. Werner von, electrical engineer
  armature invention, V 184, 185, 188
  dynamo–electric generator, V 185–6, 187, 196
  electric railway, V 233

  gravity meter, invention, VII 1463
  insulation, cables, V 222–3
  lamps, electric, V 209, 216
  rotor (disc-armature), V 193
  telegraphs, V 219
— Frederick, industrial scientist
  industrial organization, V 807, 809, 810
  open-hearth process, V 57–8, 499, 818
  regenerative furnace, V 680, 681
  steel production, V 70, 108, 804
Siemens & Halske, V 189, 192, 233, 346
Sigismund, IV 529
Sigl, G., V 707
Sikorsky, Igor I., V 404; VII 795, 823–4
Silliman Jr., Benjamin, V 109
Silver & Gay, V 645–6
Simmons, Joseph, V 398
Simms, William, IV 607
Simon, E., VI 557
— H., VI 237, 676
— Thomas, III 341
Simoni, D., VII 796
Simpson, James (1799–1869), IV 502; V 561–3, 565
— James Young (1811–70), V 312 and n.
— Thomas (1710–61), V 353
— — (c.1785), IV 500
Sinan, IV 498
Sinclair, Sir John, III 15, 135
Singer, Isaac M., V 589–90, 819; VI 679
— Sewing-machine Company, V 591
Singleton, T., V 579
Sisson, Jonathan, IV 382, 390, 603–4
— manufacturers, V 137
Siviard, St, III 198
Sixtus IV, Pope, III 288
— V, Pope, III 289, 311–12
Skaife, Thomas, V 727
Skey of Dowles, Worcestershire, IV 244
Skinner, H., V 584
Slaby, A. C. H., VII 1233
Slater, W. A., VII 945
Slichter, S. H., VI 103
Small, James, IV 2
Smalley, J., V 576
Smeaton, John, engineer
  blowing-cylinders, IV 103
  boring-mill, IV 421–2
  bridge, Pont-y-Ty-Pridd, III 436
  cast-iron, use, IV 451
  civil engineering, IV 443–4, 447, 467–9, 629
  dilatometer invention, IV 396
  drainage mills, III 305
  instrument-maker, III 436, 647
  lighthouse, Eddystone Rock, V 463
  Newcomen engine, improvements, V 124
  precision, machinery, IV 162

Smeaton, John, engineer (*cont.*)
  pyrometry, IV 350
  steam-engines, IV 166, 179–81, 186, 421–2
  water-wheels, IV 152–5, 199, 202–9
  windmills, III 100, 102–3, 104; IV 157–9
  wind-pressures, table, V 502
Smedt, E. J. De, IV 540
Smiles, Samuel, IV 672
Smirke, Sir Robert, IV 449
Smith (*c.*1843), IV 654
— A. (*fl* 1850), V 581
— Adam (1723–90), economist
  craft definitions, discussion, II 252
  factory management, discussion, VI 78, 80, 86
  free trade theories, III 87
  patents, discussion, IV 150
— Angus (1818–86), V 559
— E. Baldwin, I 305
— Francis Pettit, V 147
— George (1800–68), V 832
— — Albert (*c.*1909), VII 1306
— — P. (*c.*1912), VII 1288
— H., VII 794
— J. (*fl* 1834), V 573
— James (*c.*1707), of Whitehill, IV 172
— — (*fl* 1811–34), of Deanston, IV 7, 284, 290
— K., VII 806
— R. (*fl* 1919), VII 806
— R. H. (*fl* 1863), V 579
— T. & W., IV 590
— W. M. (*fl* 1890), V 338–9
— William (1769–1839), IV 491, 621
— Willoughby (*fl* 1873), VII 1257
Smollett, Tobias George, IV 507
Smuts, J. C., VII 796
Smyth (*c.*1800), IV 5–6
— Piazzi (1819–1900), V 46
Snell, Willebrord, III 544; IV 606; V 441
Sneller, Z. W., IV 14
Snodgrass, Gabriel, IV 578–9
— Neil, IV 283
Snoek, J. L., VII 1198
Snook, H. C., VII 1331
Snow, C. P., VI 268
— George Washington, V 467
— John, V 312, 561–2; VII 1334
Soane, Sir John, IV 449, 475
Sobrero, A., V 285
Soddy, F., VI 222
Soemmering, S. T. von, IV 653–4, 656
Soiland, A., VII 1332
Solomon, king of Israel
  bronze casts, Jerusalem, I 633
  chronology, I table F
  ivory art, I 664, 675, 676
  metal-work, I 661
Solon, II 1, 105, 129; III 570

Solow, R. M., VI 73
Solvay, Alfred, V 241, 243
— Ernest, V 241, 242–5, 802
Solvay & Cie, V 251
Somerset, Edward (Marquis of Worcester), III 680; IV 170, 672
Somerville, Lord, III 15
Sommeiller, Germain, V 517, 527
Sommelier, IV 71
Soniter, E. H., VI 479
Sophocles, II 58
Sorbière, Samuel, IV 170
Sorby, Henry Clifton, V 281, 607
Sorel, M., V 624
Sorocold, George, IV 451
Sosigenes of Alexandria, III 576–7
Sostratus of Cnidos, II 521, 566
Sothern, John, IV 484
Soubeirau, Eugène, V 313
Soufflot, J. G., IV 474–5, 480–2; V 488
Southey, Robert, IV 559
Southwark Foundry & Machine Company, V 131
— & Vauxhall Water Company, V 128, 129, 524
Southwell, R. V., VII 945
Spalding, James, III 157
Spallanzani, Lazaro, III 25
Spangler, J. M., VII 1131
Sparks, M., VII 1119
Spedding, Carlisle, IV 258–9
— Charles, IV 95
Spence, Graeme, IV 618
— Peter, IV 251
— R., VI 262, 263, 265
Spencer (*c.*1794–1801), second Earl, IV 581
— Charles Green (*fl* 1608), V 400, 406
— Christopher Miner, V 647
— Edward, V 400
— John, V 625
— Thomas, V 563
Spengler, O., VII 163
Sperry, E., VII 842, 843, 986
— L., VII 817
Spert, Thomas, III 487
Spiers, of Ayr, III 111
Spill, D., VI 552
Spinola, Ambrogio, IV 556
Spitteler, A., VI 553
Splitgerber, D. K., IV 376
Spode (eighteenth century), IV 344, 352
— Josiah II (1754–1827), IV 353
Sprague, Frank J., V 200
— H., VII 1348
Sprat, Thomas, III 682–3; IV 668–70
Sprengel, C. K., I 450
— Hermann, V 214, 293, 294 n.
Squire, William Stevens, V 247
— Chapman & Company, V 247

Staite, W. E., V 208–10
Stalhane, O., VI 490
Stalin, J., Russian statesman
  industrialization, Russia, VI 14–15
  mining, development, VI 415–16
  nuclear weapons, VI 280
  second world war, VI 18, 22
Stampfer, S. von, V 736
Stanhope, Charles (third Earl), V 691–2
Stanier, W., VII 766
Stanley, F. A., VII 1049
— William F., V 445
Stanley brothers, V 425
Stapleton, Sir G., VI 311
— H. E., II 737
Star Films, V 750
Starley, James, V 415
Starr, J. W., V 213
Stassano, E., VI 489
Statius, II 500–1, 503
Staudinger, H. P., VI 557
Steen, Jan, II 749
Steers, Thomas, III 455, 458
— William, IV 468
Steiger, O., VII 1166, 1171, 1188
Stein, Sir Aurel, III 202
Steinbeck, M., VII 1332
Steinheil, Carl A. von, V 732
Steinmetz, C. P., V 200
Stell, Joseph, III 168
Stella, Tilemann, III 450
Stelluti, Francesco, III 632
Stephen, Alexander, V 350
Stephenson, George, engineer
  boilers, multi-tubular, V 137, 322
  haulage, steam-powered, IV 84, 165, 189
  plate-girders, study, IV 486
  railways, V 322, 804
  safety lamp, IV 96
Stephenson, Robert, engineer
  boilers, multi-tubular, V 137, 322
  bridges, IV 460–1; V 363, 492, 500–1, 503–6,
    534
  hydraulic appliances, V 534
  railways, V 322, 323, 804
Stevens, John (1749–1838), V 147, 149
— — F. (fl 1907–14), VII 913
— Sir W., VI 489
Stevenson, R. L., IV 549
Stevin, Hendrik, III 304
— Simon, engineer, mathematician
  canal-locks, III 452
  decimal arithmetic, III 61, 542
  drainage-mills, III 303–5, 307
  dredging, IV 630, 632, 641
  fortification theory, III 305, 371
  hydrostatics, studies, III 305, 313; V 535

  mechanics, IV 149
  windmills, IV 157
Stevins, of Scotland, III 25
Stewart, D. Y., IV 500
Stibitz, G. R., VII 1187
Stiefel, Ralph Carl, V 629
Stieler, Adolf, IV 614, 620, 626
Stigler, G. J., VI 38
Stillerman, R., VI 34
Stillman, J. M., III 676
Stoddard, W. O., V 51
Stodola, A., VII 984
Stoeffler, III 536
Stoehrer, E., V 181, 182
Stokes, Sir George, V 546, 673
Stone, E. H., I 494
— Henry D., IV 439–40; V 646
— R., VII 1332
Stout, W. B., VII 730
Stowe, John, II 244
— W. H., VII 1323
Strabo, Greek geographer
  asphalt (bitumen), Babylonia, I 252–3
  cooperage, art, II 136
  dye, madder plant, I 246
  glass-making, Rome, II 323
  globe, terrestrial, III 614
  gold, extraction, I 581
  mastic, I 254, 256
  metal extraction, II 45, 48
  mining, Laurion, II 3
  natron, I 260
  quarries, Carrara district, II 28
  road drainage, II 529
  rock-salt, I 257
  threshing, Britain, II 97
  vegetable oils, I 288
  viticulture, II 135
  water conservation, I 529
  water-lifting machines, II 676–7
  water-mills, II 593, 597
  wines, I 284–5
Stradanus (Van der Straat, Jan), III 350–1,
  394–5, 397–9, 707
— Johannes, II 749
Strasser, P., VII 799
Strassmann, F., VI 224, 268
Strauss, B., VI 492
Stringfellow, John, V 408
Strowger, A. B., VII 1226, 1245
Strutt, Jedediah, V 597
— William, IV 475–6
Strutts, the, IV 283
Struvé (nineteenth century), IV 93
— F. G. W. von (1793–1864), IV 602
Sturge, Edmund, V 253
Sturgeon, William, IV 653, 655; V 178, 203

Sturm, J. C., V 717
Sturtevant, Simon, III 79, 83
Suarez, Pablo, V 406
Suetonius, II 522; III 504
Sugg, IV 272
Sully, Henry, IV 409
— Maximilien de Béthune, Duc de
     canals, III 459–60
     horology, III 557
     roads, IV 521, 525
     town-planning, III 291
Summers, J., VI 496
Sundstrand, D., VII 1170
— O., VII 1170
Sutcliffe, R., VI 361
Sutton, Henry, III 623
— Thomas, V 732
Svaty, V., VI 665
Swan, Sir Joseph, chemist and inventor
     artificial silk invention, V 215
     battery improvements, V 206
     incandescent filament lamps, V 213–17, 229; VI
       285
     lighting systems, 19th cent., V 208–9
Swann, H., VII 1357
Swart, Claudius Claussøn, see Claudius
Swedenborg, Emmanuel, III 31, 51
Swinburne, Sir James, V 203; VI 555
Swinton, A. A. C., VII 1258–9
Swiss metallurgical Company, V 93, 608
Switzer, Stephen, IV 174
Sylvester II, Pope, see Gerbert
Symington, William, IV 196, 571; V 143
Symonds, Sir William, IV 584–5
Symons, G. J., IV 490; V 558
Synesius, III 603, 606
Szechenyi, Count, IV 550

Tachenius, Otto, III 681, 704; IV 253
Tacitus, I 253; II 11, 140
Taharqa, I 538 n.
Talbot, A. N., VII 945
— Benjamin, V 60; VI 479, 483
Tangye, V 131
Tank, K., VII 811
Tardin, J., IV 258
Tarelli, IV 19, 34
Tartaglia, Niccolò, III 374, 539
Tata, J. N., V 70
Taunton, Lord (Henry Labouchere), V 784
Tawney, R. H., VI 167
Taylor, C. G., VII 804
— Clement, IV 255
— F. Sherwood, II 732, 735
— F. W., engineer
     machine tools, VII 1046, 1047–8
     management system, VI 80–1

steels, high-speed, VI 453, 492; VII 1037
— G. E. (c.1872), VI 497
— George (c.1792), IV 255
— James, of Lancashire, III 162
— John, VI 552
— Michael Angelo, IV 506 and n.
— Walter, IV 437
— William, V 666
Taylor and Martineau, IV 197; V 130
Teal, G. K., VII 1119
Tedesco, N. de, V 489, 496
Telegraph Construction & Maintenance Company,
     V 221
Telford, Thomas, engineer
     bridges, IV 457, 458–9, 460, 484
     building works, IV 455
     chains, Clifton bridge, V 507
     Darby's bridge, comments, IV 456
     Grubenmann's bridge, comments, III 428
     harbour works, V 539
     pavements, IV 543, 544
     roads, IV 527, 528, 531–4, 536
     training, IV 444
     waterways, III 455; IV 571–2
Tellegen, B. D. H., VII 1102
Teller, E., VI 279
Tellier, Charles, V 48; VII 1400
Temple de la Croix, Félix du and Louis du, V
     408, 409
Templeton, IV 315
Tengnagel, Otto, IV 636
Teniers, David, the younger, II 348, 749
Tennant, Charles, IV 239, 245, 247–8
Tertullian, II 531
Terzaghi, K., VII 946, 1366
Tesla, Nikola, inventor, electric motor
     current controversy, A.C./D.C., V 200
     fan motor, VII 1128
     induction motor, V 231; VI 285; VII 1086–7
     synchronous motor, VII 1087
Tessaert, IV 256
Thaer, Albrecht Daniel, IV 38, 41
Thales of Miletus, I 105; II 606
Thénard, Louis Jacques, IV 256, 374
Theodelinde, II 451
Theodoric the Great, II 457
Theodosius I the Great, II 515, 531; III 572
Theon, of Alexandria, III 586–8, 591, 606
Thephano, II 761
Theopilus Presbyter, metallurgist
     bell-founding, III 364
     case-hardening, iron, III 34
     chemical ash, II 354
     copper, purification, II 49
     enamelling, II 458–9
     glass-making, II 315, 329, 332, 358; III 206,
       207, 208

glaze recipes, II 300, 315
gold and enamels, III 48, 49
gold mining, II 12
leather-gilding craft, II 174
niello (metallic sulphides), II 480
parchment, preparation, II 189
Schedula diversarum artium, II 63–4, 351; III
  677
Theophrastus of Eresus, Greek writer
  Babylonia, crop yields, I 550
  blue frit, I 240
  building construction, III 250
  cotton cultivation, India, II 199
  pigment, white, II 360
  ploughs, wooden, II 85
  pōros limestone, II 398 n.
  sugar, I 371
  timbers, description, II 226, 234
  vegetables, I 355, 356
  viticulture, II 131–2, 134
  walnut varieties, I 358
Thévenet (*c.*1865), IV 523
Thévenot, Melchisédec (1620–92), IV 604
Thevet, André, III 8
Thimonnier, Barthélemy, V 588; VI 679
Thiout (*c.*1741), III 345
— Antoine (1692–1767), IV 384–5, 420
Thölde, Johann, *see* 'Valentine, Basil'
Thom, John, V 241
Thomas, C. J., VI 61
— David, IV 111
— G. H., VII 800
— John, IV 348, 352
— Sidney Gilchrist, metallurgist
  industrial change, V 804, 809
  steel-making process, V 60, 810, 820; VI 462
Thompson (*c.*1793), IV 166
— Benjamin, *see* Rumford
— C. B., VI 84
— H. W., VII 1206
— R. Campbell, III 224 n.
— Silvanus P., V 200, 203, 216
Thomson, E., VII 1331
— G. P., VI 269, 270; VII 1216
— J. E. (1822–92), VI 79, 82, 89
— James (*fl* 1850s), V 525–6, 530
— Sir Joseph John (1856–1940), VI 155, 223; VII
  1092
— R. W., IV 523; V 422, 771
— William (Lord Kelvin), physicist
  compass, invention, V 456, plate 31 C
  current battle, A.C./D.C., V 200
  electric circuit theory, VII 1112
  electricity distribution, V 228
  hydrodynamic wave studies, V 387–8
  lighting installation, V 217
  Neptune Bank power station, VII 1068

  rotors, electric, V 193
  sounding-machine, hydrographic, V 453–4,
   461–2
  telegraphy, V 224, 226
  transatlantic cables, IV 674
Thorek, M., VII 1319
Thorley, J. P., I 409
Thorneycroft, V 425
Thothmes I, I 782, table F
— III, Egyptian king
  botanical collection, I 543–4, 545
  chariots, II 544
  chronology, I table F
  copper from Cyprus, I 565
  glass, painted, II 341, plate 24 D
  ivory, I 669, 673
  shadow-clock, I 112
  stone wall, construction, I 480
  weaving, evidence, I 439, plate 14
— IV, Egyptian king
  chariot, I 727
  chariot-horses, control, I 722
  ivory artifact, I 669
  weaving, evidence, I 439–40, plate 15 A
Thotnefer, I 439
Thucydides, II 529
Tiberius, II 52
Tibullus, II 28
Tice, J. T., VI 653
Ticquet, III 86
Tiernan, M. F., VII 1377, 1378
Tiglath-Pileser I, I 246, 553, table F; II 494–5
— — III, I 54, table F
Tilghmann, B. C., VI 610
Tillmann of Cologne, II 65
Tilly, Johann Tzerclaes, III 31
— family, IV 209
Titley, A. F., II 744
Titus, III 574
Tizard, H. T., VI 162
Toft, Thomas, IV 345–6
— II, IV 345
Toglietta, Guido, IV 521
Tompion, Thomas, III 557, 665, 667; IV 382
Torricelli, Evangelista, III 234, 332, 636; IV 169
Torroja, E., VII 951
Torsellus, *see* Sanudo, Marino
Toufaire, IV 105
Tourasse, IV 548
Tournachon, F., V 400
Tourville, A. H. de C. de, III 492
Towler, F., VII 1061
— J., VII 1061
Towne, H. R., VI 80
Townley, Richard, IV 490
Townsend, A. J., VI 496
— C. P., VI 519

Townshend, Charles, Viscount, III 15; IV 39
Tracy, N., I 4
Trajan, Roman emperor
  bridge structure, Orsova, I 508, 512
  chronology, II table II lvi
  roads, Petra, II 498
  streets, Roman towns, I 530
  town plan, Timgad, III 280
  water-mills, II 601
Transon, Abel, V 397
Traube, A., VII 1282, 1297
Trechsler, Christopher, III 629
Tredgold, Thomas, writer and engineer
  books, IV 165, 445, 484
  carpentry, definition, II 242
  gauge, boiler pressure, V 538
Trenchard, Sir H., VII 811
Trésaguet, Pierre, IV 527–8, 532
Trestrail, Nicholas, V 129
Trevithick, Richard, engineer
  boiler invention, V 125, 133
  Cornish pumping engine, IV 191–3, 499; V
    125, 126
  locomotives, IV 188, 189; V 322, 349, plate 20 A
  rock-boring machine, IV 71
  stationary engines, IV 165, 166, 188–94, 196,
    197; V 133, 138
  steam-turbine invention, V 150
  tunnel attempt, IV 463
Trew, John, III 451
Triewald, Marten, IV 173, 175
Troncet, G., VII 1161
Troostwijk, Adriaan P. van, V 177
Tropenas, A., VI 484
Tropsch, H., VI 540
Trotz, J. O. E., VI 497
Troughton, Edward, instrument-maker
  balances, IV 405, 406
  comparator, IV 397
  dividing-engines, IV 394
  optical instruments, IV 399, 607
Truax, C., VII 1317
Truchet, Sébastien, III 46
Trudaine, Daniel, IV 526
Tschirnhausen, E. W., Graf von, IV 338
Tsiolkovski, K., VII 859, 860
Tswett, M. S., VII 1210
Tukulti Ninurta II, I 251, 527, 583, 597
Tull, Jethro, III 15; IV 5–6, 17, 20, 38
Tunner, Peter, V 65, 67
Turgot, Anne Robert Jacques, IV 27, 526
Turing, A. M.
  ACE (Pilot) computer, VII 1194
  computers, VII 1153, 1183–4
  program concept, VII 1193, 1199
Turner, C. J., VI 263, 265
— Ipswich firm, IV 71

— John, of Lane End, IV 352
— W. E. S., II 314 n., 319 n.
Turpin, Eugène, V 296; VI 540
Turriano, III 16
Tusser, Thomas, III 134
Tutankhamen, see Index of Subjects
Tyer, Charles, V 331

Uchatius, Baron F. von, V 736–7
Ueling, E. A., VI 480
Ulam, S. M., VI 279
Ulloa, Antonio d', IV 144
Ulm, C. T. P., VII 806
Ulpian, II 325
Ulugh Beg, III 584
Unas, I 434, 537
Underwood, William (c.1817), VII 1400
——— L. (d.1928), V 42
United Alkali Company, V 803
United States Chemical Manufacturing Company,
  V 105
Unwin, W. C., VII 1368
Upson, R. H., VII 798
Urban VIII, Pope, III 312, 314
Ure, Andrew, IV 135, 274; V 828
Urie, Robert W., V 341
Ur-Nammu, I 467
Ursinus, O., VII 806
Urukagina, I 280, 282, 748
Usher, A. P., IV 149
Utton, M. A., VI 53, 54, 55, 58
Utzschneider, IV 360

Vail, Alfred, IV 660
Valdés, Diego Flores de, III 480
Valency, Charles, IV 445
Valens, II 664
'Valentine, Basil' (Johann Thölde), III 29; IV 242
Valentinian I, II 515
Valerius Maximus, I 264
Vallans, William, III 451–2
Valturio, Roberto, II 620; V 141
Van Bath, B. H. S., IV 15
Vance, A. W., VII 1352
Vancouver, George, IV 598
Van der Heyden, Jan, III 695; IV 638
Van der Straat, see Stradanus
Vandry, IV 540
Van Helmont, Johann Baptista, II 745; III 24;
  IV 37, 215, 258
Van Linschoten, III 696
Van Marum, IV 408
Van Musschenbroek, Pieter, see Musschenbroek,
  P. van
Van Robais, III 152
Van Sickle, Samuel, V 113, 122
Van Veen, J., IV 640

Van Wesel, Dominicus, IV 632, 634
Varley, S. A., V 185, 186
Varro, Marcus Terentius, III 504, 578
Vásráhelyi, IV 550
Vauban, Sébastien le Prestre de, II 728; III 372–3, 467; IV 442
Vaucanson, Jacques de, III 166, 169; IV 318, 384–5, 425
Vaughan, John, V 801–2, 804, 810
— S. W., VI 477
Vauquelin, L. N., IV 145
Vavilov, N. I., I 368; VI 328–9
Vedor, C., VII 891
Veen, J. Van, see Van Veen
Vegetius, II 121, 636, 705, 707
Veksler, V., VII 1332
Venantius Fortunatus, II 597
Venturi, Giovanni B., IV 501; V 537
Venturino, Pietro, IV 636–7
Verantius, Faustus, III 18; IV 630, 635, 641
Veranzio, F., II 627
Vergil, Polydore, III 381
Verguin, E., V 272
Vermorcken, E., IV 639
Vermuyden, Sir Cornelis, III 316, 318, 320–1
Verne, J., VII 859
Verovio, Simone, III 409
Vespasian, Roman emperor
  chronology, II table II lvi
  military engines, II 715
  Roman coin, II 489
  tunnel, *Petra Pertusa*, II 511
  water-mills, II 601
Vicat, L. J., IV 447–8, 466; V 483
Victoria, English queen
  chloroform, V 312
  'clipper' ships, IV 590
  Great Exhibition of 1851, V 817, 834
  mauve dye, V 272
  porcelain, V 662
  Turbinia, speed demonstration, VII 1025, 1027
Vidal, Raymond, V 279
Vieille, P. M. E., V 294
Vierlingh, Andries, II 682, 684; III 301–2, 305, 452
Vignola, Giacomo da, III 450
Villard, P., VI 222
— de Honnecourt, II 69, 610, 643–4, 647
Vincent of Beauvail, II 665–6
Vinci, Leonardo da, see Leonardo da Vinci
Vinne, Theodore Low de, V 712
Virgil, Roman poet
  flax cultivation, II 195
  lathe, wood-turning, II 233
  milling, II 111
  plough, horse-drawn, II 94
  viticulture, II 135

Visconti, Filippo Maria, II 657
Visscher, Anna Roemers, III 229
Vitruvius Pollio, Roman architect
  anaphoric clock, III 604–5
  blue frit (*caeruleum*), I 240
  building façades, III 250, 253; IV 469
  building methods, Greek, II 397–419
  fire, origins, I 216, 217
  hodometer, III 512, 513, 628
  manuscript discovery, III 245
  maps, Roman, III 504
  military technology, II 711–12, 714, 717
  press machine, lever-and-screw, II 117
  pulleys, II 630, 658–60
  quarrying, II 27–8, 31
  roads, IV 521
  roofs, low-pitched, III 265
  survey methods, III 511–12, 513, 543
  transport, heavy stones, II 660, 662
  town-planning, III 276
  walls, Italian process, III 247–8
  water, book on, II 671
  water-clocks, I 113; III 601, 602
  water-finding, II 663
  water-lifting, II 675, 676
  water-mills, II 596, 597, 614
  water-purity, II 674
Vivian, Andrew, IV 188
— Henry Hussey, IV 128–9; V 78, 85
Viviani, Vincenzio, III 232, 315
Vizinho, José, III 547
Voch, Lucas, IV 529
Vogel, H. W., V 729; VII 1283
Voght, von, IV 41
Voightländer, J. F., V 721
Voisin, Charles and Gabriel, V 392, 406, 409; VII 791
Volk, Magnus, V 347, 418
Volkov, V. N., VII 857
Volta, Alessandro, physicist
  electricity, chemical sources, IV 678
  telegraphy, IV 649, 652–3, 654
  voltaic pile (electric battery), IV 206–7; V 177–8
Voltaire, François-Marie Arouet de, III 466
Vroom, Hendrik Cornelisz, III 486, plate 22
Vuia, Trajan, V 409

Waddell, IV 93
Wade, George, IV 530, 532
Wagner, H. A., VII 810, 1019
— L., IV 545
Waitz, von, IV 244
Waldseemüller, Martin (Hylacomylus), III 535 *and* n., 536, 538
Wales, Prince of (George V), VII 922
Walford, C., VII 1153

Walker, John, IV 255
— O. S., VII 1044
— Robert, V 223
— Thomas, V 459–60
— W. H., V 730
Walkers, Rotherham ironfounders, IV 457–8
Wall, John, IV 354
Wallace, C. F., VII 1377, 1378
Waller, A., VII 1340
Wallerius, IV 39
Wallis, John, III 257
Walpole, Horace, IV 216
— Robert, IV 39
Walschaerts, Égide, V 335, 336–7
Walsingham, Sir Francis, III 317
Walter, John, V 696
— L., VII 1105
— de Millinate, II 620
Walton, Izaak, IV 505
Wampen, H., VI 675
Wang Ching, III 439
Warburton, J., engineer, IV 321–3
— John, potter, IV 349, 351
— Peter, IV 356
Ward, Joshua, IV 242–4
Waring, R. S., VII 1074
Warington, R., VII 1387
Warren, J. C. (1778–1856), V 312
— Captain James (c.1848), IV 461
Warwick, Earl of, III 475
Washington, George, V 4
Waterhouse, J. J., V 282
Waters, R. M., VII 1334, 1335
Watson, J. D., VII 1394
— Richard, IV 237, 259–60, 263
— T. J., VII 1181
Watson-Watt, Sir R., VI 449, 559; VII 845
Watt, Charles, V 96
— Gregory, IV 264–5
— James (1736–1819), engineer
    alkali manufacture, IV 237, 247
    biographical note, V 350
    boilers, V 137, 152
    condenser invention, IV 181–3, 421, 673; V
        124; VI 450
    industrial change, discussion, V 786, 800, 840;
        VI 52
    iron industry, IV 103–4, 116
    latent heat discovery, IV 186, 680
    machine-tools, V 640, 649, 650, 668–9
    marine trunk-engine, V 148
    optical indicator, VII 989
    pottery industry, IV 348, 352
    precision mechanics industry, III 368, 647; IV
        380
    pumping-engines, IV 77–9, 452, 499; V 524
    rotary engine, III 84

screw-propulsion, V 146
sculpturing machine, IV 436
steam engines, IV 161–3, 164, 181–95, 199,
    421–2; V 125–9
steam-power, IV 161, 165, 166, 452, 680
subtense (surveying instrument), V 445
tea-kettle myth, IV 175
textile industry engines, IV 246
— James (1769–1848), IV 264
— — & Company, V 148, 363, plate 41 A
Watts, Isaac, IV 588
Wayss, Gustav A., V 489
— & Freitag, V 489, 495
Weale, Samuel, III 157
Webb, B., VI 97
— Francis W., V 335, 337–8, 339
— J. E. (fl end 19th cent.), IV 516
— James E. (c.1969), VII 858
— S., VI 97
Webster, James Fernley, V 91
— Crown Metal Company, V 249
Wedding, T. F., IV 105
Wedgwood, John, IV 350
— Josiah, potter, IV 328
    biographical account, IV 351–4
    ceramics industry, discussion, V 658, 659
    glazes, IV 334, 335, 356
    industrial change, discussion, V 799, 800
    porcelain, IV 341, 344, 350, 352–3
    'pyrometric cone' invention, IV 350–1
    water-power, IV 348
— Thomas, IV 350, 354; V 718
Wedgwood Company, V 662
Wedgwoods, the, IV 347
Wegener, A., VII 1454
Wehnelt, A., VII 1093
Weid, E. von der, VII 568
Weidenreich, F., I 35
Weigel, von, III 11–12
Weigert, Carl, V 281
Weild, W., V 579
Weir, Lord, VI 287; VII 771
Weis, J. P., VI 682
Weiss, José B., V 406; VII 808
Weizmann, C., VI 542, 543
Weldon, Walter, V 237
Wellman, G., V 573
— S. T., VI 479, 485
Wells (c.1791), IV 580
— E. C., VII 829
— Horace (1815–48), V 312
— John (c.1700), IV 467
— — I. (fl 1819), V 694
Wells & Company, V 212
Welsbach, Carl Auer von, chemist
    gas mantle, invention, IV 274; V 209; VII
        1083, 1128

lamp filaments, development, V 98–9, 218
  Nernst lamp, VII 1083
Wenceslaus Hagecius, *see* Hajek, Václav
Wendel, de, IV 105
Wenham, Francis H., V 410
Wenström, Jonas, V 191, 193
Wenzler, Johannes B., IV 263
Werner, Johann, III 535–6; V 458
Wesel, D. Van, *see* Van Wesel
Westinghouse, George, engineer
  air-brake system, trains, V 344, 345
  current controversy, A.C./D.C., V 200–1
  electric motors, V 231–2
  electric power, V 199, 201
  vertical steam-engines, V 134
Westinghouse Company, V 134
Westly (Lawson and), IV 293, 315
Westman, A. E. R., VI 631
Weston, Sir Richard, III 457; IV 15–16
Wharton, J., VI 83
Wheatstone, Charles, electrical engineer
  electric generator, V 185–6, 187
  telegraphy, IV 654, 657–60; V 218–20, 224
Wheeler, J. A., VI 225
— W. H., V 541
Whewell, William, V 546
Whieldon, Thomas, IV 348, 351, 353
Whinfield, J. R., VI 561, 650
Whipple, Squire, IV 486
Whiston, William, IV 621
White, E., VII 857
— H., VII 1331
— John, IV 242–3
— M., VI 453, 492; VII 1037, 1048
Whitehead, A. N., V 776; VI 162
Whitehouse, Cornelius, IV 271
Whitehurst, IV 350
Whitney, Eli, IV 435, 438; V 652–3, 818
Whitten, A., VII 806
Whittle, F., VII 819, 990, 1018–19
Whitworth, Sir Joseph, manufacturer
  American methods, comment, V 639
  biographical notes, IV 418, 428, 431; V 637–8,
    801
  lathe, IV 432, 434
  machine-tools, IV 418, 431–5; V 636
  micrometer screw, VII 1050
  planers, IV 433–5; V 651, 818
  scholarships offer, V 791–2
  screw-cutting tools, IV 432, 433; V 652
  'Street Sweeping Machine', VII 1131
Whitworth & Company, V 642, 801
Whitworths, the, IV 572
Whytock, R., V 583
Wibault, M., VII 810, 834
Wickham, Sir Henry, V 774
Wicks, Frederick, V 684

Wieck, of Chemnitz, IV 297
Wiggin, Henry, & Company, V 88
Wigner, E. P., VII 252
Wilberforce, William, V 830
Wilcke, J. C., IV 621
Wilcox, Stephen, V 138
Wilde, Henry, V 184–7, 193, 194, 196, 202
Wilde & Company, V 184
Wilfley, Arthur R., V 74
Wilkes, M. V., VII 1193, 1199
Wilkinson, David, IV 425
— Isaac, IV 161
— John, ironmaster, IV 428
  accuracy claims, IV 162
  boring machines, III 367; IV 103–4, 183, 421–
    3, plate 27 A; V 640, 649
  bridge, cast-iron, IV 455
  machine-tools, IV 418
  ships, iron hulls, V 350
— W. (*fl* 1872), VI 497
— William (*fl* 1775–80), IV 101–2, 105
Willans, P. W., V 135–6; VII 1070
Willats, Richard, V 723, 724
Willcocks, Sir William, V 543
William (*c*.1631), Landgrave of Hesse, III 358
— I, the Conqueror, II 128–9, 648
— III, English king, III 267; IV 672
— IV (*c*.1561), Landgrave of Hesse, III 658
Williams, E. G., VI 558
— Edward, V 447
— F. C., VII 1193, 1196
— F. N., VII 1331
— H. B., VII 1341
— L., VII 1324
— Thomas (eighteenth century), III 435
— — (*d*.1802), IV 64
— Trevor I., VI 48, 57
Williams & Robinson, V 135
Williamson, E. D., VI 588
Willibrord, St, III 197
Willson, T. L., VI 523, 541
Wilm, A., VI 439
Wilmanns, G., VII 1298
Wilson, Allan B., V 588–9; VI 679
— E., V 232
— P. N., IV 201, 206
— Thomas, IV 84 *and* n.
— W., VI 12, 151
Winchcombe, John, III 151
Winckler, J. H., IV 648
Windsor, Herbert, Lord, III 434
Winkler, Clemens, V 247; VI 517
— E., V 501, 502
Winlaw, William, IV 9
Winsor, Frederic Albert, IV 252, 264, 268
Winstanley, Henry, V 463
Winthrop, John, III 685

Winzer, Friedrich Albrecht, *see* Winsor
Winzler, Zachaus Andreas, IV 264
Wise, John, V 400
Wiseman, B. K., VII 1333
Witherbee, W. D., VII 1330
Witsen, Nicolaas, III 6
Witt, Nicolaas Corneliszoon de, III 312
— Otto, N., V 274, 281
Witte, Count Sergei Yulievich, V 115
Wohler, A., VII 985
Wöhler, Friedrich, chemist
  aluminium, isolation, V 90, 248
  calcium carbide, preparation, VI 523
  organic compounds, discovery, V 268
  research contribution, V 788, 821
  sulphuric acid process, V 247
Wöhlwill, Emil, V 96
Wolcott, Alexander, V 721, 722
Wölfert, K., VII 796
Wolff, D., of The Hague, III 229
Wollaston, William Hyde, IV 145; V 246, 717–18, 720
Wolseley Sheep Shearing Machine Company, V 432–3
Wolsey, Thomas, III 245
Wood, Aaron, IV 354
— Coniah, IV 278
— F. W., VI 497
— H. W., VII 1278
— John (1707–64), IV 470
— — (1727–82), IV 470
— R. W., VII 1295
— T. N., VII 1158
— Thomas, IV 314
Woodcroft, Bennet, IV 303
Woodford, IV 631
Woodhead, Sims, IV 503
Woodward, J., IV 37
Woolf, Arthur, IV 191, 193–4; V 125–6, 133
Woolley, Sir Leonard, I 199, 200, 396, 469
Woolrich, John Stephen, V 633
Worcester, Marquis of, *see* Somerset, Edward
Worden, A., VII 868
Wordsworth, IV 293–4
Worsdell, Thomas W., V 338
Wortley, Henry B., V 379
Woulfe, Peter, V 257, 296
Wratten, III 129
Wren, Sir Christopher, architect
  biographical notes, III 246, 256, 257–8; IV 668–9
  brick-work, Hampton Court, III 262
  building stone, II 34; III 260
  domes, III 254–5, 258–9
  London, rebuilding plans, III 297–8; IV 443
  roofs, II 442; III 264, 265–7
  timber and joinery, II 442; IV 472

Wright (c.1790), of Manchester, IV 287
— (fl 1851) (Cubitt and), IV 462
— Edward (c.1558–1615), III 464, 543, 551
— — (fl 1696–1701), IV 101
— M. D., VII 865
— Thomas, IV 603
Wright brothers, Orville and Wilbur, aircraft pioneers
  airships, comparison, VII 796
  engines, VII 1010
  flying achievements, VII 789–91, 838
  gliding experiments, V 406, 407, 411
  innovation, discussions, VI 45; VII 699
  pioneer flights, V 391–3, 396, 410, 411–12, plate 23 B
  rocket flights, comparison, VII 861
Wurtz, Charles Adolphe, V 113
Wyatt, John, III 162; IV 151
Wyke, John, IV 421
Wynn Timmins & Co., III 116
Wynn-Williams, C. E., VII 1188

Xanthudides, S., I 199, 202, 411
Xenophon, II 496, 557, 697
Xerxes, I 715, table F

Yarranton, Andrew, III 458
Yarwell, John, III 634
Yeager, C. E., VII 830
Yeivin, S., I 633 n.
York, H., VI 279, 280
Young, A. (c.1965), VI 114
— A. D. (c.1942), VII 824
— Arthur (1741–1820), Secretary, Board of Agriculture
  agricultural education, III 15
  farming improvements, IV 21, 29, 39
  mechanization, midlands, IV 153
  roads, comments, IV 530
  seed-drilling, IV 5
  threshing machines, IV 10
— J. H. (fl 1840), V 685, 686
— James (1811–83), V 104, 105, 118, 779
— Thomas, IV 164, 443, 482–4; V 353
Yuaa, I 270
Yvart, IV 33

Zachaire, Denis, II 741–2
Zachow, O., VII 726
Zacuto, Abraham, III 547
Zahn, Johann, V 717
Zeidler, O., VI 324
Zeiss, Karl, V 673, 722, 732; VII 1354
Zeller, Anton, of Aschhausen, III 32–3

Zeppelin, Count F. von, V 403; VII 797, 799
Zer, I 434
Ziegler, J. M., IV 614, 626
— K., VI 559
Ziehl, Franz, V 282
Zimmerman, Samuel, III 64
Zinn, W. H., VI 252
Zobel, O. J., VII 1114
Zonca, Vittorio, architect and engineer
  bearing-metals, III 41
  canals, II 685–6; III 441 n. 2, 451 *and* n.
  fulling-stocks, illustration, II 216

rolling-mills, III 342
  textile machinery, II 207, 645; III 172
Zopyrus of Tarentum, II 699
Zoréz, Ferdinand, V 472
Zoser, Egyptian king
  chronology, I table F
  Nile, famine claim, I 537
  stone buildings, I 475–8, 482, plate 19
  vineyards, I 283
Zosimos of Panopolis, I 280–1; II 140, 733, 736
Zuse, K., VII 1193
Zworykin, V. K., VII 1105–6, 1260

PART 3
INDEX OF PLACE-NAMES

INDEX OF PLACE-NAMES

Aa, river, Flanders, III 453
Aachen, *see* Aix-la-Chapelle
Abbeville, France, II 644, 649; III 152, 703
Abbey Mills, *see* London
Abbiate, near Milan, Italy, III 444, 446
Aberdeen, Scotland, IV 247, 592; V 422, 566; VII 1403
— Maryland, U.S.A., VII 1187, 1189
Aberystwyth, Cardiganshire, Wales, VI 311
Abingdon, Berkshire, England, III 456
Abu Dhabi, United Arab Emirates, VII 1461
Abu Ghalep, Egypt, I 362
Abusir, Egypt, I 482, 597 n., map 6; II 665
Abydos, Egypt, I map 6
    cubit, standard measure, I 779
    hunting techniques, I 165
    hut, ivory carving, I 305
    iron artefact, I 597 n.
    ivory-carvings, I 668
    textiles, ancient, I 434
Abyssinia, III 530; V 555
Accra, Ghana, VII 925, 1366
Accrington, Lancashire, England, V 578
Achères, Seine-et-Oise, France, V 420
Acre, Palestine, III 159
Actium, Greece, II 571, 573
Acton Town, *see* London
Adda, river, Italy, III 446, 448-9
Aden, III 5, 106
Adige, river, Italy, III 309
Admont, Styria, II 69
Adriatic Sea, III 525; IV 35; V 400
Afghanistan, I 359, 367, 368; V 103
Africa
    charts, west coast, III 544
    cotton species, III 159
    flints, 'knapping', III 356
    food production, V 24
    knitting techniques, III 183
    maps, medieval, III 514, 518-19, 523, 530
    ore mines, 1870s, V 812
    peace-keeping, European, VI 2
    rubber-bearing plants, V 752, 773
    ships, coastal trade, III 477
Agade, Mesopotamia, I 725
Aggtelek, Hungary, I 359, 370
Agrigento (Acragas), Sicily, II 401, 403, 678
Agrilesa, Crete, II 545
Ahenny, Co. Tipperary, Eire, II 544
Aichbühl, Württemberg, I 310-11, map 3

Airdrie, Lanarkshire, Scotland, V 350
Aire, river, Yorkshire, England, III 458; IV 572
Aix-en-Provence, France, V 558
Aix-la-Chapelle (Aachen), Germany, II 55, 423, 761; III 72, 196
Ajustrel, Spain, II 53
Akeman Street, England, III 278
Alabama, U.S.A., VI 483
Alaca, Cappadocia, I 594, map 4
Alalakh (Atshana), Syria, I 523 n., 655, 658, 673, 679, map 4
Alaska, U.S.A., V 533; VI 357, 414
Alatri, Italy, II 530, plate 40 A
Albania, III 579
Albany, New York State, U.S.A., IV 544, 553, 555; V 143
Alberta, Canada, IV 551; VI 321, 335
Albi, France, II 440
Alcalá, Spain, I 511
Alcantara, Spain, II 508, 510
Aldermaston, Berkshire, England, VI 281, 282
Alençon, Orne, France, IV 339
Aleppo, Syria, I 665, map 4; II 199; III 600
Alexandria, Egypt
    alchemy, practice, II 735-6, 738
    aqueduct, II 670
    building, Roman period, II 410
    chemical technology, II 350-1, 738
    cartography, III 505-6, 507, 531
    copper, trade, II 53
    craft changes, seventh century, II 761
    glass techniques, II 322; III 210
    hydrometer, fourth century, III 22
    maps, I map 6; III 507
    Mehmondieh canal, IV 555
    observatory instruments, III 586-7, 604-5
    pharos (lighthouse), II 521-2, 566; III 275; V 463
    ships, trading, III 473
    town-planning, II 528-9; III 275
    water-clocks, development, I 113
Alexandria, Virginia, U.S.A., IV 536
Al Firdan, Egypt, VII 890
Algeciras, Spain, VII 951, 953
Algeria, North Africa, I 72; V 258; VI 5
Algiers, N. Africa, IV 585
Alishar, Cappadocia, I 373, 518, map 4
Aljubarrota, central Portugal, II 583
Alkmaar, Holland, III 312, 679; IV 560; V 441
Alleghenies, U.S.A., IV 549

Allier, river, France, III 463
Almaden, Spain, IV 134
Alps, mountains, Europe
 balloon, first crossing, IV 400
 bridge structures, III 417
 canal project, 1786, IV 553
 chemical industry, VI 521
 farming techniques, IV 29
 fertilizer trade routes, VI 510
 hydro-electric power, VI 217
 height, barometric, IV 605
 Italy, northern boundary, III 308
 passes, IV 529, plate 36B
 roads, IV 529–30, 536
 tunnels, IV 516
 war, 1914–18, VI 11
Alsace, France
 building-materials, V 665
 computers, design, VII 1163
 farming techniques, IV 29, 31, 41–2
 fertilizer source, V 10
 madder (*Rubia*) cultivation, V 260
 industrialization, discussion, V 808
 logging, oil boreholes, VI 394
 silk dressing machine, IV 322
 textile industry, V 574, 577
Alsace-Lorraine, France, VI 2
Alston Moor, Cumberland, II 65
Altamira, Spain, I 147, 148, 151, plate 2
Altare, near Genoa, III 217
Altenberg, Germany, IV 124–5
Alton, Hampshire, England, III 116
Altona, W. Germany, VII 1373
Alum Bay, Hampshire, England, VII 1232
Alzau, stream, Languedoc canal, III 464, 467
Amalfi, Italy, II 454
Amarna, El-, *see* El-Amarna
Amazon, river, South America, V 754
Amberg, Bavaria, II 75
Ambleside, Westmorland, England, III 277
American Gulf Coast, VI 377, 402
Amiens, France, II 435; III 544, 703; IV 260,
 297, 304
Amoy, Fukien, China, III 696
Amphlett Islands, I 44, 406, 407
Amstel, river, Holland, III 293–5
Amsterdam, Holland
 coffee, cultivation source, III 6
 dikes, II 685
 drainage engineers, III 305, 312
 dredging, IV 638–40
 dye industry, III 695
 ethylene polymer discovery, VI 558
 glass-furnace, III 700
 printing, III 402
 sewerage system, VII 1384
 shipping, access, III 485

 town-planning, description, III 293–5 (map)
 tunnels, VII 905
 water-supply, IV 498; V 554
 waterways, IV 556, 560, 631
 windmills, II 89; IV 156
Amwell, Hertfordshire, England, III 453
Amyklae, Greece, II 499
Anaconda, Montana, U.S.A., V 81
Anatolia (Asia Minor), III 695; V 266
Anau, Turkestan, I 367, 501, 502, 721, map 5,
 table F
Ancona, Italy, III 311
Andalusia, Spain, III 152; IV 33
Andaman Islands, Bay of Bengal, VI 21
Andernach, Germany, III 16
Anderton, Cheshire, England, IV 571
Andes, mountains, South America, V 353
Anglesey, Wales, IV 64, 126; V 76, 500
Anjou, France, IV 29, 31, 33
Ankerlöken, Oslo, Norway, VI 522
Annonay, near Lyons, France, V 395
Antarctic, IV 58, 60–1, 598, 619
Antikythera, island, III 617–18
Antinoë (Antinoöpolis), Egypt, III 195
Antioch, Syria
 enamelling art, II 465
 Islamic expansion, 17th century, II 761
 Roman road, II 504
 street-lighting, II 531; III 282
 town-planning, Hellenistic, III 275
 water-wheels, II 614
Antipodes, the, IV 27
Antrim, Ireland, III 162
Antwerp, Belgium, map IV 557
 cotton industry, III 159
 dredging, IV 636
 farming techniques, IV 13, 16
 money market, 17th century, III 714
 oil-tankers, design, V 382
 printing, III 391, 401–2
 technical schools, V 714
 tunnels, VII 905
 water-supply, V 554, 566
 waterways, III 452; V 556–8
 woollen yarns, V 795
Anyang, China, I 722
Anzio, Italy, II 520
Aornos, Mount, unidentified site near Taxila, II
 701
Aosta, Italy, II 511; III 280
Apamaea, Syria, II 637
Apothecary Island, Leningrad, U.S.S.R., VII
 1322
Appleby Frodingham, Lincolnshire, England, VI
 470, 476, 483
Apulia, Italy, V 20, 21
Aqar Quf, Mesopotamia, I 468, map 7

Aquileia, Italy, II 321, 323, 520; III 286
Aradac, Yugoslavia, I 421
Araña, Cueva de la, Bicorp, Valencia, I 275
Aranjuez, Spain, IV 651
Ararat, *see* Urartu
Arboga, river, Sweden, III 455
Arctic, the, V 400; VII 853, 1024
Arctic Circle, III 506, 509, 553; IV 58
Arcueil, near Paris, France, III 462
Ardagh, Ireland, II 464
Ardeer, Scotland, V 288, 289, 292, 293, 821
Ardennes, France, IV 29
— the, France-Belgium, V 16, 17; VI 19, 23
Ardrossan, Ayrshire, Scotland, VII 899
Arendal, Norway, VII 755
Arezzo, Italy, I 390, 632; II 271, 274
Argentina, V 2, 3, 8, 9, 48, 49
Argonne, France, VI 255
Arisaig, Inverness-shire, Scotland, VI 245
Arizona, U.S.A., V 77, 79
Arkansas, U.S.A., V 4
Arlberg Pass, Alps, IV 529
Arles, France, II 68, 598–9, 671, 679
Arlesey, Hertfordshire, England, V 484
Armagh, Northern Ireland, V 344
Arminghall, Norfolk, England, I 314
Armorica, France, IV 29, 33
Arno, river, Italy, III 309–10, 428, 448
— valley of the, III 218
Arpachiyah, Mesopotamia, I 365, 484 n., map 7
Arqueros, South America, IV 134
Artemisium, Euboea, II 484
Artois, France, IV 31, 498
Aryn, imaginary place on the equator, III 521
Ashur, Assyria/Mesopotamia, I map 7
  bitumen in building, I 255, 550
  lentils, cultivation 7th century, I 362
  ivory-work, 13th century B.C., I 671
  roads, c.700 B.C., II 494, 495
  water-supply, I 550
Asia, III 159, 164, 378, 380, 518–19; V 265
Asia Minor, *see* Anatolia
Assam, India, V 21; VII 895
Assur, Mesopotamia, III 271
Astyra, Troad, II 42
Aswan, Egypt, I 238, 568–70, map 6; III 506; *see also* Syene
Atchana, Turkey, II 316
Athens, Greece, I map 7
  Antikythera, machine relic, III 618
  building-construction, II 398–403
  calendar, III 568, 572
  cubit measure, I 777
  furniture, II 233
  ivory-work, I 674–5
  metal-work, I 661–2; II 470, 487
  mines, I 584; II 25, 27

  pottery, I 401, 406, 408–9; II 259–62, 267
  power, water-mills, II 602
  roads, II 499, 529
  statues, gold and ivory, I 661–2
  town-planning, III 272–3, 281
  water-supply, II 663, 665, 667
Atlantic Ocean (North Atlantic)
  aeronautics, V 400; VII 799, 806
  airline travel, VII 829
  cartography, IV 619, 621
  Challenger, expedition, VII 1447
  diesel power, VII 1006
  fish preservation, IV 50
  floating-docks, V 541
  France, canal projects, III 459–60
  grain transport, 1850s, V 5
  Great Britain, voyage 1845, IV 476
  mapping, III 522, 525
  oceanography, studies, VII 1447, 1450, 1454
  Panama Canal, IV 562; VII 913, 915–16
  petroleum (oil) cargo, 1861, V 121
  ports and harbours, VII 921, 925
  radio, VII 838, 1232, 1233, 1252
  railways, VII 785, 788
  Ridge, Mid-Atlantic, VII 1454
  St Lawrence Seaway, VII 920
  shipbuilding, late middle ages, IV 575, 590, 592
  ships, 16th century, III 472, 474–5, 476–7
  ships, 20th century, VII 693, 743
  ships, iron and steel, 1860s, V 62, 362, 365, 377–8, 380
  steam-ships, V 144, 145, 147, 149, 153, 362
  submarine cables, IV 660–1; V 225; VII 1251, 1252
  whaling, IV 57, 62
Atshana, *see* Alalakh
Attock, India, VII 883, 885
Åtvidaberg, Sweden, IV 129
Auchinleck, Ayrshire, Scotland, IV 262
Aude, Gaul, II 60
— river, France, III 459, 465
Aue, Saxony, IV 338
Auge, river, France, IV 31
Augsburg, Bavaria (Germany)
  cartography, III 535; V 445
  diesel engine manufacture, VII 1004
  hydraulic engineering, II 666
  instrument-makers, III 585, 621–2
  machine-tools, III 341
  metallurgy, II 75
  printing, III 380–1, 406, 535
  water-mills, II 611
Auron, river, France, III 450
Australia, VII 789
Austria, III 161, 341, 407
Autun, France, II 507; III 504; V 118
Auvergne, France, IV 32, 169

Auxerre, France, III 504
Avaricum (Bourges), France, II 716
Avebury, Wiltshire, England, I 491
Avignon, France, III 287, 418
Avon, river, Gloucestershire, England, IV 468; V
    507; VII 960
— — Somerset, England, III 457
— — Warwick, England, III 457
Ayrshire, Scotland, IV 238, 532; V 34
Azerbaijan, Iran, I 261
Azores, islands, III 531, 544, 549; IV 62

Baalbek, Syria, II 30, 409–10, plate 2A
Babylon(ia), I plate 18
    bridges, I 715
    building methods, I 457, 468, 471
    calendar, III 559, 566–8, 571, 574
    irrigation system, I 554–5
    kingdom, 2500 B.C., I 56
    maps, I 251, map 7; II 496
    roads, I 713; II 494, 496 (map)
    town-planning, III 270–1
Bács-Földvar, Yugoslavia, IV 550
Bactria, I 631, 632, map 5
Badakhshan, Russia, I 55, map 5
Badari, Egypt, I map 6, table E
    basketry, I 418–19, 420, 421
    building materials, I 304, 456, 473
    copper culture, I 601
    glazing, II 315
    houses, c.4000 B.C., I 304
    implements, I 511, 514, 516, 518, 601
    pottery, I 387
    textiles, c.4000 B.C., I 431, 432
Baden, Germany, IV 610; V 329
Baetica (Western Andalusia), II 8
Bagatelle, near Paris, France, V 392
Baghdad, Iraq (Mesopotamia), I map 7
    cartography, medieval, III 521, 529
    ceramics, II 286–7, 303
    observatory, III 521, 584
    roads, II 497
    water-mills, II 608, 614
Bagram (Kapisa), Afghanistan, II 324
Bahrain, Persian Gulf, I 664
Baiae, near Naples, Italy, II 665
Baikal, lake, I 301
Bakewell, Derbyshire, England, IV 21
Baku, Azerbaijan, U.S.S.R.
    fuel-oil (mazout), V 117
    natural gas, burning, V 102, 104
    oil derricks, 1869, V 110
    oil seepages, V 102–3
    oil-refining, V 114, 115
    rotary drilling, discussion, VI 381
    well logging technique, VI 394
Balabish, Egypt, II 151, 158, 161

Balawat, Mesopotamia, I 646, map 7
Balboa, Panama, IV 562
Balkans, the, III 8; IV 529; VI 1, 7
Ballinderry Crannog, Ireland, II 255
Balreji, Sind, I 406
Baltic Sea
    canals, Gotha route, IV 554
    canals, North Sea, III 444, 455; IV 560–2
    (map); VI 6; VII 913
    fish preservation, IV 45, 48, 54
    islands, III 89
    map, c.1400, III 532
    rockets, Peenemünde, VII 863
    shipping, III 499
    windmills, discussion, IV 156
Baltimore, Maryland, U.S.A., IV 592, 660; V 331,
    348; VII 1077, 1340
Baluchistan, south-west Asia, I 359, 385, map 5;
    III 505
Bamberg, Bavaria, III 405
Bankside, see London
Baradello (lake-village), I 357
Barbados, West Indies, V 23, 102, 526
Barbegal, near Arles, II 598–9, 602, 639
Barkaer, Denmark, I 312
Barking, Essex, England, IV 52, 53, 510; V 10;
    VII 1031
Barnhart Island, Canada, VII 918
Barnsley, Yorkshire, England, V 676; VII 785
Barrow, Lancashire, England, VI 478, 497
Barvas, Lewis, I 391
Barygaza (Broach), India, II 53
Basle (Basel), Switzerland
    chemical industry, IV 233; VI 507, 564
    publications, III 538; IV 233
    repoussé work, II 468
    roads, VII 872
    spinning and weaving, III 168
Basque country, III 186
Basra, Iraq, II 498, 614
Bassa Valle Padana, Italy, III 309
Bassae, Greece, II 400, 401
Bath, Somerset, England
    Avon, river, navigation, III 457
    cartography, IV 622
    quarrying, II 32
    Roman towns, discussion, III 277
    stone building, III 260
    town-planning, IV 470; VII 958, 960
    woollen-manufacture, III 152
Bathgate, West Lothian, Scotland, V 104, 105
Batley, Yorkshire, England, V 570
Battersea, see London
Battle Creek, Michigan, U.S.A., VII 1278
Bavaria, IV 41, 135, 610
Bavian, Mesopotamia, I 469–70, map 7
Bayonne, France, II 648; III 476; IV 540

Bear Island, Barents Sea, IV 58
Béarn, France, IV 29
Beaulieu Heath, Hampshire, I 313
Beaumont, Texas, U.S.A., VI 366
Beaune, France, II 137
Beauvais, France, III 241
Beaver Falls, Pennsylvania, U.S.A., VI 497
Bedford, England, V 137
Bedford Level, *see* Great Level
Bedworth, Warwickshire, England, V 826
Beemster, meer, Holland, III 306, 321
Beer, Devon, England, II 34
Beeston, Nottinghamshire, England, VII 950
Beit el-Wali, Egypt, I 483, map 6
Belfast, Northern Ireland, V 254, 574, 772
Belgian Congo, I 506; VII 788
Belgium, IV 557 (map)
Belgrade, Yugoslavia, III 279
Belgrade forest, Istanbul, IV 498
Bellefontaine, Ohio, U.S.A., IV 539
Belleville, France, II 691
Belmont brook-reservoir, Lancashire, England, IV 495, 501
Belper, Derbyshire, England, IV 475–6
Belt cave, Caspian Sea, I 351
Benares, India, V 608
Bendigo, Victoria, Australia, V 47
Benediktbeuern, Bavaria, IV 360–1
Bengal, India, III 159; IV 308, 610; V 440
Beni Hasan, Egypt, I map 6
    boats, river use, I 734
    building materials, I 477
    caravans, pack-asses, I 706
    chemical arts, I 241
    cosmetic arts, I 291
    foods, preservation, I 263, 265
    horticulture, I 544, 545
    Neolithic farming, I 43, 73
    pets, I 330, 331
    pottery, I 388
    textiles, I 437, 438
Benojan, Malaga, Spain, I 148
Bereguardo, near Pavia, Italy, III 446
Beresina, river, Russia, IV 554
Bergen (Mons), Belgium, IV 557
— Norway, III 533; V 49
Bergen-op-Zoom, Holland, III 316, 318; V 441
Bergerac, France, III 16
Bergermeer, the, Holland, III 302
Berkshire, England, IV 19
Berlin, Germany
    air transport, VII 800
    buildings, fireproof test, V 482
    canal locks, III 450, 454
    ceramics, machinery, V 669
    chemical industry, III 682; V 242, 270
    dental industry, VII 1322

    dyes, IV 249; V 270
    electric motors, V 232
    ocean technology, VII 1448
    photography, VII 1282, 1283, 1288
    printing, VII 1269
    railways, VII 768
    roads, VII 872
    rocket tests, VII 863
    steel, experiments, IV 108
    technical schools, 1876, V 714; VI 139
    turbojet engine, VII 1019
    wartime, *c.*1945, VI, 23, 24
    water-supply, V 554
— New Hampshire, U.S.A., V 251
Bermondsey, *see* London
Bermuda, V 541; VII 700
Berne, Switzerland, II 531; V 440
Berri, France, III 152
Bersham, Denbighshire, England, IV 101, 421, 455
Bersheh, El-, *see* El-Bersheh
Berwick-on-Tweed, Northumberland, England, III 284; IV 460; VII 887, 893
Beth-Asbea, 'town of Byssos', Palestine, I 249
Bethlehem, Pennsylvania, U.S.A., VI 471
Beth-Marē, Lebanon, II 575
Bettis, U.S.A., VI 253
Bettystown, near Drogheda, Eire, II 473
Beveland, Holland, IV 559
Bezdán, Yugoslavia, IV 550
Béziers, Hérault, France, III 467
Bhita, India, II 528
Bideford, Devonshire, England, III 418; IV 346
Bidford-on-Avon, Warwickshire, England, III 457
Bielethal, Saxony, Germany, VII 732
Bienne, lake, I 361, 369, 370, map 3
Bignor, Sussex, England, II 446
Billancourt, Paris, France, V 433
Billingham, Co. Durham, England, VI 59
Billingsgate, *see* London
Bill Quay, Durham, England, IV 238
Binasco, Lombardy, III 446
Bingley, Yorkshire, England, IV 568
Birdlip, Gloucestershire, England, II 460, 463
Birdsboro, Pennsylvania, U.S.A., VI 473, 474
Birka, Sweden, II 313; III 202
Birkenhead, Cheshire, England, V 351
Birmingham, Alabama, U.S.A., VI 479
— England
    brass manufacture, V 607, 618, 624
    canals, IV 564–5 (maps)
    ceramics, IV 352; V 669
    chemical industry, IV 243–4
    domestic appliances, VII 1144, 1147
    electricity supply, VII 1068
    gas lighting experiments, IV 263, 266
    gas manufacture, Murdock's, IV 266

Birmingham, England (*cont.*)
  glass, IV 360; V 677, 681
  'Lunar Society', IV 673
  machine-tools, IV 419
  mail-coach system, IV 531
  marine engines, V 143, 363
  metal articles, V 624, 630
  metals, non-ferrous, IV 119, 123, 130, 133, 138,
    145, 146
  motorway, 1954, VII 876
  New Street station, design, V 474
  plating processes, metals, IV 633
  radio relay system, VII 1255
  railways, V 323
  rubber-proofing, V 769
  sewage treatment, VII 1394, 1395
  spinning inventions, III 162
  steam-power, IV 165–6, 183, 352; V 136, 143
  steel, IV 108; V 59, 63
  technical colleges, V 714
  telegraphy, IV 657
  telephone exchange system, VII 1246
  town-planning, VII 965
  tube-making, V 626
  water-power, discussion, IV 153
  water-supply, V 553, 558; VII 1368, 1369
Bir Sgaoum, Algeria, II 639
Birs-Nimrod (Borsippa), Mesopotamia, I 468,
  map 7
Biscay, Bay of, IV 122
Biscay coast, France, IV 49
Bisenzio, river, Italy, III 314–15
Biskupin, Poland, I 323
Black Country, England, IV 113
Blackfriars, *see* London
Blackley, Manchester, England, V 274; VI 565
Black Patch, Sussex, England, I 560, 561, 562
  (map)
Blackpool, Lancashire, England, IV 516; V
  835
Black Sea, III 523; IV 35, 554, 587; V 372
Blackwall, *see* London
Blaenau Ffestiniog, Merionethshire, Wales, VI
  213; VII 1060, 1073
Blaencorrwg, Glamorganshire, Wales, III 140
Blakeney, Norfolk, England, II 439
Bleasdale, Lancashire, England, I 313
Blenheim, Bavaria, III 347
Blois, France, II 649; III 254
Bloomfield, Staffordshire, England, IV 162,
  183
Bloomington, Illinois, U.S.A., IV 545
Bochum, Rhine–Westphalia, Germany, V 788
Böckingen, Heilbronn, I 355, 357
Bockum, Hanover, I 303
Bodman (lake-village), I 357, map 3
Boesinghe, near Brussels, Belgium, III 454

Boghazköy (Hittite Hattusas), I 457, 489, 757,
  map 4; II 494
Bohemia
  ceramics, V 662
  cobalt minerals, IV 335
  glass-making, III 206, 221–2, 229
  mining, III 82; IV 64, 65, 156
  oleum production, V 247
  railway locomotives, V 339
  tinplate manufacture, III 689
  windmill, description, IV 156
Bolivia, I 223; III 1; VI 190
Bologna, Italy
  alchemical manuscripts, II 738
  bronze age tools, I 613
  canal locks, III 447, 450
  cartography, Roman period, III 504
  ceramics, II 269
  farming techniques, 16th century, IV 34
  Hero's 'Pneumatics' translated, IV 672
  iron-smelting, I 595
  land drainage, III 312, 314
  printing, III 409
  radio, Marconi's experiments, VII 1232
  railways, modern signal box, VII 783
  twisting-mills, 13th century, II 205, 206
Bolsena, Italy, II 376
Bolton, Lancashire, England, IV 279, 496; V 131,
  565, 823; VI 214
Bombay, India, V 70; VII 1394
Bonn, Germany, II 671; V 115, 587; VII 872
Bons, France, II 511
Boonton, New Jersey, U.S.A., VII 1378
Bordeaux, France, II 137; III 318, 493; IV 540;
  V 773
Borneo, I 226, 227; VI 548
Bornholm, Denmark, I 369
Borsippa, *see* Birs-Nimrod
Borum Eshoj, Denmark, I 442
Boscoreale, Italy, I 223; III 613
Bosphorus, Turkey, VII 887
Boston, Lincolnshire, England, III 515; V 545;
  VII 846, 1323, 1357
— Massachusetts, U.S.A.
  buildings, V 478
  canning industry, VII 1400.
  education, discussion, VI 139, 141
  medical technology, VII 1328, 1331, 1348
  paving, streets, IV 543
  printing, V 709
  sewer-outlet, tunnel, V 521
  telegraph, visual system, 1800, IV 647
  telephony, development, VII 1226, 1227, 1230
Botany Bay, New South Wales, Australia, V 574
  n.1
Botel-Tobago Island, I 118
Botesdale, Suffolk, England, I 601

Bothnia, Gulf of, Baltic, III 533
Bouchet, Seine-et-Oise, France, V 284
Boughton Mounthelsea, Kent, England, II 34
Boulogne-sur-Mer, France, II 522; V 483; VII
    1232
Bourges, France, II 716; III 291, 450
Bourget-du-Lac, *see* Le Bourget-du-Lac
Bourgneuf, *see* La Baie
Bournemouth, Hampshire, England, VI 168; VII
    1232
Bournville, Birmingham, England, V 830; VII
    962, 963
Bow, *see* London
Bowness, Cumberland, England, II 31
Box, Wiltshire, England, II 34
Boxmoor, Hertfordshire, England, II 134 n.
Brabant, Low Countries, III 301–3, 320; IV 15,
    19
Brachay, near Joinville, France, IV 261
Bradford, Yorkshire, England
    chemical industry, IV 244
    textile industry, IV 298; V 574, 577; VI 671
    water-supply, V 552
    woollen industry, V 810
Bradford-on-Avon, Wiltshire, England, V 764
Bradley, Shropshire, England, IV 104, 428
Brampton, Cumberland, England, II 31
Brandenburg, Germany, (Prussia), I 591; III 449–
    50; IV 610
Brandon, Suffolk, England, I 134–5; III 356
Brasilia, Brazil, VII 958, 980
Braunschweig, W. Germany, VII 1291
Brazil
    brasil-wood source, V 264
    food plants, discoveries, VII 4, 6
    hammocks, III 484
    hunting weapons, I 160
    rubber, sources, V 752, 753, 755, 828
    sugar cane, V 23
    waterworks, technology, V 564
Breac Moedoc, Scotland, II 169
Breadalbane, Perthshire, Scotland, VII 900
Breda, Holland, IV 15
Brenta, river, Italy, III 441 n. 2, 447, 451
Brentford, Ontario, Canada, VII 1380
Brescia, Italy, II 60, 75
Breslau, Silesia (Poland), IV 539; VII 1322
Brest, France, IV 582, 633, 635, 646; VII 847
Breteau, near Briare, France, III 460
Briare, on R. Loire, France, III 460
Bridgeport, Connecticut, U.S.A., VII 1223
Bridgnorth, Shropshire, England, IV 244
Brie, central France, V 16
Brieskow on the Oder, Germany, III 454
Briey, France, II 69
Brighton, Sussex, England
    electricity distribution, VII 1074

electric power station, V 198, 199
electric road-carriages, V 418
motor-carriages, 1886, V 429
railways, development, V 330, 343–4, 347
Swan lamps, V 217
Brigmilston, Wiltshire, England, I 604, 605
Brigue, Valais, Switzerland, IV 530
Brill, Buckinghamshire, England, II 296
Brimscombe, Gloucestershire, England, IV 305
Bristol, England
    alchemical furnace, II 741
    Avon river, navigation, III 457
    buildings, roofing materials, V 468, 470
    ceramics, IV 336, 340, 341, 343, 344, 349
    chemical industry, V 246
    fish preservation, IV 49
    gas, lighthouse proposed, IV 261
    glass furnace, III 226
    metal industries, IV 101, 119, 126, 130, 133
    pitch and tar, ship use, III 688
    port construction, IV 468
    roads, 19th century, IV 532
    road vehicles, V 423; VII 734
    ship building, V 361, 362
    soap-making, III 703
    Stratford corn-mill, IV 209–10
    telephones, dialling, VII 1247
    water-supply, V 553, 555
Bristol Channel, England, VI 218; VII 1232
British Columbia, Canada
    dug-out boats, I 738
    fishing methods, I 160, 176–7
    houses, ancient techniques, I 312, 318
    railways, VII 769
    sledges, I 707
    village habitats, I 178, 179, 184–5
Brittany, France
    farming, 18th–19th century, IV 29, 31, 32, 33
    mechanization, union dispute, VI 97
    pottery, I 388, 389, 407
    ports, 'Catalan Atlas', III 525
    shipbuilding, discussion, III 478
    tin mines, II 46
    windmills, III 89, 95–7, 99, 102, 104
Brixham, Devonshire, England, IV 52
Brixworth, Northamptonshire, England, II 425
Brno, *see* Brünn
Broach, *see* Barygaza
Broken Hill, New South Wales, Australia, IV 75;
    V 75; VII 418–19, 421, 423–4
Brooklyn, New York, U.S.A., VII 1358
Broomielaw, Clyde, Scotland, IV 468
Broseley, Shropshire, England, III 80, 84, 688;
    IV 162, 183
Brücken, Germany, II 76
Bruges, Flanders, Belgium, III 303, 453, 473; IV
    558–9, 629

Brünn (Brno), Czechoslovakia, V 254; VI 205
Brunsbüttel, N. Germany, IV 561
Brunswick, Germany, III 160
Brussels, Belgium
  cartography, IV 620
  dredging, developments, IV 636
  electrochemical processes, IV 251
  metals, V 609
  printing, III 380
  textile machinery, VI 665
  waterways, III 450, 451 (map), 452–3; IV 548, 557–8 (map)
Bubastis (Zagazig), Egypt, I 484 n., map 6
Bucharest, Rumania, V 116
Buchau, Württemberg, I 316, 355, map 3
Buckingham, England, V 421
Budapest, Hungary, V 115, 187; VII 770
Buenos Aires, Argentina, V 48, 333
Buffalo, New York State, U.S.A., IV 555
Buildwas, Shropshire, England, IV 457
Bulgaria, III 579
Burford, Oxfordshire, England, III 152
Burghausen, W. Germany, VI 556
Burgos, Spain, III 201
Burgundy, France, IV 29, 31, 32
Burma
  dyestuffs, V 266
  fire-making implements, I 218, 226
  Gokteik Viaduct, VII 880
  oil strata, IV 112
  wartime, 1942, VI 21, 23
Burma Road, Pennsylvania, U.S.A., VII 814
Burslem, Stoke-on-Trent, Staffordshire, IV 344, 348, 351
Burton-on-Trent, Staffordshire, England, V 273, 804
Bury, Lancashire, England, III 169; IV 187, 195, 303
Bury St Edmunds, Suffolk, England, II 426, 446
Buryet, Siberia, I 301
Bush Valley, Ireland, V 347
Butte, Montana, U.S.A., V 79; VI 418, 420, 496
Buxton, Derbyshire, England, II 169
Byblos (Jbail), Lebanon, I 762, map 8, plate 31A
Byzantium (anc. city), see Constantinople (Istanbul)

Caballos, Cueva de los, Spain, I 161
Cabeira, Pontus, II 593
Cadiz, Spain, I 257; III 477; V 258
Caen, Normandy, II 33, 428, 577
Caernarvon, Wales, III 284; V 129
Caerwent, Monmouthshire, Wales, III 277
Caffagiolo, Italy, IV 329 n.
Cairnpapple, West Lothian, I 491
Cairo, Egypt, I map 6: see also Fostat
  basketry, 3400 B.C., I 420
  building, stone materials, I 477

  cartography, V 442
  ceramics, II 286, 304
  cubit rods, measuring, I 778
  Delta barrages, construction, V 543
  globe, III 613
  instrument-making, III 584
  ivory-work, I 676
  solar power experiment, VI 221
  spinning and weaving, I 427; II 199; III 179, 182
  town-planning, III 269
Calabria, Italy, II 199; V 20
Calah, see Nimrud
Calais, France, II 527, 536, 726; III 74; IV 33; V 622
Calcutta, India
  bridges, VII 882, 885
  clipper ships, IV 590; V 62
  rubber (Hevea) seedlings, 1873, V 774
  steam-ships, V 362
  water-supply, V 563; VII 1366
Calder, river, Yorkshire, England, III 458
Cales, Apulia, II 268
Calicut, India, III 178
California, U.S.A.
  cardiac surgery, VII 1357
  charts, sailing-routes, IV 620
  Fageol Safety Coach, 1921, VII 728–9
  hydraulic engineering, V 532; VI 416
  knives, stone bladed, I 69
  lamp-oil, V 116
  magnetic measurements, VII 1462
  metals, extraction, IV 135, 141; V 95
  mining, IV 65–7; V 532, 825; VI 411, 416
  photography, V 739
  pioneer settlement, V 3, 5, 825
  radiology, development, VII 1332–3
  tribal economies, I 174, 176
  water-supply, V 558
Calydon, Greece, II 403
Camborne, Cornwall, England, V 129
Cambrai, France, III 178; IV 555
Cambridge, England
  arrow-head, bronze-age, I 511
  building materials, II 34
  electricity, generation, V 180
  King's College, construction, II 432–3, 436
  roads, II 532
  skin-boat specimen, II 582
  stained-glass windows, III 241–2
  windmills, III 103
  wireless telegraphy, V 227
  wooden vessels, bronze-age, I 517
— Massachusetts, U.S.A., VII 1294
Cambridgeshire, England, I 232; III 152, 315; V 10
Camden Town, see London

Camlachie, Scotland, V 241
Cammazes, France, III 467
Campagna, Italy, IV 34
Campeche, Mexico, V 263
Campine district, Belgium, IV 557
Campsie, Stirlingshire, IV 251, 256
Canada, VI 256, 418, 420
— Great Lakes, IV 553 (map)
Canary Islands
    banana cultivation, V 23
    cartography, III 531, 544
    dyestuffs, III 694; V 258, 265
    shipbuilding, III 477
    sugar-cane, III 6
    windmills, III 89–90
Canberra, Australia, VII 958, 959
Cane Wood, Hampstead, London, IV 492
Canopus, Egypt, III 564
Canosa, Apulia, II 268, 342, 558, 559
Canterbury, Kent, England
    cathedral, 12th century, II 436, 690
    fabrics, medieval, III 196
    leather craft, II 178
    Saxon sun-dial, III 596–7
    stained-glass, III 241
    town-planning, III 277
    water-supply, II 690, 691
    windmills, II 623
Canton, China, III 696; VI 16
— river, China, IV 328
Canvey Island, Essex, England, VI 408
Cape Breton, Nova Scotia, IV 50
Cape Finisterre, France, IV 590
Cape Horn, South America, IV 465; V 362
Cape of Good Hope, South Africa, IV 591; V
    362, 441, 442
Cape Verde Islands, Atlantic Ocean, IV 62
Cappadocia, Turkey, I 212, map 4
Capua, Italy, II 53, 500, 501 (map), 531, 654
Carambuc (Peking), III 380
Carcassonne, France, III 241, 283, 369, 459, 465
Carchemish, Mesopotamia, I 395–6, 675, 758,
    map 4; II 167
Cardross, Dunbarton, Scotland, V 406
Caribbean region, Central America, V 264
Caribbean Sea, VI 20; VII 915
Carillon Rapids, river Ottawa, Canada, IV 552
Carinthia, II 60, 67, 70; V 339
Carlisle, Cumberland, England, II 437; V 333; VI
    564; VII 786, 1388
— Pennsylvania, U.S.A., IV 105
Carmel, Mount, Palestine, I 31, 503, 520, map
    1A, 8
Carnarvon, Wales, II 237
Carney's Point, New Jersey, U.S.A., VI 508
Carnon, Cornwall, England, I 563
Carolina, U.S.A., V 261

Carrara, Italy, II 24, 28, 406
Carreghofa, Wales, IV 428
Carrhae, see Harran
Carrington, Cheshire, England, VI 557
Carron, Falkirk, Scotland, IV 105
Carshalton, Surrey, England, IV 207
Cart, river, Scotland, VII 890
Cartagena, Spain, II 5, 44
Carthage, II 516–17, 568, 660, 670, 699, table I
Casablanca, Spain, IV 554
Caspian Gates, III 505
Caspian Sea, IV 554; V 111, 121, 130; VII 788
Casquets, rocks, Channel Islands, III 499
Cassandra, Greece, II 3
Cassel, Germany, III 343, 658
Castel Durante (Urbania), Italy, II 285; IV 329 n.
Castelli, Italy, IV 329 n.
Castelnaudary, Aude, France, III 464
Castile, Spain, III 152, 170, 480, 703
Castille, II 356
Castle Eden, Durham, England, II 340
Castleford, Yorkshire, England, V 674
Castle Green, Bristol, England, IV 340
Castleton, Derbyshire, England, V 768
Castor, Northants, England, II 429
Catalonia, Spain, II 56
Catasauqua, Pennsylvania, U.S.A., IV 111
Catskills, New York, U.S.A., VII 1371
Caucasus, U.S.S.R., VI 21; VII 812
Caughley, Shropshire, England, IV 344, 354
Cauvery, river, India, V 542
Cawnpore, on Ganges, India, V 542
Celle, Hanover, IV 41
Central America, V 23, 264, 265, 773
Centuripe, Sicily, II 268
Cerne, West Africa, II 321
Cesse, river, Languedoc, France, III 466–7
Ceylon (Sri Lanka), V 2, 21–2, 24, 305, 752, 774
Chacewater, Cornwall, England, IV 179
Chagar Bazar, Mesopotamia, I 212, 213, 594, 721,
    map 7
Chalais-Meudon, Seine-et-Oise, France, V 402
Chalcis, Palestine, II 504
Chalk River, Ontario, Canada, VI 241, 262, 263,
    265
Châlons-sur-Marne, France, III 179, 290, 441
Chalon-sur-Sâone, France, IV 556; V 719
Champagne, France, II 69; IV 27; V 16
Champigneulles, France, I 559, 560, 562 (map)
Champlain, Lake, N. America, IV 553
Chanda, central India, V 70
Chandu-Daro, Indus valley, I 717, 719, map 5
Ch'ang-au, China, III 439
Channel Islands, IV 32
Channel Ports, England, VI 11
Chantilly, Oise, France, IV 338
Chaourse, Aisne, France, II 468

Chard, Somerset, England, V 408
Charente-Inférieure, *see* La Terrière
Charleroi, Belgium, IV 548, 557
Charleston, Virginia, U.S.A., IV 545
Charleville, Ardennes, France, III 290
Charlottenburg, Germany, V 64; VI 139, 147
Charlton Rode, Norfolk, England, I 612
Charnwood, Leicestershire, England, II 32
Chartres, France, II 444, 445, 644, 645, 649; III 241
Châtellerault, Vienne, France, III 430–1
Chatham, Kent, England
  cements, manufacture, IV 448
  concrete constructions, 1835, V 487
  paddle-boats, V 142
  shipbuilding, IV 576, 582; V 373
  watermills, IV 202
Chatsworth, Derbyshire, England, IV 476
Chattanooga, Tennessee, U.S.A., VI 479
Chaumont-en-Vexin, Oise, France, III 319
Chauny, Aisne, France, IV 555
Cheam, Surrey, England, II 297, 310
Cheddar, Somerset, England, V 34 n.
Chedworth, near Cirencester, England, II 446
Chellaston, Derbyshire, England, II 37
Chelles, France, I 131, map 2
Chelmsford, Essex, England, VII 1254
Chelsea, *see* London
Cheltenham, Gloucestershire, England, V 54
Chemnitz, Saxony, Germany, II 66; III 28; IV 178, 297; V 598, 788, 794
Chenab, river, India, VII 883, 897
Chenôve, Burgundy, II 647
Chepstow, Monmouthshire (Gwent), Wales, III 458; IV 461–2; V 514; VII 887, 888
Cher, river, France, III 450
Cherbourg, Manche, France, IV 465–6
Chertomlyk, south Russia, II 470–1, 555, 556
Cheshire, England
  chemical industry, IV 236, 239; VI 507
  'cottage' workshops, 1806, V 817
  cotton industry, V 810
  education, developments, V 779
  farming techniques, IV 19
  railway system, VII 783
Chesnoy, near Montargis, France, III 462
Chester, Cheshire, England, II 532; III 277–8; IV 460; V 336
Chesterfield, Derbyshire, England, II 445
Chesterford, Essex, England, II 218
Chesters, Northumberland, England, II 415
Chew, river, Somerset, England, IV 209
Chiangsu, province, China, III 439
Chiaramonte Gulfi, near Syracuse, II 343
Chiaravalle, Italy, III 309
Chicago, Illinois, U.S.A.
  air transport, VII 699
  bridges, V 513
  buildings and architecture, VII 930, 933
  buildings, design changes, V 62, 467, 478, 479, 481; VII 931
  canning industry, V 42
  car (automobile) industry, VII 715
  ceramics, V 668
  domestic appliances, VII 1133
  education, developments, VI 150
  fire-protection, buildings, V 481
  management, experiments, VI 85
  meat-packing industry, V 4, 42
  medical technology, VII 1318, 1324, 1329, 1331
  navigational equipment, VII 851
  nuclear reactors, VI 236, 252
  plutonium extraction, research, VII 260
  printing machinery, VII 1277
  railways, VII 703
  roads, IV 545
  steam turbines, VII 1026, 1031
  steel industry, V 62, 64
  telephony, development, VII 1223–4, 1230
Chicama valley, Peru, I 421, 436
Chichester, Sussex, England, II 505–6; III 265, 267, 278
Chiddingfold, Surrey, England, II 443; III 207
Chignal Smealey, Essex, England, II 439
Chile, V 24, 77, 244; VI 325, 357, 414, 521
Chiltern Hills, England, III 146; IV 20
China Sea, IV 592
Chinese Turkestan, III 195, 201–2
Chiozza, Italy, V 412
Chislehurst, Kent, England, V 226
Chiu Chen, Indo-China, I 393
Choisy-le-Roi, Seine, France, IV 360, 367
Choukoutien, China, I 23–5, 30, 37 (map)
Christmas Island, VI 282
Cincinnati, Ohio, U.S.A., VII 949, 1224, 1322
Cirencester, Gloucestershire, England, II 32; III 152, 277
Cissbury, Sussex, England, I 559, 562 (map)
Civita Lavinia, Italy, III 596
Civitavecchia, Italy, II 520; III 311
Clacton-on-Sea, Essex, England, I 28, 29, 141
Clairvaux, France, II 650
Clayton Tunnel, Sussex, England, V 332
Cleigh, Argyll, Scotland, I 605
Clermont, Puy-de-Dôme, France, II 763
Cleveland, Ohio, U.S.A., IV 540–1; V 664; VI 494, 658; VII 1022, 1321
Cleveland district, Yorkshire, England, V 68, 811
Cleveland Hills, Yorkshire, England, V 801
Clifden, Ireland, VII 1248
Clifton, Bristol, England, V 507
— Yorkshire, England, IV 607
Clinton, Massachusetts, U.S.A., VI 239, 245, 246

Clintonville, Wisconsin, U.S.A., VII 726
Clipsham, Rutland, England, II 35
Clos de Vougeot, Burgundy, II 647
Clough, County Antrim, Ireland, I 636
Clunie, Scotland, VII 900
Clwyd, Vale of, Wales, I 32
Clydach, Glamorgan, Wales, V 89
Clyde, river, and Clydeside, Scotland
  bridges, welded steel, VII 878
  chemical industry, VI 507
  docks, construction, IV 468
  electricity generation, VII 1068
  harbours, V 539–40
  industrialization process, V 815, 816
  iron smelting, IV 109; VI 478
  metal corrosion, discussion, VI 450
  shipbuilding, IV 548, 592; V 350–2, 372, 373,
    810–11
  steam-power, marine engines, V 143, 149
  town-planning, VII 965
Coalbrookdale, Shropshire, England
  bridges, IV 455, plate 30A, 457, 458
  canals, IV 569
  ceramics, 1830s, IV 355
  ironmasters, IV 235
  iron ore, smelting process, IV 99–102, 105–6
  social change, V 830
  steam-engines, IV 162, 179, 188–91
Coalport, see Coalbrookdale
Coatsville, Pennsylvania, U.S.A., VI 495
Coblenz, Germany, III 284
Cobridge, Staffordshire, England, IV 349, 353
Coburg, Bavaria, II 749
Colchester, Essex, England, II 32, 323, 515; III
  277–8; IV 648
Coleford, Gloucestershire, England, V 65
Collingham, Yorkshire, England, II 33
Collyweston, Northamptonshire, England, II 34
Colne valley, Hertfordshire, England, V 553
Cologne, Germany
  aqueduct, II 671
  ceramics, medieval, II 287, 299
  electric fires, development, VII 1134
  fabrics, III 196
  glass analyses, II 313
  glass-making, II 323, 339, 340
  iron trade, middle ages, II 76
  metal-work, ornamental, II 454
  mills, II 607, 620; IV 212
  plough-model, Roman period, II 86
  town-planning, III 279, 284; VII 974–5
Colombia, South America, IV 134; VII 1409
Colon, Panama, IV 562; VII 913
Colorado, U.S.A., VI 415
Columbia, Ohio, U.S.A., VII 731
Columbia, river, U.S.A., VI 239, 240, 261
Columbus, Ohio, U.S.A., V 663

Commachio, Italy, III 314
Como, and lake, Italy, II 60; III 448
Comorin, Cape, India, IV 610
Compostella, north Spain, II 526
Comstock Lode, U.S.A., VI 418
Condé, France, IV 646
Conflans St-Honorine, France, IV 556
Congo, VII 788, 802
— river, VII 1453
Connecticut, U.S.A., IV 58, 536–7
Connel Ferry, Argyll, Scotland, VII 880, 881
Conques, France, II 462, 468, 474
Constance, Baden, Germany, II 195, 691; III 386,
  389; V 667
— lake, Europe, I 517, map 3; III 427; V 403
Constantinople (anc. Byzantium), see also Istanbul
  agricultural implements, II 99
  alum industry, II 368
  building-construction, c.900, II 422–3
  ceramics, medieval, II 288
  crusades period, II 761–7
  dyestuffs, II 367, 368
  fabrics, III 196
  glass-making, II 328
  harness design, horses, II 553, 556
  international tension, 1901–15, VI 7, 9
  Islamic influence, II 761–7
  metal-work, II 452, 454, 467
  military methods, II 718, 720; III 361, 363
  printing, 18th century, III 405
  quarrying, c.10th century, II 29–30
  roads and land travel, II 497
  shipbuilding, II 577
  silk-weaving industry, II 197
  town-planning, III 279–80
  water-mill, 5th century, II 600
  water supply, II 664
Conway, Wales, III 284
— river, Wales, V 366
Copais, lake in Boeotia, Greece, II 678
Copan, Honduras, I 125
Copenhagen, Denmark
  ceramics, V 661, 666, 670
  fermentation industry, V 303
  magnetic deflection observation, IV 654
  marine diesel engines, VII 1006
  moler, building material, V 666
  radiology, medical, VII 1329
  rocket flights, VII 859
  veterinary school, 1773, IV 35
Coplay, Pennsylvania, U.S.A., V 485
Corbridge, Northumberland, England, II 409
Corby, Northamptonshire, England, VI 414
Corcelettes, Switzerland, I 722, map 3
Cordel, near Treves, Germany, II 313
Cordouan, France, II 524
Corfe Castle, Dorset, England, II 36

Corinth, Greece, I map 4, table F
  aqueducts, discussion, II 670
  ceramics, from c.700 B.C., II 259, 260, 288, 295
  glass-making, II 328, 332
  pottery, I 392, 401, 406, 409
  roads, II 498
  town-planning, III 275
Cork, Ireland, VII 1407
Cornete, Italy, I 518
Cornwall, England
  agricultural vehicles, III 140–1, 144
  atmospheric engines, IV 177
  ceramics, IV 340, 351, 352
  chemical industry, V 245, 284
  copper, IV 126; V 76
  Cornish beam-engine, V 126–30
  explosives, V 284
  gas lighting, IV 252, 262
  mining, IV 74, 77–80; V 76; VI 420
  pumping engine, IV 188, 499
  steam-engines, IV 179, 183–4, 188–91, 193–4;
    V 126–30
  stone building material, III 261; VII 922
  tin, IV 123, 124
  tinplate, III 689
  water supply, IV 499
  wool production, III 152
Corsica, Mediterranean sea, V 345
Cortaillod (lake-village), I 508, 512
Corvo, Azores, II 568
Cos, Aegean Sea, II 566–7
Costa Hill, Orkney Islands, VI 216
Coteau du Lac, Canada, IV 552
Cotswolds, the, England, III 145, 147, 257, 260–1
Couillet, Belgium, V 243
Courpalet, Rigole de, Orleans canal, III 463–4
Coventry, Warwickshire, England, III 703; IV
  176, 319; V 414–16, 428, 826
Cowdray Manor, Sussex, England, II 522
Coxside, Plymouth, England, IV 340
Craig Lwyd, Penmaenmawr, N. Wales, I 44, 562
Cranbrook, Kent, England, III 95
Crécy, France, II 721, 726
Crete
  ceramics, II 262
  food and drink, II 62
  glass-making, II 317, 318, 321
  vehicles, Kamaresa, II 545
  water supply, II 663
  windmills, II 618
  world war, 1939–45, VI 19, 20; VII 814
Crewe, Cheshire, England, V 63, 331, 335
Crofton, Yorkshire, England, IV 259
Croisac, Picardy, IV 239
Cromberg, II 75
Cromer, Norfolk, England, IV 49
Crossness, Kent, England, IV 510

Croton valley, U.S.A., V 554, 559
Croydon, Surrey, England, IV 499; VII 1378
Cuba, III 3, 8; V 264; VI 2; VII 473
Cues, Rhineland, III 585
Culebra and hills, Panama, IV 562; VII 911
Culross, Fife, Scotland, IV 260
Cumae, Campania, II 323
Cumberland, England, III 63, 140, 152; IV 251;
  V 60; VI 482
— Maryland, U.S.A., IV 537
Cumbernauld, Dunbartonshire, Scotland, VII 972,
  980
Cuxhaven, W. Germany, VII 864
Cyprus
  cast iron, 2nd century, II 7
  ceramics, II 274, 276–7
  copper, II 48, 50–1, 55
  cotton production, II 199; III 159
  glass-making, II 318–19, 341
  sugar-cane, III 6
Czechoslovakia, IV 120; VI 11, 18, 256

Dagenham, Essex, England, III 422; VI 528
Dahlen, W. Germany, VI 268
Dahlgren, Virginia, U.S.A., VII 1186
Dahshur, Egypt, I 308, 597 n., 657, 699, map 6
Dakhla oasis, Egypt, I 239, 262, 528, 534, map 6
Dakota, North and South, U.S.A., V 5, 6
Dale Dike, Yorkshire, England, IV 497
Dalmarnock, Scotland, IV 249
Dalmation coast, Adriatic, III 514
Dalmuir, Dunbartonshire, Scotland, IV 238
Dalsland, Sweden, I 513
Dalswinton lake, Scotland, V 143
Damascus, Syria, I map 8; II 759 (map)
  ceramics, medieval, II 301
  cotton production, III 159
  crusaders and traders, II 760, 761, 765
  glass-making, II 328
  iron trade, c.700 B.C., I 598
  ivory-work, I 665
  military weapons, III 350
  steel industry, II 57; III 350
  tool-making, III 115
  town-planning, III 275
Damastion, Epirus, II 5
Dammarie, near Montargis, France, III 462
Damme, Flanders, III 442; IV 558–9, 629
— Oldenburg, III 657
Dannemora, Sweden, V 55
Danube, river, IV 35, 548, 550, 555; VII 702
Danzig, II 76; III 167
Daoulas, Brittany, III 178
Darjeeling, India, VII 882
Darling Harbour, Sydney, Australia, V 47
Darlington, Durham, England, IV 292, 460; V
  333; VII 881

Darnley, Renfrewshire, Scotland, IV 247
Dartford, Kent, England, IV 202; VI 615
Dartford Tunnel, Kent, England, VII 905
Dartmouth, Devonshire, England, III 481, 487;
  IV 173
Daventry, Northamptonshire, England, VII 1255
Davis, California, U.S.A., VII 1407
Davis Strait, Greenland, IV 58
Dayton, Ohio, U.S.A., V 391; VII 789, 1010
Deal, Kent, IV 647
Dean, Forest of, England, II 10, 61, 70; III 31,
  78
Dee, river, England and Wales, IV 568
Deir el-Bahri, Egypt, I map 6
  agriculture, I 544
  basketry, I 422, 423
  incense arts, c.1500 B.C., I 295
  ships, c.1500 B.C., I 55
  textiles, c.2000 B.C., I 435, plates 12A, 13
  wood-work art, I 701
Deir el-Gebrawi, Egypt, I 733, map 6
Dejbjerg, Jutland, II 538, 540, 545, 547, 548, 551
Delaware and river, U.S.A., IV 549; VII 1371
Delfshaven, Holland, III 442; IV 640
Delft, Netherlands
  dredging methods, IV 638-9, 640
  education, technology, V 788
  pottery ware, IV 329, 336-7, 345
  textile manufacture, VI 657-8
  windmills, II 620; III 304
Delhi, India, VII 883, 1377
Delos, Greece, II 404, 446; III 274
Delphi (Mount Parnassus), Greece, I 262; II 484,
  499
Delvenau, river, Germany, III 444
Dendra, south Greece, II 481
Deniliquin, New South Wales, Australia, V 47
Denmark, III 97, 99, 217, 579
Deptford, see London
Derby, England
  buildings, fire-proofing, IV 475
  ceramics, IV 338, 341-2, 344
  machine-tools, IV 433
  silk-mill, IV 311
  steel, buildings, IV 475; V 477
  water supply, V 555
Derbyshire, England, III 80, 260; IV 212, 345; V
  597; VI 296
Dergmeer, the, Holland, III 302
Deruta, Italy, IV 329 n.
Derwent Valley, Cumberland, England, VII
  1369
Dessau, Anhalt, Germany, IV 622; VII 955
Detroit, Michigan, U.S.A.
  motor-vehicle industry, VII 715, 721, 730
  nuclear reactors, VI 255
  tunnels, VII 905

  typewriter invention, V 689
  tubeless tyres, VI 563
— river, U.S.A., VII 905
Deux-Ponts, Zweibrücken, Bavaria, IV 135
Devon, England
  agricultural vehicles, III 140-1, 146
  ceramics, IV 349, 351
  farming techniques, IV 19
  stone building, III 261
  wool industry, III 152
Devonport, Devon, England, IV 582
Dewsbury, Yorkshire, England, V 570, 810
Dhanauri, India, V 542
Dhuis, tributary, river Marne, France, V 555
Dieppe, France, I 665; II 141
Dieps and Tjaalingermeer, Holland, III 302
Digoin, on river Loire, France, IV 556
Dijon, France, III 379; V 563
Dilolo, Congo, VII 788
Dinard, France, VI 218
Dira Abu'n-Nega, Thebes, Egypt, I 370, map 6
Dixmude, Belgium, IV 629
Djemila, Tunisia, II 639
Dodona, Greece, II 557
Dogger Bank, North Sea, IV 52
Dolaucothy, South Wales, II 10
Doliche, Syria, I 598, map 4
Dolkheim, East Prussia, II 559
Dolphin Holme, Lancashire, England, IV 295
Domnarvet, Sweden, VI 473
Domodossola, Piedmont, Italy, IV 530
Don, river, Russia, see Tanais
— — Scotland, VI 495
— — Yorkshire, III 458
Donawitz, Austria, VI 489
Doncaster, Yorkshire, England, IV 107, 300; VI
  586
Donegal, Ireland, III 162
Donetz coal basin, Russia, V 70
Donnerupland, Jutland, II 83
Dorchester, Dorset, England, I 491; II 671
— Oxfordshire, England, III 241
Dordrecht, Holland, IV 156
Dorset, England
  building construction, III 260; IV 450
  button-making, V 826
  ceramics, IV 349, 351
  railways, V 331
  wool industry, III 152
Dortmund, Germany, IV 519; VI 97
Doubs, river, France, V 142
Dounreay, Caithness, Scotland, VI 248, 249
Dover, Kent, England
  breakwater, construction, IV 466
  building-construction, II 427
  Castle clock, mechanism, III 651, plate 32A
  cables, wire, V 622

Dover, Kent, England (*cont.*)
    cartography, IV 607
    harbour, III 317, 459; VII 738, 920, 922–3,
        924
    lighthouses, Roman period, II 523
    radio, experiments, VII 1232
    town-planning, III 278
    telegraphy, IV 658–60; V 622; VII 1232
    watermills, II 648; IV 199
— Straits of, IV 604, 607–8, 660
Dowlais, Glamorgan, Wales, V 611, 612; VI 480
Dowles, Worcestershire, England, IV 244
Down, Co., Ireland, I 400
Downs, the, roadstead off Kent coast, England,
    IV 590
Dreros, Crete, II 469, 470
Dresden, Saxony, Germany
    camera industry, VII 1290, 1291
    ceramics, II 307; IV 338
    engineering design methods, V 501
    gas-lighting, 1799, IV 261
    instrument-making, III 629
    tinplate industry, 17th century, IV 104
Dublin, Ireland
    ceramics, II 307; IV 341
    chemical industry, IV 239; V 254
    concrete, *c*.1750, IV 449
    filtration, water-supply, V 563
    map, traffic-flow, IV 623
    metal-work, II 475
    railway station design, V 333
    shipbuilding, II 581
    telegraphy, submarine, V 225
    waterways, III 455; IV 572
Dublin Bay, Ireland, IV 608
Duck Creek, Ohio, U.S.A., V 108
Duddul, *see* Hit
Dudley, Worcestershire, England, III 226; IV
    112
Dugga (ancient Thugga), near Tunis, II 530
Duisberg, W. Germany, VII 884
Dulruth-Superior, Minnesota, U.S.A., VII 917
Dumfries, Scotland, VI 548
Dümmer, Oldenburg, I 364
Dunbar, Pennsylvania, U.S.A., VI 476
Dundee, Scotland, III 176; IV 58, 251; V 570
Dundyvan, Scotland, IV 111–12
Dungeness, Kent, England, V 183, 210; VII 922
Dunkirk, France
    agriculture, 19th century, IV 33
    cartography, IV 665
    dredging, IV 629, 635
    wartime, 1940, VI 19, 168
    waterways, III 453
Dunnose, Isle of Wight, England, IV 607
Dunston, Staffordshire, England, VII 1031, 1071
Dura-Europos, Mesopotamia, II 363; III 275

Durham, England
    atmospheric engine, IV 177
    chemical industry, V 245
    coal-mining tools, IV 81, 85; VI 361
    fabrics, burial, II 756, 763; III 196–7
    metal-work, 7th century, II 457
    railway, VII 771
    roads, II 525
    rocketry, VII 860
    St Cuthbert's tomb, II 456, 756, 763
Durnberg, Austria, I 443
Dürrenberg, Austria, II 166
Dursley, Gloucestershire, England, II 32
Düsseldorf, W. Germany, VII 872
Dvina, river, Russia, IV 554
Dyers Cross, Surrey, England, II 326
Dystrup Mose, Denmark, I 209

Eagley, river, Lancashire, England, IV 495
Earls Barton, Northamptonshire, England, II 237
Easby, Yorkshire, England, II 437
East Africa, V 23
East Anglia, England
    building materials, III 261
    farming techniques, IV 17, 19
    farm tools, III 140, 145
    fish preservation, IV 47
    hydraulic engineering, V 544–5
    industrial change, V 810, 811, 816
    watermills, IV 200
    windmills, III 98, 101, 103, 107
Eastern Flevoland, Netherlands, VII 907
East Goodwin, English Channel, V 228
East Indies
    borax supply, III 228
    'Catalan Atlas', III 525
    food and drink, III 4, 6
    trade, 16th century, III 679
    rubber industry, V 752, 773, 774
    zinc, exports, 16th century, III 690
East Rand, South Africa, VI 422
East River, New York City, U.S.A., VII 881
East Texas oilfield, U.S.A., VI 405
Ebbw Vale, Monmouthshire, Wales, IV 113–14;
    VI 475, 479, 496
Eddystone Rock, off Plymouth, England, V 463
Eddyville, Kentucky, U.S.A., V 53
Edfu, Egypt, I 538, 539, 665 n., map 6
Edinburgh, Scotland
    bridges, VII 887
    cannon, 15th century, III 361
    chemical industry, IV 237, 243–4; V 254
    civil engineering, IV 444, 486
    computers, VII 1151, 1179
    drainage-works, IV 509
    education, agriculture, IV 41
    fabrics, bleaching process, III 176

machine-tools, IV 429–30
photography, V 723
railways, V 332; VII 769
roads, 19th century, IV 539
shipbuilding, V 362
type-founders, V 683
water-supply, IV 490, 500
weaving process, 1850s, V 583
Edmonston, Midlothian, Scotland, IV 177
Egham, Surrey, England, VII 1379
Egmondermeer, Holland, III 302
Egtved, Denmark, I 442, 518, plate 17A
Eider, river, Schleswig-Holstein, IV 560–1
Eilean an Tighe, North Uist, I 390, 397, plate 8
Eindhoven, Netherlands, VII 1198
Eire, VII 1407, *see also* Ireland
El Alamein, Egypt, VI 22
El-Amarna, Egypt, I map 6, table F
    balance, *c.*1350 B.C., I 783
    basketry, *c.*1500 B.C., I 422
    buildings, I 306, 473–4
    chemical arts, I 240
    glass and glazes, II 315, 318, 320, 332, 336, 341
    ivory-work, I 669
    leather, II 163
    loom remains, I 433
    sculpture, art, I 242
    steel-tools, I 649
El-Amrah, Egypt, I 304, map 6
Elan, river, central Wales, V 553, 558
Elath, Jordan valley, I 633
Elba, island, Mediterranean Sea, IV 647, V 59
Elbe, river, Germany, III 444, 455; IV 548; VI 23; VII 864
Elberfield, Germany, V 140
El-Bersheh, Egypt, I 200, 437, 700, map 6
Elbing, Poland, II 583; IV 539
Elburz, Iran, I 261
Eldora, Iowa, U.S.A., V 481
Elephantine, Egypt, I 539, 663, map 6
Eleusis, Greece, II 499
El-Hejra, Arabia, I 766
Ellis, Greece, II 499, 529
El-Jisr, Palestine, I 671–2, map 8
El-Kab, Egypt, I 260, map 6
El Mina, West Africa, III 547
El-Omari, Helwan, Egypt, I 366
El Salvador, VII 1409
El-Ubaid, Mesopotamia, I map 7, table E
    building techniques, I 306, 456, 465
    metal-work, I 638, 639–41
Elugalab, Eniwetok Atoll, Pacific Ocean, VI 279
Ely, England, II 679
Emilia, Italy, II 681
Enfield, Middlesex, England, IV 440; V 639
English Channel
    aeronautics, V 400, 410; VII 791–2

canal links, IV 555–6, 561
cartography, IV 612–13 (map); V 439 n. 2
'Catalan Atlas', III 525
Cherbourg harbour, IV 465–6
navigational equipment, VII 847–8
oil exploration, VII 1464
paddle-steamers, 1888, V 146
radio transmission, 1899, VII 1232
shipbuilding, iron ships, V 351
ships, *c.*1865, IV 592, plate 43A
telegraphy, IV 659
tidal power, VI 217
Eniwetok Atoll, Pacific Ocean, VI 279
Enkhuizen, Holland, IV 635
Enkomi, Cyprus, I 624, 677, map 4; II 318, 481
Ennsmündung, on river Danube, IV 550
Entwistle, Lancashire, England, IV 496
Ephesus, Asia Minor, II 341, 671, 760; III 275, 282
Epiphania (Hama), Syria, II 199
Epirus, Greece, I 359, map 4
Epsom, Surrey, England, VII 1226, 1246
Erbil, Mesopotamia, I 533–4, map 7
Erech (Uruk; *mod.* Warka), Mesopotamia, I map 7, tables E, F
    buildings, brick and stone, I 461–3
    canals, I 549
    metals, extraction, I 582
    potter's wheel, *c.*2000 B.C., I 199, 200
    roads, *c.*300 B.C., II 494
    vehicles, *c.*3500 B.C., I 205, 206, 710
    weapons, stone, I 496, 510, 512
    writing, *c.*3500 B.C., I 745–6, 747, 748
Erfurt, Thuringia, I 359
Eridu, Mesopotamia, I 461, 462, 463–4, 526, map 7
Erie, Lake, Canada–U.S.A., IV 552; VI 465; VII 917, 918
Erith, Kent, England, VII 1073
Ermine Street, England, III 278
Erösd, I 309, map 4
Ertebølle, Denmark, I 303, 495–6, 499, 507
Erzgebirge, the, Bohemia and Saxony, III 28; IV 335, 338
Esbly, France, VII 892
Eshnunna (Tell Asmar), Mesopotamia, I 466–7, 670, map 7; II 528
Eskilstuna, river, Sweden, III 452, 455
Esna, Egypt, I 477, map 6
Essen, Rhineland, Germany, IV 108; V 57
Esslingen, Germany, II 138
Essex, England
    building practices, III 265; V 467
    farming and tools, III 143; V 11, 13
    road traffic census, 1922, VII 872
    sewage system, VII 1384
    timber, shipbuilding, III 487
    wool industry, III 152

Estonia, I 166
Estremadura, Spain, III 152
Ethiopia, III 5, 505, 522; VI 17
Euboea, Greece, II 52
Euphrates, river, III 270, 438
Évreux, France, III 241
Exe, river, England, III 451, 456 (map)
Exeter, Devon, England
  ceramics, 13th century, II 293
  mail-coach system, 1784, IV 531
  sewage system, VII 1394
  ship design experiments, V 386
  town-planning, III 277
  waterways, III 451, 456
Ezinge, Holland, I 322

Fabriano, near Ancona, Italy, III 415
Faenza, Italy, II 311 n.; III 678; IV 329 n., 337
Faeroe Islands, North Sea, I 442; IV 57; VI 357
Fairfax, Virginia, U.S.A., IV 536
Fairford, Gloucestershire, England, III 242
Fairhaven, Massachusetts, U.S.A., IV 58
Fairlight, Sussex, England, II 34
Falkirk, Scotland, IV 208
Falmouth, Cornwall, England, II 577
Falun, Sweden, III 342
Fano, Italy, III 414
Fara, Mesopotamia, I 748, 750, map 7
Farnham, Surrey, England, I 303, 394, 497; V 422, 430
Faversham, Kent, England, II 339; V 284
Fayum, Egypt, I map 6, table E
  basketry, I 418–19, 423
  buildings, Neolithic, I 304
  flax cultivation, I 354, 372–3
  glaze, blue, II 315–16
  gypsum source, I 241
  metal implements, I 602
  plant cultivation, I 354, 365, 367–8, 372–3
  pottery, I 376, 377, 387
  reaping-stick, Neolithic, I 541, 542
  survey tools, III 512
  textiles, I 431, 448
  tools, non-metallic, I 501, 506, 511–12, 514, 516
  water-wheel, A.D. 113, II 671, 676
Federsee, Württemberg, I 303, 310, 316, 355, 359, map 3
Felixstowe, Suffolk, England, VII 796
Felling, Durham, England, IV 238
Fens, The, England, III 315–22, 459; IV 20, 32; V 544–5
Fenton, Staffordshire, England, IV 351
Ferguson Island, I 406
Fernando Po, West Africa, V 22
Ferrara, Italy, III 312, 314, 382, 678
Ferriby, Yorkshire, England, I 738, 739
Ferro (Hierro), Canary Islands, III 531; IV 597

Ferrol, Corunna, Spain, IV 582
Ferté-, see La Ferté-sous-Jouarre
Finavon, Scotland, I 322
Findon, near Aberdeen, Scotland, IV 54
Finisterre, Cape, Spain, III 548
Finland, I 166, 452; III 553; VI 10; VII 473
Finmark, Norway, VI 357
Finow, river, Germany, III 454
Fintelle, Flanders, III 441 n. 2
Firth of Forth, Scotland, V 62
Fishguard, Pembrokeshire, Wales, VII 920, 922, 923–4, 925
Fiume Morto, Italy, III 314
Fiume (Reika), Yugoslavia, IV 528, 553
Flanders, France-Belgium, VI 11; VII 847
Flax Bourton, Somerset, England, IV 210
Fleet Ditch, see London
Flint, Wales, III 284
Flintshire, Wales, IV 177
Florence, Italy
  architecture, Renaissance, III 245–50, 254, 258
  bridges, III 418, 424, 428–30
  canals, III 448
  cannon, 1326, II 726
  ceramics, IV 329 n., 337
  drainage engineering, III 309–10
  dyestuffs, II 365; III 694; IV 234
  glass spectacles, III 230
  paintings, Palazzo Vecchio, III 707
  printing, III 406
  roads, IV 544
  silk-culture, II 198
  spinning processes, II 201, 202, 207
  town-planning, III 280
Florida, U.S.A., VII 1447
Flushing, Holland, III 486
Foggia, Italy, II 510
Fogliari, see Phocoea
Folkestone, Kent, England, II 419, 446
Font-de-Gaume, Dordogne, France, I 151, 300
Fontéchavade, Charente, I 22, 29, 34, 37 (map), table C
Forest of Dean, Gloucestershire, England, V 55, 65
Forli, Italy, IV 329 n.
Formby, Yorkshire, England, VI 296
Fort Benton, Montana, U.S.A., V 666
Fortezza, Crete, II 321
Forth, river, Scotland, VII 890, 898
Fort Madison, Iowa, U.S.A., VII 890
Fortunate Islands, see Canary Islands
Fostat or Fustat (old Cairo), II 199; III 179, 182
Foyle, Lough, Ireland, II 582; IV 608, 610
Franche-Comté, France, IV 29
Frankfurt-am-Main, Germany, II 75, 531; V 226, 332, 346, 489
Frankfurt-on-Oder, Germany, III 454; VII 872, 1291

Freiberg, Saxony, Germany, II 11, 23, 65, 66; IV 139
Freibourg, Switzerland, III 538; IV 459
Fréjus, Var, France, V 118
French Riviera, *see* Riviera
Freschines, France, IV 38
Fresné la Mère, Falaise, France, I 609
Fresquel, river, France, III 464–5
Friedrichshütte, Silesia, IV 105
Friesland, Holland (Netherlands)
  drainage-mills, III 305
  drainage-techniques, II 681–3; III 301, 305; VII 907–8
  farming techniques, IV 15, 26; V 18
  milk processing, V 18
  windmills, IV 156
Frisian Islands, North Sea, IV 560
Friston, Suffolk, England, III 94
Frodingham, Lincolnshire, England, V 60
Froges, Isère, France, V 93
Froghall, Staffordshire, England, IV 569
Fucinus, lake, Italy, II 679
Fuente de Guarrazar, near Toledo, Spain, II 451, 455
Fundy, Bay of, Nova Scotia, VI 217, 218
Furness, Lancashire, England, IV 162; V 60
Fusina, Italy, III 441, 442
Fustat, *see* Fostat

Gagarino, Russia, I 301
Galapagos Islands, I 3; VI 357
Galgate, near Lancaster, England, IV 326
Galgenberg, Rhineland, II 295
Galicia, Poland–U.S.S.R., V 103, 110, 111, 113, 115; VI 9
Gallipoli (Gelibolu), Turkey, II 625; VI 9
Galloper shoal, off the Naze, Essex, England, V 451
Galloway, Scotland, VI 200
Gambia, West Africa, V 23–4
Ganges, river, India, III 506; V 440; VII 882, 883
Garonne, river, France, III 459, 466
Gascony, France, III 283; IV 32
Gaspé, Quebec, Canada, IV 50
Gatchina, Russia, IV 529
Gateshead, Co. Durham, England, IV 189, 448; V 138, 217; VII 1069
Gattow, E. Germany, VI 268
Gatun, Panama Canal, VII 915
Gaul, III 206, 509
Gaza, Palestine, I 410(22), maps 6, 8
Gazonne, Étang de la, France, III 462
Gebel-el-Arak, Egypt, I 516, 667, 669, 672, map 6; III 110
Gebel Zabara, Egypt, I 526
Gedding, Suffolk, England, III 101

Geneva, Switzerland, IV 539; VI 243; VII 700
Genoa, Italy
  building construction, III 248
  glass industry, II 327; III 217
  lighthouse, I 139, II 523
  printing, colonies, III 380
  railways, VII 770
  roads, II 526
  shipbuilding, III 478
  soap-making, III 170, 703
  textiles, III 170
  trade, II 766
  water-supply, V 556
Georgia, U.S.A., V 341
Georgian Bay, Lake Huron, Canada, IV 552
Georgsmarienhütte, near Osnabrück, Germany, V 68
Gerar, Palestine, I 597; II 85–6
Germany, VII 1025, 1026
Germigny-des-Prés, near Orleans, France, II 762, plate 44
Gerona, Spain, II 641
Gerthe, Bochum, W. Germany, VI 528
Gerumsberg, Sweden, I 442, 443
Gerzeh, Egypt, I 431, map 6, table E
Gezer, Palestine, I 201, 239, 248–9, 274, 762, 764, map 8
Ghana (1957), *see* Gold Coast
Ghassul, Palestine, I 419, 432, 433, map 8
Ghent, Belgium, II 526; IV 13, 16, 556–7, 559–60; V 116
Giant's Causeway, Northern Ireland, V 347
Gibraltar, III 505, 523; IV 580
Giessen, Hesse, Germany, V 269
Gihon, near Jerusalem, III 510
Gilbert Islands, Pacific Ocean, VI 21
Giovi Incline, Italy, VII 770
Girsu, *see* Lagash
Giza, Egypt, I 241, 482, 597 n., 714, map 6
Gjeithus, Norway, I 150
Glace Bay, Canada, VII 1248
Glamorganshire, Wales, III 139, 146–7, 434
Glasgow, Scotland
  aeronautics, V 404
  blast-furnaces, IV 109
  bridges, IV 454; VII 878
  buildings, cast-iron use, V 470
  chemical industry, IV 238–9, 247–9, 253, 673; V 250
  civil engineering, IV 468
  hydraulic power, V 535
  industrialization, 18th century, V 815
  machine engineers, IV 162
  Mechanics Institute, IV 445
  printing, V 699, 714
  railways, V 348; VII 783
  road-vehicles, V 422

Glasgow, Scotland (*cont.*)
  rubber-proofing, V 761
  shipbuilding, IV 593; V 143, 361
  soap-making, III 705
  standardization, V 801
  steam-engines, IV 181; V 134, 143
  textile industry, IV 285, 300, 315; V 583
  town planning, VII 965, 972
  water-supply, IV 489, 497; V 553
  waterways, V 540
Glastonbury, Somerset, England, I 318–21, 322,
    323, 404–5, plate 4; V 263, 423
Glatz (now Klodzko), Silesia, IV 520
Gleiwitz, Upper Silesia, IV 105, 235
Glencrutchery, Isle of Man, I 410(18)
Glendon, Pennsylvania, U.S.A., VI 476
Glodzko, *see* Glatz
Gloucester, England, II 479; III 172, 241, 277,
    688; V 323; VII 895
Gloucestershire, England, III 9, 137, 152; IV 19,
    104, 304
Gnathia, south Italy, II 268
Godalming, Surrey, England, II 445; V 198
Godavari, river, India, V 542
Gokstad, Norway, II 237, 579–80
Goldberg, Neresheim, Württemberg, I 311, 322
Gold Coast (now Ghana), West Africa, V 22, 93;
    VII 922, 925, 926
Golden Horn, Constantinople, III 433
Golling, near Salzburg, Austria, V 251
Goodwin Sands, Kentish coast, England, IV 467
Goole, Yorkshire, England, VII 890
Goslar, Harz Mountains, Germany, II 11, 611;
    III 691–2, 706; IV 71
Gosport, Hampshire, England, IV 106, 658
Göta, river, Sweden, III 455
Göteborg (Gothenburg), Sweden, III 455; V 333
Gotha, Germany, IV 620
Gothenburg, *see* Göteborg
Götland, Sweden, I 369; VII 1082
Göttingen, Germany, II 532; V 269, 458; VII
    828, 829, 1019
Gouda, Holland, III 442
Gournia, Crete, I 387, map 4
Govan, Glasgow, Scotland, IV 244
Governolo, on river Mincio, Italy, III 442
Graff-Hendriks polder, Holland, III 301
Gragnano, near Naples, Italy, II 520
Granada, Spain, I 359
Grand Banks, North Sea, VI 350
Grand Congloue, island near Marseilles, II 569
Grand Pa's Knob, Rutland, Vermont, U.S.A., VI
    216
Grand Pressigny, France, I 500–1, 505, 509
Grand Rapids, Michigan, U.S.A., VII 1380
Grange, Limerick, Ireland, I 195
Grateley, Salisbury Plain, England, V 330

Gravelines, *see* Grevelingen
Gravesend, Kent, England, VII 901
Great Barrier Reef, Australia, IV 575
Great Lakes, Canada–U.S.A.
  Erie canal, IV 555
  grain transport, V 5
  hydro-electric power, VI 195
  land transport, VII 702
  mapping, IV 597
  St Lawrence Seaway, IV 551–3; VII 920, 921
  shipbuilding, VII 740, 750
Great Level (Bedford Level), England, III 316–21
Great Ouse, river, eastern England, V 545
Great Plains, North America, V 3
Great Wall, China, VI 15
Greenburg, Pennsylvania, U.S.A., V 491
Greenford Green, Middlesex, England, V 271
Greenland, III 532–3, 705; IV 58; V 92
Greenock, Renfrewshire, Scotland, IV 181, 501,
    592; V 362, 540
Greenstead, Essex, England, II 425
Greenwich, *see* London
Grenelle, Paris, France, IV 498; V 211, 555
Grenoble, Isère, France, IV 539; VI 508
Grevelingen (Gravelines), Nord, France, III 453;
    IV 629
Grignon, France, IV 41
Grimaldi caves, Italy, I 168, map 2
Grimes Graves, East Anglia, England, I 44, 235,
    515, 558–61, 562 (map)
Grimsby, Lincolnshire, England, IV 51, 52, 54;
    VI 343
Groningen, Holland, II 681; III 303; IV 15
Grote Wiericke channel, Holland, III 304
Grozny, Russia, V 110
Guadalcanal, Seville, Spain, II 11
Guadalentin, river, Spain, IV 498
Guadalquivir, river, Spain, IV 551
Guatemala, Central America, V 258; VII 1409
Gubbio, Umbria, Italy, IV 329 n.
Guernsey, Channel Islands, IV 543
Guiana, South America, I 164–5, 166; V 23
Guildford, Surrey, England, III 456–7
Guinea, West Africa, III 547, 680
Guisborough, Yorkshire, England, IV 250
Gulf of Mexico, VI 390; VII 1453
Gulf of St Lawrence, Canada, VII 917
Gulf Stream, IV 619; VII 1447
Gurob, Fayum, Egypt, I 419, 420, 422, 433, map
    6; II 318
Gut Gunern, Silesia, IV 28
Guyana (1966), *see* British Guiana
Gytheion, Laconia, II 30

Haarlem, Holland
  dredging machine, IV 639
  linen manufacture, 1755, III 175

precision instruments, IV 408
printing, III 377, 383, 392
windmills, II 619–20
Haarlemmermeer, Holland, III 307
Hadhramaut, *see* Wadi 'Amd
Hadley Base, Pennsylvania, U.S.A., VII 868
Hadra, near Alexandra, Egypt, II 268
Hague, The, *see* The Hague
Haida, British Columbia, Canada, I 178
Hailsham, Sussex, England, III 129
Haine, river, Belgium, IV 557
Halaf, Mesopotamia, I 201, 365, 367, 460, maps 4, 7, table E
Haldon Hill, Exeter, England, I 312
Halifax, Yorkshire, England, III 152; IV 268; V 105, 582
Hall, Tyrol, Austria, III 216, 341
Halle, Saxony, IV 233
Hallstatt, Austria, I map 3
    glass, *c.*600 B.C., II 322
    horse-sandal, Roman, II 515
    leather relics, Iron Age, II 166
    metal implements, I 617–18, 619
    mining tools, Iron Age, I 567
    weapons, *c.*900 B.C., II 455
Hama, Syria, I 201, map 4; II 199, 614
Hamairan, Persia, I 262
Hamburg, Germany
    air transport, VII 800
    buildings, fire-protection, V 482
    dredger design, IV 632–3
    glass-making, III 225
    railways, VII 768
    shipbuilding, IV 594
    wartime bombing, VII 816
Hamilton, Ohio, U.S.A., VI 478
Hammersmith, *see* London
Hampshire, England, III 139, 152; IV 19; V 554
Hampstead, *see* London
Hampton, Middlesex, England, IV 494
Handsworth, Yorkshire, England, IV 107
Hanford, Columbia River, U.S.A., VI 239–40, 243, 245, 252, 257–8, 261–2
Hankow, Hupeh, China, V 70
Hanley, Staffordshire, England, IV 344, 347
Hanover, W. Germany
    calculating machine, VII 1163
    mapping, IV 610
    optical glass, IV 361
    railways, V 340
    road surfaces, IV 538, 540
    telegraphy, IV 644
Han-Yang, Hupeh, China, V 70
Happisburgh, Norfolk, England, II 664
Harappa, Indus valley, I map 5
    bronze age weapons, I 613
    carts and wagons, I 207, 717, 719

circle principle, I 195
door-socket, *c.*2500 B.C., I 194
food plants, I 360 n., 366
grain-stores, I 52
potter's wheel, I 202
pottery, I 385
script, I 760
tools, non-metallic, I 512
town plan, I 52; III 270
wheel, vehicles, I 207
Haro, river, Punjab, VII 882
Harran, Mesopotamia, I 282, 371, map 4
— (Carrhae), Syria, II 736, 738
Harrow Hill, Sussex, England, I 561, 562 (map)
Hartford, Connecticut, U.S.A., IV 439; V 312, 467, 639, 818; VII 731
Hartlepool, Co. Durham, England, VI 634
Hartley, Northumberland, England, IV 92
Harty, Kent, England, I 612
Harwich, Essex, England, IV 51, 448
Harz Mountains, Germany
    mining, III 82; IV 70–1, 78, 93
    railways, V 345
    rockets, VII 865
Haskell, New Jersey, U.S.A., VI 508
Haslar, Hampshire, England, VII 747
Hassfurt, Bavaria, V 502
Hassuna, Mesopotamia, I map 7, table E
    building construction, I 306, 456, 459
    door-socket, 4500 B.C., I 193
    tools, non-metallic, I 502, 507, 512
Hastings, Sussex, England, II 587, 720; IV 466
Haswell, Durham, England, IV 84
Hatfield Chase, Yorkshire, England, III 320
Hat-Nub, Egypt, I 477, map 6
Hattusas, *see* Boghazköy
Havana, Cuba, VII 1251
Havel, river, Prussia, III 450–1, 454–5
Haverhill, Suffolk, England, III 101
Hawaii, North Pacific, V 23; VII 813
Hawara, Egypt, II 231
Haweswater, lake, Westmorland, England, VII 1368
Haworth, Yorkshire, England, VII 1391
Hayange, Moselle, France, IV 111, 115
Hayle, Cornwall, England, V 128–9
Hebburn-on-Tyne, Northumberland, England, IV 93, 97; VII 1080
Hebrides, Scotland, I 376–7, 381–2, 390, 397; 406–8; V 450
Heckington, Lincolnshire, England, III 96
Heeshugh, Schleswig-Holstein, II 165
Heidelberg, Baden, Germany, I 23, table C; IV 273, 656; V 821
Hejra, El-, *see* El-Hejra
Helbon, Damascus, Syria, I 282, map 8
Heliopolis, Egypt, I 359, map 6

Helmshore, Lancashire, England, IV 326
Helsinki, Finland, IV 497; VII 977
Helvoet, Holland, IV 635
Hendon, Middlesex, England, V 147; VII 800
Henrichemont, Cher, France, III 290–1
Herculaneum, near Naples, Italy, II 225; III 237, 596
Herend, Hungary, V 662
Herefordshire, England, III 152
Hermus, river, Asia Minor, III 704
Hertford, Hertfordshire, England, VII 950
Hertfordshire, England, III 152, 265; IV 492; VII 964
Hertogenbosch, Holland, IV 558
Hesse, Germany, IV 527, 610, 644
Het Gein, Holland, III 441
Hierakonpolis, Egypt, I map 6
    ancient society, c.3200 B.C., I 50, 51
    building construction, I 305
    ivory-work, II 161
    leather clothing, II 161
    metal-work, I 639, 641, 666; II 484
    script, Egyptian palette, I 755
Hierro, see Ferrot
High Island, Hong Kong, VII 912
Highlands of Scotland, VII 1072
Hildesheim, Germany, II 454
Himalayas, mountains, India, IV 610; V 442; VII 874
Hindu-Kush, mountains, central Asia, III 505
Hintlesham, Suffolk, England, I 616
Hiroshima, Japan, VI 23–4, 230, 235, 276–7; VII 816
Hissar, Persia, I 212, 721, map 5
Hit, Mesopotamia, I 251 (map), 252–3, 258, map 7; II 375
Hjälmar, lake, Sweden, III 452, 455
Hjedding, Denmark, V 19
Hjortspring, South Jutland, II 165, 578–9
Hobbacott Down, Cornwall, England, IV 569
Hoboken, New York, U.S.A., V 149
Hoechst-am-Rhein, Germany, V 248
Hoerde, Germany, VI 479
Hohenheim, Germany, IV 41
Holbaek, Zealand, I 167
Holborn, see London
Holderness, Yorkshire, England, I 315
Holland (Netherlands), IV 557 (map)
Hollandsch Diep area, Holland, IV 558
Hollywood, California, U.S.A., VII 1302, 1313
Holmegaard, Denmark, I 161–2, 499
Holstein, Denmark, IV 26, 41; V 20
Holt, Denbighshire (Clwyd), Wales, I 394; II 300, 305
Holtenau, N. Germany, IV 561
Holyhead, Anglesey, Wales, IV 459–60, 531; V 336, 365–6; VII 899, 1251

Holy Land, the, III 473
Homestead, Pittsburgh, U.S.A., VI 478
Homs, Syria, I 529, maps 4, 8
Honan, China, I 772; III 439
Honduras, VII 696
Honfleur, France, II 657
Hong Kong
    dam construction, VII 1367, 1369
    hydrofoil service, VII 745
    wartime, 1942, VI 21
    water-supply, VII 907, 911 (map), 912
Hooge Mierde, Holland, I 314
Hoogly, river, India, VII 882, 883, 885
Hook Norton, Oxfordshire, England, II 184
Hoorn, Holland, IV 632; V 18
Hopedale, Massachusetts, U.S.A., V 586
Hopewell, Virginia, U.S.A., VI 508
Hotham Carr, Yorkshire, England, I 608
Hot Lane, Cobridge, Staffordshire, England, IV 349
Hounslow Heath, Middlesex, England, III 554; IV 607; V 447
Houston, Texas, U.S.A., VII 1391
Hradish, Moravia, II 69
Huai, river, China, III 439
Huaian, China, III 440
Huancavelica, Peru, IV 139
Huang-Ho, river, China, III 438
Huddersfield, Yorkshire, England, V 523, 810
Huddleston, Yorkshire, England, II 32, 33, 384
Hudson, river, U.S.A.
    aqueduct, VII 1371
    canal links, IV 549, 553 (map), 555
    outflow, ocean floor, VII 1453
    paddle-steamers, V 143, 145
    tunnelling methods, V 519–20; VII 1371
Hudson Bay, I 498
Hull (Kingston-upon-Hull), Yorkshire, England
    brick and tile industry, II 305, 438
    building construction, II 438
    cement production, V 486
    ceramics, II 305; IV 349–50
    dredgers, IV 640, 641
    fishing industry, IV 52, 54, 58
    railway retarders, VII 786
    telephone system, IV 227 n.
    town-planning, III 284
    trade and transport, II 527; IV 349–50
    trade unions, VI 98
Humbeek, Flanders, III 453
Humber, the, England, III 260
Hungary, III 82, 579, 706
Hunsbury, Northamptonshire, England, I 322
Huntingdon Beach oilfield, California, U.S.A., VI 392
Huntingdonshire, England, III 315
Hurlet, Renfrewshire, Scotland, IV 250–1

Huron, lake, N. America, IV 552
Hyde, Cheshire, England, IV 84
Hymettos, mount, Greece, II 25–7

Iberian peninsula, III 95, 100
Iceland, III 532; IV 49, 54; VI 357; VII 1454
Idaho, U.S.A., VI 335
Idbury, Oxfordshire, England, II 37
Idria, Italy, IV 134–5
Iffley, Oxfordshire, England, III 456
Ifield, Sussex, England, V 133
Iglau, Czechoslovakia, II 66
Igornay, Seine-et-Loire, France, V 118
IJ (or Y), river, Holland, III 293–4; IV 640; VII 905
Ijmuiden, Holland, IV 560
IJsselmeer, Holland, V 545; VII 907, 908, 909
IJsselmeer Polders, Netherlands, VII 907
Île de France, IV 31
Illinois, U.S.A., V 4, 36
Inchgarvie, Firth of Forth, Scotland, V 503
Inchinnan, Renfrewshire, Scotland, VII 890
India, VI 6, 221
Indian Ocean, III 521
Indiana, U.S.A., V 4; VI 544, 559
Indo-China, V 774
Indonesia, V 752
Indre, France, I 149
Indret, island near Nantes, France, IV 102
Indus, river, I map 5; III 438, 505; VII 879, 883, 886
International Rapids, St Lawrence River, Canada–U.S.A., VII 918
Inverness, Scotland, V 323
Iowa, U.S.A., V 4, 5
Ipswich, Suffolk, England, IV 3, 71; V 490
Iran, VI 186, 191
Iraq, I 160, 198, 306
Ireland, Dublin area, IV 623 (map)
Irenopolis, II 57
Irgenhausen, Switzerland, I 436, map 3, plate 12B
Irkutsk, Siberia, IV 529
Ironbridge, Shropshire, England, IV 455
Iron Gates rapids, Romania, IV 550
Iroquois Dam, Canada, VII 918
Irrawaddy, river, Burma, VI 392
Irwell, river, Lancashire, England, III 458; IV 563, 571
Ischia, Italy, VI 220
Isle of Dogs, see London
Isle of Man, Irish Sea, III 140; VII 1251
Isle of Sheppey, Kent, England, VII 891
Istanbul, Turkey, VII 1366; see also Constantinople (anc. Byzantium)
Istria, Italy, IV 34
Italy, II 501 (map)
Iwo-Jima, volcano islands, Pacific Ocean, VII 814

Jamaica, West Indies, IV 413; V 23, 93, 264
Jammu, India, VII 874
Jan Mayen Island, Arctic Ocean, IV 58
Janville, Oise, France, IV 555
Japan, III 5, 378, 524, 579, 696–7
Jarmo, Iraq, I map 7, table E
   basketry, c.5000 B.C., I 418
   food plants, I 365, 367, 368
   house-building, c.5000 B.C., I 306
   pottery, I 376
   tools, non-metallic, I 502
Jarrow, Co. Durham, England, II 756; III 595; IV 95, 238; VI 480
Java, East Indies, I 226; V 21, 23, 258, 773, 774
Javel, Paris, IV 247 and n.
Jbail, see Byblos
Jemdet Nasr, Mesopotamia, I 745, 747, map 7, tables E, F
Jena, Germany, II 749; V 673; VII 951, 1322
Jerash, Palestine, III 280
Jericho, Palestine, I map 8, table F
   basketry, traces, I 419–20
   building techniques, I 306
   chariots, I 725
   ivory-work, I 672
   plant cultivation, I 372
   potter's wheel, c.2000 B.C., I 201, 388
   pottery, I 376, 387, 391, 420
   tools, non-metallic, I 501, 502
   water-supply, I 530
   wood-work, I 684, 685
Jersey City, New Jersey, U.S.A., V 519; VI 522; VII 1378
Jerusalem, Palestine, I map 8, table F; III 520 (map)
   apothecaries, I 292
   Byzantine influence, II 761
   cotton industry, III 159
   Jewish calendar, III 573–4
   metal-work, 950 B.C., I 661
   military technology, A.D. 70, II 715
   Pool of Siloam, III 510
   salt, food preservative, I 259
   water conduit, c.700 B.C., III 510
Jhelum, river, India, VII 883
Jisr, El-, see El-Jisr
Joachimsthal, Bohemia, II 11, 13 and n.; III 28
Johannesburg, South Africa, V 828; IV 424; VII 782
Johnstown, Pennsylvania, U.S.A., VI 477
Jordan, Palestine, III 280
Jotapata (Tel Jefat), Palestine, II 715
Jülich-Berg, Western Germany, IV 520
Jumièges, Normandy, II 426
Jutland, Denmark, I 156, 496, 514, 518; V 20

Kaar-el-Banaf, Egypt, II 212
Kafr Ammar, Egypt, I 434, map 6

Kahun, Egypt, I 433, 542, map 6; III 269
Kaifeng, China, III 439
Kalmuk steppes, I 716, 717, 719
Kalvträsk, Sweden, I 708
Kamaresa, Crete, II 545
Kampen, Holland, IV 636
Kamperdiep, the, Holland, IV 636-7
Kanam, Kenya, I 22, 25, 37 (map)
Kansas, U.S.A., V 5; VI 189, 401
Kansas City, Kansas, U.S.A., VII 1226
Kapisa, see Bagram
Kara Hissar, Armenia, I 262, map 4
Karhof Cave, Westphalia, I 367
Karlsborg, Sweden, VI 612
Karlstadt, Yugoslavia, IV 528
Karmir Blur, Armenia, I 684, 685
Karnak, Egypt, I map 6
  building methods, I 457-8, 479-80, 483
  leather, uses, II 164
  plant cultivation, I 543-4, 545
  quarrying methods, c.3000 B.C., I 570
  water-clocks, c.1400 B.C., III 601
  water-control, I 539
Kashira, U.S.S.R., VII 1082
Kashmir, Vale of, India, VII 874
Katrine, Loch, Scotland, V 553
Kattegat Sound, off Sweden and Denmark, IV
  554, 560
Keeweenaw, Michigan, U.S.A., VI 413
Kehl, Rhineland, Germany, V 514
Keighley, Yorkshire, England, V 777
Kelso, Roxburghshire, Scotland, IV 459
Kempten, W. Germany, VII 1005
Kempton Park, see London
Kendal, Westmorland, England, III 152; VI 206
Kennet, river, England, III 457
Kent, England
  bridges, VII 891
  building methods, III 261, 265; V 467
  cartography, IV 611, 614
  cement industry, IV 448, 449
  ceramics, IV 345
  dilatometer invention, IV 397
  dye stuffs, III 694
  farming techniques, IV 4, 17, 19
  maps, IV 614, 621
  port construction, IV 466
  roads, IV 543; VII 872
  shipbuilding, III 487
  water-mills, IV 200
  wool industry, III 152
  windmills, III 101
Kentish Town, see London
Kentucky, U.S.A., IV 548; V 4; VII 1377
Kephalos, Greece, II 58
Kerch, Crimea, II 545; III 200
Kerkmeer, Holland, III 302

Kerma, Nubia, I 422, 435, 699
Keswick, Cumberland, England, VI 212
Ketley, Shropshire, England, IV 569
Kew Gardens, Surrey, England, V 774; VI 338
Key West, Florida, U.S.A., VII 1251
Khafaje, Mesopotamia, I map 7
  chemical arts, 2500 B.C., I 242
  ivory-work, I 681(3)
  plants, cultivation, I 354, 362, 372-3
  transport methods, I 705
  wheeled vehicles, 2500 B.C., I 205
Kharga oasis, Egypt, I 262, 528, 534, map 6; II
  666
Khartoum, Sudan, I 501
Kherbet-Agoub, Algeria, II 639
Khorsabad, Mesopotamia, I 469, 470, 472, 675,
  map 7; II 54
Khotan, Turkestan, II 197
Khurasan, Persia, III 191
Khyber Pass, India, VII 883
Kidderminster, Worcestershire, England, V 581
Kiel fiord, Baltic, IV 560-1
Kilbowie, Dunbartonshire, Scotland, VI 685
Kilmarnock, Ayrshire, Scotland, V 581
Kimberley, South Africa, V 828
Kimpech, see Campeche
Kingsberg, Norway, II 11
King's Cross, see London
King's Lynn, Norfolk, England, II 440, 462, 463,
  469; V 544
King's Norton, Birmingham, IV 568
Kingston, Ontario, Canada, IV 552
Kingston Down, Kent, England, II 456
Kingston-on-Thames, Surrey, England, V 739
Kingston-upon-Hull, see Hull
Kirjath Sepher, Palestine, I 248
Kiruna orefield, Sweden, VI 200, 417
Kish, Mesopotamia, I 206, 208, 209, 717, 753,
  map 7; II 328
Kittis, Finland, III 553
Kitty Hawk, North Carolina, U.S.A., V 391, 407;
  VII 789, 838, 861, 1010
Kivik, Sweden, I 213, 726
Kizzuwadna, Armenia, I 595, map 4
Klagenfurt, Austria, II 546
Klondike, Canada–U.S.A., VI 411
Knapsack, W. Germany, VI 524
Knidos, Asia Minor, I 237, map 4
Knockadoon, Limerick, Ireland, I 312
Knole, Kent, England, II 256
Knolls, Schenectady, New York, U.S.A., VI 253,
  255
Knossos, Crete, I map 4
  building methods, I 485-8
  drainage system, I 535; II 528; IV 504
  ivory work, I 672, 673
  plants, food use, I 361

pottery, I 387, 388, 398, 400, 403
roads, I 714; II 494, 528
script, Minoan, I 758
town-planning, II 528
water-pipes, I 535
wheels, c.1500 B.C., I 212, 214
Kocsi, Hungary, II 547
Kodok (Fashoda), Sudan, I 732
Kohala, India, VII 884
Köln-Lindenthal, Neolithic village, I 308, 309
Kolyma, river, Siberia, U.S.S.R., VI 416
Königsberg, near Chemnitz, Saxony, IV 178
Korea, III 378; VI 6; VII 1023
Korpilahti, Finland, I 413
Kostenki, Russia, I 302
Kostromskaya Stanitsa, Russia, I 324–5
Kotri, India, VII 879, 880
Kouklia, Cyprus, II 450, 460
Kouphonisi island, I 247, map 4
Kowloon, Hong Kong, VII 911
Krain bei Strahlen, Germany, IV 28
Krasnoe Sormova, U.S.S.R., VI 498
Krasny Octyabr, U.S.S.R., VI 498
Krefeld, Rhineland, Germany, IV 320; V 587; VI 664
Kreimbach, Rhenish Bavaria, I 620
Kromwater, Holland, III 302
Kronstadt, Russia, IV 179
Kuala Lumpur, Malaysia, VI 417
Kuangsi, province, China, III 439
Kuban, river, I 588, 611, map 4
Kul Oba, Kerch, Russia, I 646–7, map 4
Kuortane, Finland, I 707, 708
Kurdistan, I 520, map 7; VII 894
Kuyunjik, Mesopotamia, I 452–3, 710, 741, map 7; see also Nineveh

La Baie (now Bourgneuf), France, IV 50
Labrador, Canada, IV 58; VII 917
La Caunette, Languedoc, France, II 65
La Chapelle, France, V 211
Lachine, Canada, IV 551
Lachine Rapids, St Lawrence River, Canada, VII 918
Lachish (Tell ed-Duweir), Palestine/Israel, I map 8
glass-ware, c.1600 B.C., II 319
ivory-work, I 674
potter's wheel, I 201, 388
pottery, I 385, 388, 390
script, inscriptions, I 762, 766
water-supply tunnel, I 530
Laconia, Greece, II 58
Ladoga, lake, Russia, IV 554
Ladbroke Grove, see London, England
Ladysmith, Natal, South Africa, V 397
La Ferté-sous-Jouarre, France, III 16

Lagash (Girsu, Tello), Mesopotamia, I map 7
bitumen industry, I 253
cubit measure, I 778, 780
ivory-work, I 665 n.
metal-work, I 582, 583, 640–1
syballic signs, I 748, 749
Lagonza, Italy, I 359
Lagos, Nigeria, V 23
Lake District, England, III 261; IV 496; V 554
Lakewood, California, U.S.A., VII 1357
La Madeleine, France, I 134
Lambaesis, Algeria, II 530
Lambeth, see London
Lambro, river, Italy, III 446
La Motte d'Apremont, Haute Saône, France, I 443
La Mouthe, Dordogne, France, I 235
Lampertsloch, Alsace, V 102
Lampy, stream, Languedoc Canal, III 464, 468
Lancashire, England
bridge construction, VII 898
building methods, III 260, 264
ceramics, IV 352
chemical industry, IV 239; V 245, 301; VI 507
clothing manufacture, VI 681
cotton industry, III 159, 162; V 628, 810, 825
education, technical, V 779
industrialization, discussions, V 815, 816–17, 825
metals, refining, IV 64, 115, 126
metal-working, V 628
mining, IV 64
railways, V 333
steam-engines, IV 166, 177
steel industry, V 62
spinning machinery, III 162; IV 277, 287, 315; V 577, 578
textile industry, V 574, 576, 579, 580, 588, 809, 810–11
water-power, V 301
water-supply, IV 501
windmills, III 89
Lancaster, England, IV 315; VII 1271
Landes, France, III 183
Lane Delph, Staffordshire, England, IV 347
Lane End, see Longton
Langley Field, U.S.A., VII 1021
Langres, France, II 647
Languedoc, France, III 15, 152, 283, 319, 459, 464–8
Lansberg, Bavaria, IV 135–6
Lanskron, Austria, IV 261
La Perche, France, II 74
Lapland, I 413, 436; IV 596
La Providence, Réhon, France, VI 458
La Quina, Charente, I 31
La Rance, France, VI 218

Larderello, Italy, VI 219, 220
La Rochelle, Bay of Biscay, France, III 679
Lascaux, France, I 147, 331, 744, 745
La Tène, Lake Neuchâtel, II 166, 170, 557
La Terrière (Charente-Inférieure), I 391
Latium, Italy, II 270
Latvia, VII 1292
Laudot, river, near Toulouse, France, III 468
Lauenburg, Germany, III 443
Laumes, Côte d'Or, France, I 610
Laurel Hill, New York, U.S.A., V 85, 96
Laurence, Massachusetts, U.S.A., VI 658
Laurion, Greece, I 558, 584, map 4; II 1–4, 43–5, 110
Laussel, Dordogne, France, I 152, 153
Lavenham, Suffolk, England, III 152
Lawton, Cheshire, England, IV 565
Layer Marney, Essex, England, II 439, 440
Lea, river, Essex, England, III 451, 456; IV 492; V 554
Leafield, Oxfordshire, England, VII 1249
Leatherhead, Surrey, England, VI 595
Lebanon, V 259
Le Bourget-du-Lac, Savoy, I 359, 361, 369
Le Creusot, France, IV 105; V 66
Lee, see London
— river, England, VII 1368
Leeds, Yorkshire, England
    architecture, III 257
    chemical industry, V 242
    education, technical, V 714
    engineering industry, V 801; VII 1061
    industrialization, discussion, V 828
    machine-tools, IV 433
    road-steamer, V 421
    textile industry, IV 292, 299, 323, 326; V 570, 810
    waterways, III 458
Leek, Staffordshire, England, IV 345
Leghorn (Livorno), Italy, II 523; III 287
Le Havre, France, IV 555; V 48, 351; VI 97
Lehringen, Germany, I 30
Leicester, England, V 598; VII 1388
Leicestershire, England, I 382; III 152; IV 569
Leiden, Holland
    canal-locks, III 452
    chemical industry, IV 233, 243
    clock by Salomen da Castor, III 663
    dredgers, IV 632, 633, 641
    farm-produce factories, V 18
    land drainage, III 303
    scarlet dye discovery, III 695
    weaving looms, III 167
    windmills, IV 156–7
Leighton Buzzard, Bedfordshire, England, V 487
Leipzig, Germany, V 683; VI 527; VII 800, 1269, 1289

Leith, Edinburgh, Scotland, V 539
Lek, river, Holland, III 442
Le Manoir, France, VII 895
Le Mans, Sarthe, France, V 392, 423
Le Moustier, Dordogne, I 31, maps 1A, 2
Lena, river, Siberia, U.S.S.R., VI 415
Lendersdorf-Krauthausen, Germany, IV 115
Lengyel, Hungary, I 369
Leningrad (Petrograd), U.S.S.R., VI 10, 12, 21, 280, 328; VII 1322; see also St Petersburg
Leoben, Austria, II 76; V 64, 664
Leominster, Herefordshire, England, III 154
Lepanto, Gulf of Corinth, III 350, 473
Le Petit-Gentilly, Seine, France, V 395
Lepini, mountains, Italy, III 311
Leptis Magna, north Africa, II 519–20
Le Puy, Haute Loir, France, II 764
Lers, river, near Toulouse, France, III 464–5
Les Baux, Bouches-du-Rhône, France, V 91, 666
Lesbos, Aegean Sea, II 58
Les Cascades, Canada, IV 552
Les Cottes, Vienne, I 147
Les Eyzies, Dordogne, France, I 26
Lesparre, Gironde, France, III 319
Letchworth Garden City, Hertfordshire, England, VII 964
Leubingen, Saxony, I 314
Leuna, E. Germany, VI 527
Levallois-Perret, France, I 133, map 2
Levant, the, III 159, 217, 472, 702
Lewisham, see London
Lexington, Massachusetts, U.S.A., VII 1226
Liberia, VII 696
Libya, I 150
Lichterfelde, Germany, V 233
Lickey Hills, Midlands, England, V 323
Liége, Belgium
    axe, cooper's, III 118
    blast-furnaces, mechanization, II 655
    canal system, IV 556, 558
    coal-mining, II 68; III 72, 74–5; IV 81
    iron industry, II 68, 74, 76; IV 105 and n.
    tools, coal-getting, IV 81
    zinc production, IV 132–3
Liepe, on tributary of Oder, Germany, III 454–5
Liffey, river, Ireland, IV 449
Liguria, Italy, III 218
Lilla Edet, Sweden, III 455
Lille, France, III 453; IV 31, 645–6; V 685, 724
Lillo, Holland, IV 559
Limehouse, see London
Limmat, river, Switzerland, III 428
Limmer, Hanover, IV 540
Limoges, France, II 464, 467; III 241; IV 340, 528
Limousin, France, IV 32
Limpsfield, Surrey, England, II 285, 296

Linares, Spain, II 8, 9
Lincoln, England
    aqueduct, II 671
    ceramics, II 293
    drainage canal, II 680
    glass windows, mosaic, III 241
    stone houses, II 426, 446
    town-planning, III 277
    water-treatment, V 567
Lincolnshire, England, III 101, 146, 152, 315–16;
    IV 18, 200; VI 465
Lindisfarne, Northumberland, England, II 429,
    762
Linz, Austria, VI 489
Lipari Islands, II 368
Lisbon, Portugal, III 480, 544, 549; V 395; VII
    887
Lisht, Egypt, I 668, 699, map 6; II 318
Lisieux, France, II 37
Lissue Rath, Co. Antrim, north Ireland, II 255
Little Falls, New Jersey, U.S.A., VII 1378
Little River, Virginia, U.S.A., IV 536
Little Woodbury, Wiltshire, England, I 317–18
Livermore, California, U.S.A., VI 279
Liverpool, Lancashire, England, general; see also
        Liverpool, place names
    building materials, V 470
    cartography, survey, IV 610
    ceramics, IV 336, 344, 349, 354
    chemical industry, IV 238–9, 241, 247
    docks, III 458; IV 468
    education, discussion, V 795
    industrial development, V 801, 810
    navigation aids, V 455
    palm-oil imports, 1772, V 23
    photography, V 728
    population movements, V 825
    printing, V 701
    railways, V 323
    roads, IV 541
    sanitation operations, IV 506, 509, 514
    shipbuilding, V 62
    shipping, IV 594; V 540
    social effects, technology, V 828, 829
    steel industry, V 62
    street repair levy, 1329, II 532
    telegraphy, IV 657
    town-planning, VII 965
    water-supply, IV 497; V 553, 557, 559
    waterways, V 540; VII 916
Liverpool, place names
    Crown Hill, V 468
    Edge Hill, V 468
    Lime Street, V 468, 474
Livoli, river, Italy, III 311
Livorno, see Leghorn
Lixhe, near Liége, Belgium, III 586

Lizard, The, Cornwall, England, V 192
Llanbrynmair, Montgomeryshire, Wales, III 141
Llandudno, North Wales, V 835
Llanfairfechan, Carnarvonshire, Wales, II 507
Llantwit Major, Glamorgan, Wales, II 415
Llyn Cerrig, Anglesey, Wales, II 539, 540, 551
Lobito Bay, Angola, VII 788
Lobsann, Alsace, France, IV 539; V 118
Locarno, Switzerland, II 323
Lochaber, Inverness-shire, Scotland, VII 900
Loch Cuaich, Inverness-shire, Scotland, VI 212
Loch Katrine, Scotland, IV 497
Loch Lomond, Dunbartonshire, Scotland, VI
    202
Loch Sloy, Dunbartonshire, Scotland, VI 201
Lodi, Lombardy, IV 34
Logelbach, Colmar, Haut-Rhin, France, V 114,
    115, 119
Loing, river, France, III 460, 462–4
Loire, river, France, III 459–60, 462–4, 543; IV
    29, 50, 556, 641
Loire valley, III 95
Lombardy, Italy, III 439, 444–6, 449; IV 34–5;
    V 20
London, England, general; see also London, place
        names, and Index of subjects
    aircraft, VII 800, 812
    alcoholic drinks, III 11, plate 2A
    architecture, VII 930, 939, 964
    beverages, introduction, III 5–6
    bridges, III 418; IV 453–6, 458; V 513; VII
        885, 887, 889, 892, 894
    building methods
        fire-proofing, IV 446–7, 475; V 482–3
        Gothic period, II 444
        materials, III 264; IV 446–7, 449, 451; V
            477, 482–3, 486, 490
        Renaissance period, III 245, 252, 254, 256–9,
            295–9
        Roman Wall, II 31, 408
        twentieth century, VII 930, 939, 948
    cartography, III 531, 544, 553; IV 394, 602,
        607, 615
    ceramics, IV 349; V 668
    chemical industry, IV 234, 237, 244, 251–2; V
        254, 272, 301; VI 568
    cinematography, VII 1302, 1306
    civil engineering, IV 443
    coal, II 68; III 77, 80
    coinage machinery, III 340
    computers, VII 1153, 1154, 1174–6, 1188, 1194
    dividing-engines, IV 394
    domestic appliances, VII 1127, 1129, 1132,
        1143
    dredging, IV 641
    dyestuffs, V 272; VI 568
    education, technical, IV 445; VI 152, 162

London, England, general (*cont.*)
  electricity
    nineteenth century, V 209, 214, 217, 224, 226–7
    -supply, V 178, 199–201, 202; VII 1068, 1071–2, 1074–5
    twentieth century, VI, 285, 294, 295; VII 1068, 1071–2, 1074–5
  fabrics, III 203
  fertilizer factory, V 254
  Fire of, 1666, III 264, 295–8; IV 443
  fish industry, IV 44, 54
  food-supply, V 32–3, 41, 49, 50; VII 1402
  gas, coal-, IV 268–9, 271–2; VII 1127
  gas turbine, VII 1020
  glass-making, III 221–2; IV 360; VI 584
  harbour facilities, V 540
  horses, bridle-bits, II 559
  hydraulic power, V 535
  industrial change, discussion, V 801, 828
  instrument-making, III 621–2; VII 1215, 1322
  invention, chemical industries, III 679, 681, 702
  leather, medieval period, II 176
  lens-making, III 234; IV 360
  lighting systems, electric, V 209, 214, 217
  machine-tools, IV 431
  maps, III 296, 297; IV 496; VII 969
  medical, VII 1322, 1335, 1346, 1347
  metals, extraction, IV 105
  paper-making, III 416
  photography, V 721, 722, 725, 746; VII 1282, 1285, 1302, 1306
  printing industry, V 687, 694, 706, 709, 710, 714; VII 1269, 1272
  radio broadcasting, VII 1255
  railways, V 323, 333, 348; VI 295; VII 771, 779
  roads, II 505–6, 508–9, 532–3; IV 525, 530–1, 540–6; VII 871–2, 876
  road-vehicles, III 145; V 429; VII 728, 731
  Roman, III 277–9
  safety-lamp, mines, IV 95–6
  sanitary engineering, IV 505, 507, 509–10, 514, 516–19; VII 1383, 1387
  shipbuilding, II 576; IV 579, 590; V 353, 361, 366; VI 745
  smoke-control, VII 1480
  social effects, technology, V 828, 831
  space technology, VII 865
  steam-engines, 19th century, V 129, 142
  steam-power, IV 166, 172, 177, 181, 183, 189, 191, 197
  steam turbine, VII 1031, 1032
  stone implements, I 27
  surveys, III 543, 544
  telegraphy, IV 647, 657–60; V 224

  telephony, V 226–7; VII 1222, 1225, 1242, 1246
  textile machinery, II 202; III 167, 203
  timekeepers, III 553, 663, 667, 671
  tools, tradesmen's, III 116
  town-planning, III 278, 283–4, 295–8; VII 964–5, 968–9 (map), 979, 981
  tunnels, V 520; VII 902–3
  turbine power, VII 1020, 1031, 1032
  water-power, IV 207; V 301; VI 213
  water-supply, IV 491–5, 498–9, 502–3; V 552, 553–4, 561–3; VI 213; VII 1368, 1372–3, 1377, 1379
  waterways, IV 548
London, England, place names
  Abbey Mills, IV 517
  Acton Town, VII 779
  Bankside, VII 1032
  Barking, VII 1031
  Battersea, I 512; II 460, 462, 468; IV 143, 354, 494; VII 1071
  Bermondsey, VII 1174
  Billingsgate, IV 53; V 212; VII 1403
  Blackfriars, III 586
  Blackwall, IV 590, 658; V 182, 210
  Bow, III 695; IV 338, 341–2, 344, 354, 492
  Camden Town, IV 492, 657
  Carlton House Terrace, IV 447
  Charles Street, IV 545
  Chelsea, IV 338, 341–2, 344, 493–4, 501; V 562
  Coleman Street, VII 1225
  Commercial Road, IV 544
  Cripplegate, V 482
  Deptford, III 485, 491; IV 208–9, 251, 254, 494, 575; V 199–200, 201; VII 1068, 1074, 1075
  Euston, IV 657
  Fleet, river, IV 510
  Fleet Ditch, II 168
  Fleet Street, IV 509–10
  Fulham, IV 331
  Green Park, IV 493
  Greenwich, III 497, 521, 665; IV 443, 494, 589, 592, 597–8, 607, 618; V 372, 487; VII 901
  Greenwich Observatory, V 448, 458
  Hammersmith, IV 651–2
  Hampstead, IV 492
  Holborn, VII 1246
  Holborn Viaduct, V 212
  Houndsditch, IV 40
  Hyde Park, IV 451, 476; V 474, 698, 772; VI 281
  Hyde Park Corner, VII 872, 892
  Isle of Dogs, VII 901
  Islington, III 543

Kempton Park, VII 1372
Kennington Park, V 829
Kensington Gardens, IV 493
Kentish Town, IV 492–3; V 330
King's Cross Station, VII 768, 769
Ladbroke Grove, VII 887
Lambeth, III 238; IV 191, 336, 344, 446, 494;
  VI 595
Lee, IV 494
Leicester Place, V 471 n. 1
Lewisham, IV 494
Limehouse, IV 157
Lower Thames Sreet, V 471
Maiden Lane, V 200, 202
Mill Hill Park, VII 779
Millwall, V 352, 362, 504
New Cross, IV 464
New River, III 464, 543; IV 179, 492
Old Ford, IV 494
Oxford Street, IV 545
Paddington, IV 657–8; VII 924
Pall Mall, IV 269
Piccadilly, IV 423, 474, 493, 495
Pimlico, IV 493
Plumstead, IV 503; V 566
Poplar, IV 467
Port of London, IV 484
Printing House Square, V 696
Regent Street, IV 474
Rotherhithe, IV 463, 494; VII 901–3
Royal Adelaide Gallery, V 402
St Margaret's parish, Westminster, IV 269
St Pancras, IV 458; VII 930
Salisbury Court, IV 172
Savoy Palace, IV 542
Silvertown, V 247
Smithfield, II 165; IV 542
Soho, V 561
Somerset House, VII 1402
South Kensington, V 372
Speaker's Green, Westminster, VII 1384
Spitalfields, IV 318–19; V 830
Stepney, IV 318–19, 492; VII 901
Strand, IV 268, 493
Stratford, V 301 n. 1, 318
Teddington, IV 216; VII 440, 1194
Thames Embankment, V 212
Threadneedle Street, IV 539–40
Tottenham Court Road, IV 545
Tower Street, IV 594 n.
Tower Subway, V 347, 516
Trafalgar Square, IV 449
Vale of Health, Hampstead, IV 492
Vauxhall, IV 170; V 398
Villiers Street, Strand, IV 493
Wapping, IV 467, 492
West End, IV 447

Westminster, III 80, 425, 686; IV 269, 543; VI
  15
Whitechapel, IV 464
Whitehall, IV 493
Wood End, V 238
Woolwich, see Woolwich, London
Longdendale valley, England, IV 496; V 553
Longdon-on-Tern, Shropshire, England, IV
  566–8
Long Island, U.S.A., III 89
Longport, Staffordshire, England, IV 348
Long Sault rapids, river Ottawa, Canada, IV 552
Longton (Lane End), Staffordshire, England, VI
  626
Longton Hall, Staffordshire, England, IV 338,
  341–2, 344
Loose Howe, Yorkshire, England, I 313
Lora Falls, Argyllshire, Scotland, VII 880
Lorraine, France
  farming methods, V 16
  glass-making, III 237, 240; IV 367
  industrial change, V 808, 809, 820
  steel industry, V 60, 820
  wartime, 1914, VI 8
Los Alamos, New Mexico, U.S.A., VI 272–5, 277,
  279; VII 858, 1332
Los Angeles, California, U.S.A., VII 972, 1382
Los Millares, Almeria, Spain, I 511
Loudon, Virginia, U.S.A., IV 536
Loughborough, Leicestershire, England, V 598,
  602
Louisiana, U.S.A.
  chemical industry, VI 508, 516
  dental equipment, VII 1325
  drilling, gas and oil, VI 382, 407
  oceanography, VII 1453
  sugar-cane cultivation, V 23
  waterways, IV 548
Louisville, Kentucky, U.S.A., IV 548; VII 1376,
  1377
Louvain, Belgium, III 539, 586, 621; IV 252, 260
Low Countries, VI 22
Lowell, Massachusetts, U.S.A., V 529, 573; VI
  204
Lowestoft, Suffolk, England, IV 341, 344
Loyang, China, III 439
Lübeck, Germany, II 64, 76, 531; III 444; IV
  560
Lucca, Italy, II 206–7; III 199, 280
Lucknow, India, VII 1377
Ludlow, Shropshire, England, V 312
Ludwigshafen, W. Germany, V 248; VI 526, 556
Lule, River, Sweden, VI 200
Lund, Sweden, III 533
Luristan, Persia, I 616, 617, map 7; II 474
Lusatia (Lausitz), Germany, I 359
Lusoi, Arcadia, Greece, II 416

Luxembourg, V 511, 820; VII 932
Luxor, Egypt, I 283, 537, map 6
Luz, Shechem, I 249
Lwow, Poland, V 116
Lydd, Kent, England, V 296 n. 2
Lyme Regis, Dorset, England, IV 466
Lynmouth, Devon, England, VI 212
Lynton, Devon, England, VI 212
Lyons, France
  chemical industry, VI 508
  footpaths, asphalt, IV 539
  gas-lighting, IV 261
  photography, V 745; VII 1296
  printing, III 396 n.
  telegraphy, IV 644
  veterinary school, IV 32
  viticulture, II 137
  water-supply, II 671
  waterways, IV 550, 556
  weaving methods, III 165, 189, 191; V 587
Lys, river, France–Belgium, III 453; IV 555

Maadi, Egypt, I 365, 368, map 6
Maastricht, Holland, III 196, 202; IV 558
Mable, river, France, III 291–2
Macao, China, IV 328
Macclesfield, Cheshire, England, IV 565
Macedonia, II 48
Machairus, Peraea, I 263, map 8
Madeira, river, Brazil, V 774
Madeira Islands, Atlantic Ocean, III 6, 544
Madras, India, I 27; IV 610; V 541
Madrid, Spain, III 341; IV 505, 651; VII 822
'Magdala of the dyers', Palestine, I 248, 259,
  map 8
Magdeburg, Saxony, Germany, III 31; IV 170; V
  764
Maginot Line, France, VI 18, 19
Maggiore, lake, Italy, III 444
Magnitogorsk, U.S.S.R., VI 468
Maiden Castle, Dorset, England, I 321–2
Maidenhead, Berkshire, England, III 441
Maidstone, Kent, England, II 32, 34; III 702; IV
  503, 658; V 567
Main, river, Germany, IV 554; V 502; VI 507
Maine, France, IV 29, 33
Mainz, Germany
  aqueducts, Roman, II 671
  ballons, 1795, V 397
  metal-work, II 454
  printing, III 377, 381–5, 388–90, 393, 404
Majorca (Mallorca), II 358, 597; III 544; IV
  329 n.
Makapan cave, Transvaal, I 20, 28, 37 (map)
Malabar, India, II 53
Malaig, U.S.A., VI 245
Malapane, Upper Silesia, IV 105

Mälar, lake, Sweden, III 452, 455
Malaya (now Malaysia, West)
  agriculture, V 2
  dyestuffs, V 266
  fire-making, I 219, 221, 226
  mining, metals, IV 66; VI 415, 416
  rubber industry, V 221, 773, 774, 828
  Sakai blow-gun, I 164
  torches, tree-resin, I 234
  wartime, 1942, VI 21
  whaling, IV 62
Malmesbury, Wiltshire, England, III 152
Malpas tunnel, Languedoc canal, France, III 467
Malperget, Sweden, VI 200
Malplaquet, Belgium, III 350
Malta, I 712, 713, 767; II 199, 499
Mal'ta, Siberia, I 413
Malton, Yorkshire, England, II 32
Malvern, Worcestershire, England, VII 1196
Malvoisine, France, III 544
Manchester, Lancashire, England
  car industry, VII 704
  chemical industry, IV 247, 252–3; VI 537, 548
  dyestuffs, V 273
  education, technical, V 778, 801; VI 152
  electric-power, V 229
  hydraulic-power, V 535, 548
  industrial change, discussions, V 810, 828
  machine-tools, IV 428, 431; V 636
  microscopy, VII 1351–2
  petroleum discoveries, V 104
  printing, V 714
  railways, V 323, 333; VII 899
  rubber industry, V 762, 769
  sewage disposal, VII 1389, 1391, 1393
  shipbuilding, V 351–2
  steam-engines, V 133, 137
  steel industry, V 63
  telegraphy, IV 657
  textile industry, III 152, 168; IV 280, 283,
    287–8, 298, 300, 326; V 573; VI 672
  town-planning, VII 965
  tunnels, VII 899
  water-supply, IV 490, 496, 500; V 553, 558,
    559; VII 1368
  waterways, IV 563
  wire-manufacturere, V 621
  wool industry, III 152
Manchuria, VI 6, 15
Manitoba, Canada, V 6
Mannheim, Baden, Germany, V 427; VI 517
Mansfield, Nottinghamshire, England, IV 286
Maracaibo, lake, Venezuela, VII 1456
Marägha, Persia, III 584
March, Cambridgeshire, England, VII 786
Marche-sur-Meuse, France, II 68
Marcus Hook, Pennsylvania, U.S.A., VII 1021

Mareotis, lake, Egypt, I 257, map 6; III 275
Margate, Kent, England, III 100
Mari, Mesopotamia, I 549, 552, 594, 598, 671,
    map 7
Mariana Islands, Pacific Ocean, VII 814
Marianas Trench, Pacific Ocean, VII 1451, 1453
Marib, Yemen, I 530
Markerwaard, Netherlands, VII 907
Marlborough, Wiltshire, England, III 504
Marlow, Buckinghamshire, England, III 441
Marly (Marly-la-Machine), Paris, France, IV
    154-5, 451, 500
Marly-le-Roi, Versailles, France, V 332
Marmariani, Greece, I 370, map 4
Marne, river, France, III 441; V 474; VI 8; VII
    892
Marseilles, France
    alphabet, Phoenician, I 767
    bridge, VII 889
    chemical industry, V 242, 245
    soap-making, II 356; III 703
    telegraphy, IV 644
    trading centre, 7th century, II 762
    typewriters, V 689
    waterways, II 576
Marshall Bennett Islands, Oceania, I 740, 741
Marsham, Norfolk, England, V 425
Marsiliana de'Albegna, Italy, I 769
Martha's Vineyard, island, U.S.A., IV 647
Maryland, U.S.A., IV 536
Marzabotto, Italy, III 276
Mas d'Azil, Ariège, France, I 157, 347, 358, 753,
    map 2
Massachusetts, U.S.A.
    electrokymography, VII 1343
    photography, VII 1294
    roads, 1639, IV 536
    rockets, development, VII 861
    Rumford, Count, IV 675
    sewage disposal, VII 1388, 1391
    textile industry, IV 286, 298; V 571, 582, 588
    vehicle, goods, VII 723
Matarrah, Mesopotamia, I 367, 502, map 7
Matmar, Egypt, I 431
Mattapoisett, Massachusetts, U.S.A., IV 58
Matterdale, Cumberland, England, V 260
Matto Grosso, South America, I 171
Maubeuge, Nord, France, V 397
Mauretania, Roman province, north Africa, II 227
Mayen, near Treves, II 326
Mayfield, Staffordshire, England, V 260
Meare, Somerset, England, I 318, 320
Mecca, Arabia, III 580, 599-600
Mechlin, Belgium, II 532
Mecklenburg, Germany, II 125; III 285, 450; IV
    23
Medina, Arabia, III 580

Medina de Rioseco, Spain, IV 554
Medinet Habu, Egypt, I 209, map 6
Mediterranean Sea and region
    agriculture, II 86; IV 29, 31
    antibiotics, VII 1355
    cartography, III 509, 518-19, 522-3, 525, 530;
        IV 596-7; V 441, 442
    ceramics, IV 328, 331, 333, 335, 337
    chemical industry, III 683; IV 232, 236, 250
    cotton cultivation, III 159
    dyestuffs, V 259, 266
    glass-making, III 240
    hydrofoils, VII 745
    navigation, III 525, 544, 548
    railways, VII 770
    shipbuilding, III 471-4, 476-7, 479-80, 493
    telegraphy, IV 660
    town-planning, III 271-2
    waterways, III 459-60, 465-7; IV 556, 561
    windmills, III 95, 99 n., 100, 103
Medway, river, Kent, England, IV 462; V 514
Megara, Greece, II 108, 268-9, 498
Megiddo, Palestine, I map 8
    chariots, 1479 B.C., II 544
    ivory-work, I 665, 673-4, 676
    plough-shares, c.926 B.C., II 86
    potter's wheels, I 201
    water-supply, I 530
    weaving tools, I 433, 434, 435
Meissen, Saxony, II 307; IV 329, 337-9, 341-2;
    V 662
Melbourne, Victoria, Australia, IV 541; V 47; VII
    887, 898
Melle (Metullo), Dép. Deux-Sèvres, France, II
    489
Melos, island, I 423, map 4
Melrose, Scotland, II 551
Melsunger, W. Germany, VII 1322
Memphis, Egypt, I map 6
    agriculture, I 537
    handicrafts, I 659, 690
    vineyards, Dynasty I, I 283
    water-control, I 529, 537, 539; II 671
— Tennessee, U.S.A., VII 1411
Menai Strait, Wales, IV 429, 459-60, 531; V 363,
    366; VII 899
Mendip Hills, Somerset, England, II 3, 10, 13;
    III 703; IV 130; V 282
Ménilmontant, brook, Paris, France, IV 505
Menlow Park, California, U.S.A., VI 146
Mercurago, N. Italy, I 209, 213, 214, map 3
Merida, Spain, II 671
Merimde, Egypt, I map 6
    agriculture, c. 4400 B.C., I 360, 362, 365
    buildings, Neolithic, I 304, 473
    pottery, I 376
    tools, non-metallic, I 509, 511, 512, 514, 516

Merionethshire, Wales, III 140
Meroë, Nubia, I 597; III 509
Mersey, river, England
    bridges, VII 888, 889
    hydraulic engineering, V 540, 550
    pottery materials, transport, IV 349
    tunnel, Queensway, VII 905
    waterways, III 458; IV 558, 563, 566, 572
Mersin, Cilicia, I 456, 502, 509, 513, map 4
Merton, Surrey, England, V 242
Merwede, river, Holland, IV 612
Mesabi Ranges, U.S.A., VI 411, 463, 465
Mesopotamia, I map 7; III 270–1, 502, 521; V
    102, 103
Metz, France, II 726
Meudon, Seine-et-Oise, France, V 484
Meuse, river, France and Low Countries, III 74;
    IV 556–8, 629
Mexico, Gulf of, IV 597
— Republic of
    cochineal, culture, I 245; III 695; V 258
    dyestuffs, IV 258, 263
    food resources, III 1, 3
    fossil fuels, I 73; VI 190
    magnetic anomalies, VII 1462
    metals, extraction, IV, 138; V 95; VI 473, 474
    population growth, discussion, VII 1409
    railways, IV 345
    rubber, production, IV 752
Meydum, Egypt, I 434, 482, map 6
Michelsberg, Germany, I 399
Michigan, U.S.A., V 77, 79, 588; VI 413; VII
    1131, 1278, 1380
Mid-Atlantic Ridge, VII 1454
Middelburg, Holland, II 656; III 301; IV 630, 635
Middelkirke, Belgium, V 567
Middlesbrough, Yorkshire, England, V 512, 801,
    810; VI 479; VII 889, 1071
Middlesex, England, III 265; IV 492 (map)
Middletown, Ohio, U.S.A., VI 478
Middle West region, U.S.A., VII 781
Midlands, England, V 350, 815
Midlothian, Scotland, I 402; IV 255
Midway Island, Pacific Ocean, VI 21; VII 814
Milan, Italy, II 501 (map)
    agriculture, IV 34
    canals, III 444, 445 (map), 446–50, 459
    design, theory, V 501
    fabrics, figured, III 196, 202
    irrigation works, III 309
    metallurgy, II 53, 60, 75, 468; III 39
    polymers, VI 560
    refuse disposal, II 681
    roads, IV 544; VII 873
    survey instruments, V 446
    timepieces, III 656
    town-planning, III 285–6

Miletus, Asia Minor, I 722, map 4; II 528–9,
    671; III 272–3
Milford Haven, Pembrokeshire, Wales, VII
    898
Millares, Los, see Los Millares
Mill Hill Park, see London
Mill Rapids, Canada, IV 552
Milos, Aegean Sea, II 368
Milton, Yorkshire, England, IV 105
Milwaukee, Wisconsin, U.S.A., V 688; VII 1230,
    1277, 1346
Minworth, Birmingham, England, VII 1394
Mississippi, river, U.S.A.
    agriculture, V 5
    bridges, V 62, 510; VII 890
    dredgers, V 540
    steam-boats, IV 548; V 145
    transport services, IV 548, 551
Missouri, U.S.A., IV 66; VII 1288
Mistelbach, Austria, I 359
Mitanni, north Mesopotamia, II 541 and n. 2
Mittelwerke, Harz, W. Germany, VII 865
Mitterberg, Salzburg, Austria, I 566, 609, map 3;
    II 50
Mockfjärd, Sweden, VI 200
Moddershall Valley, Staffordshire, IV 348
Modena, Italy, III 230, 504; V 102, 103
Moeris, lake, Egypt, I 418, 545, map 6
Mogden, Middlesex, England, VII 1391
Mohenjo-Daro, Indus valley, I map 5
    basketry and mats, I 420
    bitumen, I 251 (map)
    building methods, I 466
    cotton, c.2000 B.C., I 374
    dyestuffs, I 246
    pottery techniques, I 202, 381, 406
    script, Indus valley, I 760
    textiles, c.3000 B.C., I 432
    town-planning, III 270
Moisson, France, VII 797
Moldavia, central Europe, V 41
Molgora, river, Italy, III 446
Mölln, lake, Germany, III 444
Monaco, VII 1447, 1448
Mönchberg Pass, Alps, IV 529
Monckton, New Brunswick, Canada, VII 788
Mondsee (lake–village), I 357, map 3
Mongolia, III 195; VI 16
Monkwearmouth, Durham, England, II 425; IV
    457
Monmouth, England, III 458; VII 901
Monmouthshire, England, III 697; IV 104, 115;
    V 519, 615
Monpazier, Dordogne, France, III 282–3
Monreale, Sicily, II 392
Mons, Belgium, III 72; IV 557
Montagis, Loiret, France, III 460, 462–4

Montagne Noire, France, II 69; III 467
—— Rigole de la, France, III 467–8
Montana, U.S.A., V 4, 77, 79
Mont Cenis tunnel, Alps, IV 71; V 345, 516, 517, 527
Monte Bello Islands, Australia, VI 263, 281
Monte Carlo, Monaco, IV 542
Monte Cassino, Italy, II 126, 454
Monte Chiaro, Italy, V 103
Montélier (lake–village), I 369
Monte Sant' Angelo, near Foggia, Italy, II 454
Montespan, France, I 152
Monteverdi, Lombardy, III 309
Montmartre, Paris, IV 449
Mont-Olympe, on river Meuse, France, III 290
Montreal, Canada
  nuclear research, VI 241, 262
  railways, VII 775, 784
  St Lawrence river (Seaway), IV 551–3 (map); VII 916 (map), 917–18, 921
Montreuil, Paris, France, V 750
Mont St Michel, Brittany, II 37
Monza, Italy, II 451
Moravia, Czechoslovakia, IV 264
Morecambe Bay, Lancashire, England, VI 217
Mörigen, Switzerland, I 369, 370, map 3; II 95
Morley, Yorkshire, England, V 570
Morocco, VII 5, 529
Morwellham, on Tavistock canal, Devon, IV 569
Moscow, U.S.S.R.
  aircraft, VII 712, 823
  electricity distribution, VII 1082
  industrial change, VI 6, 12
  nuclear research, VI 280
  roads, IV 529
  wartime, 1942, VI 21; VII 812
  waterways, IV 554
Moselle, river, France–Luxembourg–Rhineland, IV 19, 25
Mostagedda, Tasa, Egypt, I 368, 431; II 147, 159–60, 162
Mosul, Mesopotamia, III 179
Motala, Sweden, V 63
Muiden, Holland, II 685
Mülhausen (Mulhouse), Haut-Rhin, France, IV 297; V 196, 210
Mülheim, W. Germany, VI 559
Mulhouse, see Mülhausen
Muliana, Crete, I 729
Mull, Island, Argyllshire, Scotland, VI 632
Munich, Bavaria, W. Germany
  domestic appliances, V 830; VII 1126–7
  dredging tools, IV 642
  electrocardiograph, VII 1340
  machines, 15th century, II 653
  optical glass, IV 360, 361
  photography, VII 1285, 1294

  printing industry, V 704
  railway engineering, V 328, 333, 335
  roads, II 527
  science, discussion, IV 675
  social effects, V 830; VI 16
  sun-dial, III 597–8
  telegraphy, IV 653
  textiles, manufacture, VI 664
  water-supply, V 555
Murano, Venice, III 209, 217–18, 678; IV 362
Murciélagos, Cueva de los, Andalusia, I 421, 436, 450, plate 9
Mycenae, Greece, I map 4, table F
  building methods, I 459, 486–8; II 398, 413
  horses, equipment, I 722, 728
  ivory-work, I 673
  leather, uses, II 167
  metal-work, I 623, 660–1
  olive-tree, c.1200 B.C., I 359
  pottery, I 403
  spinning tools, II 202
  water-supply, I 531
  wheels, chariot, I 213, 214
Myrina, Asia Minor, II 277–8

Naga-ed-Der, Egypt, I 434, map 6
Nagasaki, Japan, VI 23, 252, 276, 277; VII 816
Nahorkatiya, India, VI 406
Namlonge, Uganda, VI 328
Nancy, Meurthe-et-Moselle, France, III 290
Nanterre, Seine, Loire Inférieure, France, V 91
Nantes, France, III 463; IV 102; V 16; VII 889
Nantucket, Massachusetts, U.S.A., IV 58
Nantwich, Cheshire, England, II 532
Nant-y-Glo, Monmouthshire, Wales, IV 114
Naples, Italy, II 501 (map)
  gases, Grotto del Cane, IV 215
  globe statue, c.300 B.C., III 612
  insecticides, use, VI 324
  optical lenses, III 231
  quarrying, alabaster, II 38
  roads, VII 873
  soap production, III 705
Naqada, Egypt, I 433, 434, 779, map 6, plate 31B; II 161
Narbonne, France, II 69, 70; III 468
Narni, Italy, II 508, 511
Narvik, Norway, VII 740
Naucratis, Egypt, I 260, map 6; II 131
Naurouze, near Toulouse, France, III 465, 468
Nauru, island, Pacific Ocean, VI 529
Naxos, Aegean Sea, II 25
Neagh, Lough, Northern Ireland, III 455
Neath, Glamorgan, Wales, IV 126
Nebraska, U.S.A., V 5
Nemi, lake, Italy, II 569, 572
Nene, river, England, III 316; V 544

Nesle, Somme, France, V 396
Netherlands, *see* Holland
Neuberg, Austria, VI 483
Neubrandenburg, Germany, III 285
Neuchâtel, lake, I 357, 361, 508, map 3
Neuhausen, Switzerland, V 93, 94, 608
Neuhaus-on-the-Spree, Prussia, III 454
Neuilly, Paris, IV 451, 453
Neva, river, Russia, IV 554
Nevada, U.S.A., IV 66
New Almaden, California, U.S.A., IV 135
Newark, New Jersey, U.S.A., IV 540; V 729
— Nottinghamshire, England, V 344
New Bedford, Massachusetts, U.S.A., IV 58; VII 1245
Newberghe, New York, U.S.A., VII 1380
New Britain, Connecticut, U.S.A., VII 1147
Newbury, Berkshire, England, III 151-2, 457; IV 20
New Caledonia, island, South Pacific, I 71; V 86
New Carthage (Cartagena, Spain), II 5
Newcastle-upon-Tyne, Northumberland, England
  bridges, IV 460-1
  building materials, V 488
  cables, telephone, VII 1242
  chemical industry, IV 238, 251, 256
  coal industry, III 77, 84, 85
  electricity, V 214, 216; VI 285, 296; VII 1069, 1071
  electric motors, VII 1086
  fish, preservation, IV 54
  furnaces, iron foundry, IV 102
  gas, light and heat, IV 259
  metals, extraction, IV 137
  railways, III 84, 85; IV 460-1; V 340; VII 768
  safety lamp, mines, IV 95-6
  shipbuilding, IV 590
  steam-power, IV 178, 189; V 139
  town-planning, VII 965
New Cross, *see* London
New Earswick, Yorkshire, England, VII 962
New England, U.S.A., IV 50, 58, 62
Newfoundland, Canada
  Cook's survey, IV 616, 618
  fish preservation, IV 49, 50
  ocean currents, VII 1447
  oils, fish and whale, III 705
  radio experiments, 1901, VII 1232-3
  telegraphy, V 225
New Guinea, I 44, 119, 158, 221; VI 21
New Hall, Staffordshire, England, IV 341, 344
New Hampshire, U.S.A., VII 1187
New Haven, Connecticut, U.S.A., IV 538; VII 1223
New Jersey, U.S.A.
  Arthur Kill Bridge, VII 891
  building methods, V 481

  chemical industry, VI 534
  oil industry, VI 190, 534
  punched-card machines, VII 1181
  roads, *c.*1891, IV 537
  tunnelling, methods, V 519
  water-supply, VII 1375, 1378
New Kensington, Pennsylvania, U.S.A., VI 458
New Lanark, Scotland, IV 287; V 830
New London, Connecticut, U.S.A., IV 58
Newlyn, Cornwall, England, V 439 n. 1
New Mexico, U.S.A., VII 862, 1332
New Orleans, Louisiana, U.S.A., IV 540, 548; VII 1325
Newport, Monmouthshire, Wales, IV 91; V 150, 512; VII 889
New River, *see* London
Newry, Northern Ireland, III 455
New Southgate, England, VII 1250
New South Wales, Australia, V 8, 9, 47; VII 1400
Newstead, Nottingham, England, II 232
Newton, Massachusetts, U.S.A., V 425
Newton Creek, U.S.A., V 105
Newton Park, Somerset, England, II 465
New York City, U.S.A., general; *see also* New York City, U.S.A., place names
  bridges, V 509, 513; VII 881, 884, 887, 891
  building methods, V 470, 471, 478, 481-2, 490; VII 942, 951, 953
  buildings, VII 942, 951
  canals, IV 555; VII 916
  canning industry, VII 1400
  cartography, IV 620
  chemical industry, VI 507
  clothing industry, VI 684, 688
  computers, VII 1187, 1191
  drilling gear, oil, V 108
  electronic engineering, VII 1097
  fire protection, V 481-2
  maps IV 553, 620
  medicine, VII 1330, 1339, 1341, 1342, 1347, 1351, 1357-9
  metals industry, V 623
  paddle-steamers, V 143
  photography, V 721
  printing machinery, V 684, 697, 710, 712
  roads, IV 532, 544, 545
  road vehicles, VII 723, 731
  rubber production, V 766
  sea routes, VII 916, 924
  shipbuilding, IV 476, 590; V 362, 365
  steam-power, marine, V 143, 147
  steel industry, IV 110; V 65
  telegraphy, IV 660
  telephony, VII 1230, 1238
  town-planning, VII 966
  tunnels, V 519; VII 922, 1371
  water-supply, V 554, 559; VII 1371

waterways, IV 553, 555; V 143; VII 916
weaving patent, V 581
New York City, U.S.A., place names
  Arthur Kill bridge, VII 891
  Bayonne bridge, VII 884
  Broadway, V 743
  Brooklyn bridge, V 509, 623
  Central Park, IV 538
  Collect Pond, V 142
  Croton valley, V 554, 559
  Harlem river bridge, V 513
  Hudson river tunnel, V 519; VII 922
  Lever House, VII 942
  Manhattan bridge, VII 881
  Pearl Street, V 134
  Verrazano Narrows bridge, VII 887
New York State, U.S.A., IV 537, 553, 555; V 35;
  VII 1174
New Zealand, VI 220
Niagara, U.S.A., VI 519
Niagara Falls, Canada–U.S.A., IV 552; V 201,
  252, 508, 533; VI 196; VII 918
Niagara River, N. America, IV 552; VI 196
Nias, Indonesia, I 191
Nicaea (Iznik), north-west Turkey, II 718, 760
Nicaragua, Central America, V 258
Nice, France, V 567; VII 1000
Nicea, Asia Minor, III 579
Nicobar Island, Indian Ocean, I 406
Nidrosia, Norway, III 533
Niedermendig, near Andernach, Germany, II 31
Nieuport, Flanders, III 442, 453; IV 629
Nieuwe Maas, river, Netherlands, VII 905
Nièvre, see Nivernais
Nigeria, I 407; IV 66; V 22
Nile, river, Egypt
  calendar, Egyptian, III 560–1
  canals, III 439; IV 555; V 542–4
  cartography, III 502, 519, 521
  irrigation systems, IV 555; V 542–4
Nîmes, France, II 404, 405, 671
Nimrud (Calah), Mesopotamia, I 54, map 7
  boats and floats, I 736–7
  building materials, I 470
  chariots and bits, c.900 B.C., 723, 727
  date-palms, cultivation, I 552
  ivory-work, I 664, 675, 678, 679–80
  metal-work, c.800 B.C., I 623, 624, 656, 678
  pulley-wheels, Assyrian, c.1500 B.C., I 523 n. 1
  vessel-making, glass, II 336
  water-supply, I 523 n.1, 527, 534, 550, 553
Nineveh, Mesopotamia, I 251 (map), map 7,
  table F
  battering-ram remnants, II 715
  boats and rafts, I 736–7
  brick and stone, c.705 B.C., I 452–3, 469, 470,
    471–2

  gardens, Assyrian, 7th century B.C., I 551, 553
  irrigation methods, I 524, 532, 552–3
  ivory-work, c.650 B.C., I 676, 677
  metal-work, I 627, 632
  roads, I 714
  rope-making, I 452–3
  staining technique, c.1500 B.C., I 243
  town-planning, III 271
  water-supply, 691 B.C., I 531–2, 552–3
  weapons, c.7th century B.C., I 167
Nippur, Mesopotamia, I 549, map 7
Nishapur, Persia, II 302
Nivernais (Nièvre), France, III 689
Niya, central Asia, II 164
Nogent-le-Roi, France, II 649
Nointol, France, I 562 (map), 563
Noisiel-sur-Marne, France, V 211, 474, 475
Noordoost Polder, Netherlands, VII 907, 909
Nordhausen, Saxony, IV 242
Nordland, Norway, VI 357
Nore, The, Thames estuary, England, V 463
Norfolk, England
  agriculture, IV 5, 10, 17, 19; V 13
  land reclamation, III 315–16; V 545
  windmills, III 97
  wool industry, III 152
Noricum, Roman province, I 595–6; II 56–7, 61,   456
Normandy, France
  agriculture, IV 29, 31, 32
  civil engineering, IV 426; VII 896
  glass-making, III 238, 240
  land drainage, III 319
  Mulberry Labour, VI, 22; VII 896, 922
  navigation systems, VII 847
North Africa, V 10, 16, 265, 713, 827
Northampton, England, II 532; III 162
Northamptonshire, England, III 137, 260; VI 414
North Atlantic Ocean, see Atlantic Ocean
North Cape, Norway, V 442
North Carolina, U.S.A., VII 789
North Chelmsford, Massachusetts, U.S.A., V 645
Northdale, Yorkshire, England, I 511
North Downs, England, IV 448
Northern Massim Islands, I 119
Northern Rhodesia, see Zambia
Northfleet, Kent, England, IV 449
North Lambeth, see London
North Pole, VII 797, 806
North Sea
  fishing industry, IV, 44, 51–3, 62
  Kiel canal, III 444, 455; IV 554–5, 560, 561
    (map); VI 6; VII 913
  land-reclamation, V 545
  natural gas, VI 191, 537
  oceanography, VII 1451, 1462
  ports, IV 51
  ships, III 476; IV 575; VI 6

North Sea (*cont.*)
  submarine cables, IV 660
  windmills, IV 156
North Shields, Northumberland, England, V 340
North Tees, Yorkshire, England, VII 1031, 1071
Northumberland, England, III 152; IV 10, 18, 177; V 128
Norway, III 532, 683
Norwich, Norfolk, England
  farm implements, IV 3
  madder dyestuff, V 260
  quarrying, building stone, II 35
  shipbuilding theory, V 386
  telegraphy, IV 658; V 227
  textile machinery, IV 319
  tool-making, III 115
Notodden, Norway, VI 522–3
Nottingham, Nottinghamshire, England
  alabaster source, II 38; IV 450
  ceramics, II 306; IV 345
  education, discussions, V 778, 795
  railways, VII 786
  road surfaces, IV 541
  textile machinery, V 598, 600, 602, 603
  water-supply, V 555
Nottinghamshire, England, IV 19, 541
Nova Scotia, Canada, IV 50; V 225; VI 323
Nuceria, Italy, II 513
Nun, cape, Morocco, III 544
Nuremberg, Germany
  alcoholic drinks, II 144
  carpenter's tools, II 394, 641
  cartography, III 532, 534–5, 548–9; V 458
  dyestuffs, V 261
  firearms, III 358
  glass-making, III 238
  instrument-makers, *c.*1471, III 585, 621–2
  machines, development, III 337–8, 342
  metallurgy, II 75; III 678
  photochemistry discoveries, IV 717
  printing, III 380–1, 532, 534
  roads, *c.*1456, II 526, 531–2
  spinning processes, II 208
  timekeepers, III 652, 656
  tin-plate production, II 655
  tool-making, II 394
  viticulture, II 138
  water-pipes, II 664
  water-power, II 611
Nutley, New Jersey, U.S.A., VII 1269
Nuzi, Mesopotamia, II 319
Nydam, south Jutland, II 457, 460, 578, 579
Nyssa, Asia Minor, II 671

Oakland, California, U.S.A., VII 728
Oak Ridge, Tennessee, U.S.A., VI 230–1, 233–4, 239, 252, 260–1; VII 1333

Oaxaca, Mexico, IV 139
Obourg, Belgium, I 560, 561, 562 (map)
Odelberg, Sweden, V 662
Odense, Denmark, IV 497
Odenwald, Hesse, Germany, II 30, 31
Oder, river, Germany, III 454; IV 551
Oedenburg, (Sopron), Hungary, I 443
Oeniadae, Greece, II 517
Offenbach, Hesse, Germany, V 346
Ohio, river, U.S.A., IV 459, 537; VII 884
— U.S.A., V 331; VI 295, 623; VII 789
Öhringen, Germany, I 355
Oil Creek, Pennsylvania, V 110, 120
Oise, river, France, IV 453, 555–6
Okinawa, Japan, VII 814
Oklahoma, U.S.A., VI 189, 401
Oldbury, Worcestershire, England, IV 241; V 91, 249, 251, 253; VI 520
Old Ford, *see* London
Oldham, Lancashire, England, V 523, 610, 801, 823
Olduvai, Tanganyika, I 25, 37 (map)
Olympia, Greece, I 661, 681, map 4; II 398, 470, 499
Olynthus, Macedonia, I 395, map 4; II 108, 112, 445; III 273–4
Oman, Arabia, II 53
Omari, El-, *see* El-Omari
Onega, lake, Russia, IV 554
Ontario, Canada, VII 1380
— lake, Canada–U.S.A., IV 551–2; VII 917–19
Ophir (Suppara), India, I 664
Oppau, W. Germany, VI 526, 527
Orbiel, river, Languedoc canal, France, III 467
Oregon, U.S.A., V 3; VI 418; VII 884
Orkney Islands, Scotland, I 382, 408, 421; IV 57, 616; VI 216
Orleans, France, II 608; III 225, 463; IV 31
Ormesbury, Yorkshire, England, V 58
Orne, river, France, IV 29
Orsova, Rumania, II 512
Orvieto, Perugia, Italy, II 58; IV 329 n.
Oseberg, Norway, I 356; II 237
Oslo, Norway, III 533
Osmundsberg, Sweden, IV 539
Ostend, Belgium, III 453; IV 558–60
Öster-Götland, Sweden, I 362
Ostia, Italy
  building methods, II 416, 446, plate 29A
  food imports, II 103, 134
  hydraulic engineering, II 677
  machines, *c.*79 A.D., II 639
  port of Rome, II 518
  town-planning, III 280–2 (map)
  viticulture, II 134
Ostrakina, Palestine, I 259
Otranto, Italy, III 474

Ottawa, Canada, IV 552; VII 1378
— river, Canada, IV 552
Otterspoor, Holland, III 441
Ouse, Fenland river, England, III 316
— river, Yorkshire, England, III 458; VII 890
Ovrebö, Norway, I 709
Oxford, England
  agriculture, IV 41
  Baconian College sequel, 1648, III 681
  building methods, III 257, 265-6
  coffee, new drink, 1650, III 6
  instrument-makers, III 585, 586
  printing, III 402
  stained-glass windows, III 241
  stones, quarrying, II 35, 36
  waterways, III 441
Oxfordshire, England, II 35-6; III 147; IV 20

Pachuca, Mexico, IV 138
Pacific Coast, America, V 36
Pacific Ocean
  aircraft, VII 806, 813-15
  canals, IV 562; VII 913, 915-17
  cartography, IV 598, 618
  marine works, IV 465
  navigation, Loran system, VII 847
  oceanography, VII 1454, 1462
Paclais, Loire-Inférieure, France, III 96
Pactolus, river, Asia Minor, II 42
Paddington Station, see London
Paderno, on river Adda, Italy, III 448-50
Padua, Italy, III 249, 314, 447, 451 n., 677; IV
  329 n.; VII 1344
Paisley, Renfrewshire, Scotland, IV 300, 501; V
  315, 362
Pajarito Plateau, New Mexico, U.S.A., VI 272
Pakistan, VII 886, 892, 905
Palatinate, the, Germany, IV 25
Palazzolo, Italy, III 309
Palermo, Sicily, III 198
Palestine, III 573-4; V 259; VI 543
Palestrina (Praeneste), Italy, I 770; II 513, 570
Palma Nuova, Venetia, Italy, III 286
Palmyra, Syria, I map 8; II 497-8, 501 (map)
  cartography, medieval, III 521, 522
  Chinese silk, III 195
  salt-oases, I 257
  town-planning, III 275, 280
  water-supply, I 528, 532
Pampeluna, Navarre, Spain, IV 554
Panama, Central America, IV 562; VII 696, 913
Panama Canal, VI 4, 494; VII 911, 913, 915-17
Panopolis, Egypt, II 140
Papal States, Italy, IV 529
Paphos, Cyprus, II 459
Papplewick, Nottinghamshire, England, IV 154,
  283

Para, Brazil, V 752, 774
Paris, France, general, I map 2; see also Paris,
  France, place names
  aeronautics, V 399, 400, 402; VII 797
  agriculture, IV 31, 674
  aircraft, VII 800, 823
  bicycle industry, V 414
  bridges, III 423, 425, 430
  building materials, IV 449-51; V 471, 472-3,
    488-9, 490, 496
  building methods, III 252, 254-5, 258, 293; IV
    482, 682; V 62-3, 488-9, 490
  buildings, fire-protection, IV 475-6
  cartography, III 531, 554; IV 596-7, 605-7,
    609, 615
  ceramics, IV 328; V 661, 670
  chemical industry, IV 234, 238, 255
  coffee-drinking, 1643, III 6
  dredging, IV 631
  drilling equipment, V 108
  dyestuffs, 19th century, V 272
  education, discussions, V 794; VI 140, 141
  electricity, V 179, 188, 196, 206, 229; VII 1074,
    1242
  electric-lighting, V 210, 211, 212; VI 296
  Exhibition, 1867, V 789; VI 141
  foods, manufacture, VII 1400
  gas, lighting and heating, IV 258, 260-1, 264
  instrument-making, III 585, 616; IV 398
  japanning trade, III 696
  machines, 17th century, III 333, 340, 341, 345
  machine-tools, V 641; VII 1041
  medicine, VII 1322, 1346, 1354
  metallurgy, IV 143, 234; V 62-3, 66, 91, 609
  military engines, II 723; III 358
  oceanography, IV 615; VII 1447
  photography, V 725, 739-40, 743; VII 1288
  printing, III 397 n., 398; V 691, 698
  roads and streets, II 532; III 293; IV 522-3,
    526, 539-40, 542, 544
  road-vehicles, V 429; VII 871
  rubber manufacture, V 765, 773
  sanitary engineering, IV 505, 512-13; V 489,
    490
  ships, V 351, 353
  spinning machinery, IV 297
  steamboats, V 142, 143
  steel industry, V 62-3, 66
  stained-glass, III 241
  telegraphy, IV 645-6
  telephone cables, VII 1242
  textile industry, III 178; IV 324; V 588
  timekeeping mechanisms, III 674
  tool-making, III 115
  town-planning, III 293; VII 966, 976
  wartime, VI 8, 11
  water-mills, II 608

Paris, France, general (*cont.*)
  water-supply, II 691; III 333; IV 498, 674; V 555, 565
  waterways, III 441, 463; IV 548, 555-7
Paris, France, place names
  Bois de Boulogne, V 395
  Boulevard des Capucines, V 745
  Champs de Mars, IV 259
  Champs Elysées, IV 264
  Eiffel Tower, V 62-3
  Hôtel Seignelay, IV 258, 261, 264
  Île-Saint-Louis, IV 261
  Place de la Concorde, IV 539
  Pont du Carousel, IV 539
  Pont Notre-Dame, III 333
  Pont Royal, IV 539
  Rue Bergère, IV 540
  Saint-Eugène, V 471
  Tuileries Gardens, IV 259; V 396
Parlin, New Jersey, U.S.A., VI 508
Parma, Italy, III 446; IV 34
Paros, Aegean Sea, II 24-5, 399
Parramatta, river, Australia, VII 894
Parys Mountain, Anglesey, Wales, IV 64
Pasion, Greece, II 58
Passau, Bavaria, II 75; IV 263
Passchendaele, Flanders, III 453
Passy, Paris, France, IV 178; V 555
Patleina, Bulgaria, II 288
Patricroft, Lancashire, England, IV 430; V 838
Pavia, Italy, III 446, 449
Pearl Harbour, Hawaii, U.S.A., VI 21, 227, 228; VII 813
Péchelbronn, France, IV 539; V 103, 111, 113, 118
Pech-Merle, Lot, France, I 144, 145
Pecos County, Texas, U.S.A., VI 388
Peenemünde, W. Germany, VII 863-5
Peking, China, III 440; IV 597; VI 15; VII 871
Pelusium, Egypt, I 257, 259, map 6
Pembrey, Carmarthenshire, Wales, V 85; VI 444
Pembroke, Wales, IV 587; V 373
Pembrokeshire, Wales, IV 81
Pencoyd, near Philadelphia, U.S.A., V 60
Pendleton, near Manchester, England, IV 251
Pennines, England
  buildings, 17th century, III 257, 260-1
  water-dams, IV 495, 497; V 556
  water-power, IV 151-2
  Woodhead tunnel, VII 899
Pennsylvania, U.S.A.
  anthracite coal, IV 110
  gas turbines, VII 1020, 1021
  medical products, VII 1355
  oil industry, V 106, 107, 109-10; VI 377, 380, 405; VII 740
  railway engineering, V 330, 331
  steel industry, IV 110; VII 479, 488, 880

Pentelicon, mount, Attica, II 24-6, 398
Pentuan, Cornwall, England, I 563
Per-Amen, Egypt, I 284
Perche, La, *see* La Perche
Pergamum (Perganon), Asia Minor, II 502 (map), 760
  aqueduct, design, II 668-9, 671
  ceramics, II 268
  metallurgy, II 48, 52
  parchment, II 188
  terrestrial globe, *c*.150 B.C., III 614
Périgord, France, III 34
Perm, Urals, Russia, IV 529
Perpignan, Pyrénées-Orientales, France, V 107
Persepolis, Persia, II 496
Persia, IV 328, 335, 337; V 103, 266
Persian Gulf, III 505; VII 788, 802, 1461
Perth, Scotland, V 323, 333, 555
Peru, South America
  agriculture, III 1, 3; V 24
  borax source, V 321
  cartography, III 554; IV 596; V 442
  coal resources, III 73
  knitted fabric, III 182
  metallurgy, IV 139
  oil resources, V 102
  rubber, discoveries, 1736, V 754
  shipbuilding theory, V 353
Perugia, Italy, I 769
Pesaro, Italy, III 313; IV 329 n.
Peterborough, Northamptonshire, England, II 441, 679; IV 658; VII 1183
Peterhead, Aberdeenshire, Scotland, IV 58
Peterhof, Gulf of Finland, Russia, IV 656
Petersinsel (lake-village), I 369, map 3
Petra, Syria, II 497-8
Petrograd, *see* Leningrad *and* St Petersburg
Petrolia, Ontario, Canada, V 110
Petrossa, Rumania, II 452
Petsofa, Palaikastro, Crete, I 403, map 4
Pewsey, vale of, Wiltshire, England, III 147
Phaistos (Phaestos), Crete, I 403, map 4; II 494
Philadelphia, U.S.A.
  cables, electric, VII 1077, 1242
  chemical industry, VI 507
  dredging, IV 641
  lighting, electric, V 213
  medicine, VII 1322, 1323, 1330-1, 1348
  paper manufacture, VI 610
  photography, V 725, 729, 737
  printing, VI 691, 692, 700; VII 1269
  railway engineering, V 327, 332
  roads and streets, IV 540, 543
  road vehicles, VII 717, 725
  rubber manufacture, V 766
  steam-engines, V 131, 137, 149

Philippine Islands, I 165, 219, 226; III 4, 8; VI 2, 21, 23
Philippine Trench, Pacific Ocean, VII 1452
Phocoea (Fogliari), west Turkey, II 368, 486
Phylakopi, Melos, I 423
Piacenza, Italy, V 103
Piano D'Oroto, Italy, VI 524
Picardy, France, III 91, 175, 319; IV 31
Piedmont, Italy, III 218; IV 34, 35; V 20
Piltdown, Sussex, England, I 141
Pimlico, see London
Pingsdorf, Rhineland, II 295
Piperno, Italy, III 311
Piraeus, Greece, II 402, 403, 413, 517–18, 529; III 271–2
Pir Panjal range, Himalayas, India, VII 874
Pisa, Italy, II 582; III 230, 314, 662; V 188, 320
Pitchford, Shropshire, England, III 688; V 103
Pittsburgh, Pennsylvania, U.S.A.
　electric cables, VII 1074
　glass manufacture, VI 578
　industrialization process, V 808
　metal rolling-mills, V 612
　natural gas, VI 407
　oil drilling, V 109
　radio broadcasts, 1920, VII 1254
　railways, VI 465; VII 785
　steel industry, V 53, 68; VI 465, 472, 486
Pittstown, New York State, U.S.A., V 589
Plaine, Rigole de la, France, III 468
Plataea, Greece, II 529, 698
Plattenberg, II 75
Plaue, lake, Germany, III 455
Plover Cove, Hong Kong, VII 911, 912
Plumstead, see London
Plymouth, Devon, England
　ceramics, IV 340–1, 344
　marine works, IV 466–9
　refrigeration, V 49
　shipbuilding, IV 582
　telegraphy, IV 647
　telephone system, V 226
　water-supply, II 691
Plymstock, Devon, England, I 612
Po, river, Italy, III 310, 314, 446; IV 35; V 20; VII 896
Pocket Nook, St Helens, Lancashire, England, IV 373–5
Poissy, France, VII 1400
Poitiers, France, II 452, 720, 735, 765
Poitou, France, III 319; IV 29, 32
Pola, on Adriatic Sea, Italy, V 567
Poland, III 229, 579
Poldhu, Cornwall, England, V 228; VII 1233
Polvaccio, near Carrara, Italy, II 28
Pomerania, Prussia, I 576; IV 23

Pompeii, Italy, II 501 (map)
　bread-making, II 118
　building methods, Roman, II 415, 416, 446, plate 29B
　cloth-making, II 214, 217–19
　donkey-mills, c.79 A.D., II 110, 639
　furniture, Roman, II 224–5, 230
　glass products, II 323, 416; III 237
　heating system, Roman, I 231
　leather-work, II 148
　machines, Roman period, II 110, 631, 638, 639
　painting remains, 1st century, II 363, 364
　roads, II 530, 531, plate 34B
　shipbuilding, II 570–1, 572
　surveying instruments, III 512
　tools, carpentry, II 230, 392–3
　town-planning, III 274–5
　water-supply, II 664, 665, 677
Pontarlier, Doubs, France, IV 540
Pont Crouzet, near Cammazes, France, III 468
Pontine Marshes, Italy, III 311–13, plate 19
Pontus, II 54, 58
Pont-y-Cysyllte, Shropshire, England, IV 567
Pontypool, Monmouthshire, Wales, III 688–9; IV 104, 115
Pontypridd, Wales, III 434
Poplar, see London
Popudnia, Ukraine, I 307, 309
Populonia, Italy, II 50, 52, 56, 60
Porjus, Sweden, VI 200
Portland, Cotswold Hills, England, III 260
— Dorset, IV 448–9, 451
Port Patrick, Wigtownshire, Scotland, V 225
Portrush, Northern Ireland, V 233, 347
Portsmouth, Hampshire, England
　machine-tools, IV 424, 426–7, 437; V 636
　shipbuilding, III 491; IV 576–7, 580–1, 584, 587
　telegraphy, IV 647
　water-supply, IV 553, 555
— New Hampshire, U.S.A., VI 6
Port Sunlight, Cheshire, England, V 806
Port Talbot, Glamorgan, Wales, VI 496; VII 920, 922, 927–9
Portus, Italy, II 518, 522
Potomac, river, U.S.A., V 142, 152
Potsdam, Prussia, III 225
Potters Bar, Hertfordshire, England, VII 902, 903, 904
Potterspury, Northamptonshire, England, II 296
Poughkeepsie, New York State, U.S.A., V 564
Pozzuoli (Puteoli), Italy, II 59–60, 355, 520, 530, 531
Praeneste, see Palestrina
Prague, Czechoslovakia
　glass-engraving, III 223
　instrument-making, III 621

Prague, Czechoslovakia (*cont.*)
  printing, V 706
  sewerage system, VII 1384
  timekeepers, III 658
  street lighting, 1817, V 116
  textiles, manufacture, VI 670
Prater, Vienna, Austria, VII 1129
Preston, Lancashire, England, V 333, 540; VII 876, 961
Prestonpans, Firth of Forth, Scotland, IV 244
Prety, Burgundy, II 599
Priene, near Miletos, Asia Minor, III 273–4
Prince Rupert, Canada, VII 788
Prottalangar, Iceland, VI 220
Provence, France, III 159, 319; IV 27; V 260
Providence, Rhode Island, U.S.A., VII 1227
Provo, Utah, U.S.A., VI 468
Puentes, Spain, IV 498
Puerto Principe, Havana, V 102
Puerto Rico, West Indies, V 23
Pulicat, Madras coast, India, IV 249 n. 2
Punjab, India, I 202
Punt (? Ethiopia, Red Sea coast)
  cosmetics, export, I 293
  gum arabic, I 243
  ebony, export, *c.*1500 B.C., I 685
  incense trees, I 295, 544
  ivory-work, I 663
  ships, *c.*1500 B.C., I 55, 735
Purbeck, Dorset, England, III 396; IV 543
Pushkari, Ukraine, I 301, 302
Puteaux, near Paris, France, V 242, 435
Puteoli, *see* Pozzuoli
Pyrenees, France–Spain, III 596; IV 29, 605; VI 508

Qau, Egypt, I 368, 477, 666, map 6
Quebec, Canada, VI 227
Queenborough, Isle of Sheppey, Kent, England, IV 251
Queensland, Australia, V 9, 23; VI 308, 315
Quesnoy, France, IV 645
Quiberon Bay, Brittany, III 499
Quina, La, *see* La Quina
Quinte, Bay of, Lake Ontario, Canada, IV 552

Radcliffe, Manchester, England, IV 245
Radley, Berkshire, England, I plate 8
Radstock, Somerset, England, V 331
Raleigh, North Carolina, U.S.A., VII 953
Rambouillet, France, IV 27
Rammelsberg, Germany, III 692
Ramsey Abbey, Huntingdonshire, England, II 760
Ramsgate, Kent, England, IV 467
Rana-Ghundai, Baluchistan, I 721
Rance, river, France, VI 218
Rand, the, South Africa, V 95

Ranipur, river, India, V 542
Raqqa, Syria, II 328
Rasselstein, Rhineland, Germany, IV 115
Ras Shamra, Syria, I 673, map 4: *see also* Ugarit
  craft guilds, I 665 n.
  dyestuffs, I 247
  metal weapons, *c.*1300 B.C., I 618, 619
Rathenow, Prussia, III 450
Ratisbon, Bavaria, IV 170
Ratman, river, India, V 542
Ratoo mountains, Borneo, IV 144
Ravenhead, Lancashire, England, IV 368–9, 371, 373
Ravenna, Italy, II 227, 413, 421–4, 679; III 312
Ravensbourne, river, Kent, IV 494
Ravi, river, India, VII 883
Rawalpindi, Pakistan, VII 884
Ray (Rhagae), Persia, II 736–7
Reading, Berkshire, England, III 457; V 723; VII 875, 1374, 1378
— Pennsylvania, U.S.A., V 327
Redbridge, near Southampton, England, IV 581
Redgrave Park, Suffolk, England, I 609
Redruth, Cornwall, England, IV 252, 262
Red Sea, III 439; IV 561; VII 1466
Regensburg, Bavaria, Germany, II 531; III 284
Réhon, France, VI 458
Reichenstein, Silesia, IV 121, 141
Reichraming, on river Enns, Austria, V 65
Reika, *see* Fiume
Remedello, Italy, I 505, 508, 510, 511
Remscheid, Germany, V 63
Rendsburg, Schleswig-Holstein, IV 560–1
Reno, river, Italy, III 314
Repudre, river, France, III 467
Revel, near Toulouse, France, III 464
Reykjalang, Iceland, VI 220
Reykjavik, Iceland, VI 220
Rheims, France, V 574, 669
Rheinhausen, W. Germany, VI 475
Rhine, river, west Europe
  aluminium industry, V 93
  bridges, III 427–8; VII 896
  Canal du Nord project, IV 556
  chemical industry, VI 507, 564
  dredging, IV 629
  land transport, VI 702
  navigation, IV 550–1
  paddle-steamers, 1816, IV 548
  pollution problem, VII 1480
  viticulture, IV 25
  windows-glass industry, III 237
Rhineland, Germany, III 206, 537–9; IV 345, 556; VI 17
Rhodes, Greece, I map 4
  calendar, Egyptian, III 564
  cartography, III 505–6, 508–10

military engines, 305 B.C., II 700–1, 702, 716
shipbuilding, II 569
windmills, II 625
Rhodesia (*now* Zimbabwe), VII 781
Rhondda Valley, Wales, IV 91; V 815
Rhône, river, France, IV 550; VII 879
Ribble, river, northern England, V 540
Richelieu, Indre-et-Loire, France, III 290–2
— river, Canada, IV 553
Richmond, Surrey, England, IV 243; V 431
— Virginia, U.S.A., VII 1322
Riddings, Derbyshire, England, V 104
Rideau, river, Canada, IV 552
Rietgreb, Holland, III 302
Riga, Gulf of, Latvia, IV 554
— Latvia, VII 1292
Rigi, Switzerland, V 346
Riihimaki, Finland, I 709
Rimini, Italy, II 513; III 311
Rimutaka, New Zealand, V 345
Ringkuhl, near Grossalmerode, Hesse, IV 244
Ringstead, Norfolk, England, I 636
Rio Tinto, Spain, II 7, 48, 50
Ripac, Bosnia, I 370
Ritchie County, Virginia, U.S.A., IV 540
Rivelaine, Belgium, IV 81
Rivenhall, Essex, England, III 241
Riviera, France, VII 871
Robenhausen (lake–village), I 356, 357, 359, map 3
Rochdale, Lancashire, England, V 523, 570, 574
Rochefort, Charente-Inférieure, France, IV 582
Rochester, Kent, England, IV 462; V 514, 537
— New York State, U.S.A., V 747; VII 1285, 1288, 1328
Rocky Mountains, Canada–U.S.A., IV 66; V 7, 449; VII 913
Rocky Point, New York, U.S.A., VII 1250
Rodney Point, Victoria, Australia, V 46
Rödöy, Norway, I 709
Rogny, near Montargis, France, III 460, 462–3
Rohri Channel, River Indus, Pakistan, VII 886
Roissy, France, VII 976
Romania, IV 539; V 110–11, 113, 114, 116; VII 1063
Rome, Italy
  building-methods, Roman, II 27–8, 404–7, 409–14, 417, 421–2
  buildings, Renaissance, III 245–50, 253–5, 257–8, 266
  calendars, III 566, 575–8
  cartography, III 511–12, 534
  ceramics, IV 332
  dyestuffs, V 265
  food-supply, Roman period, II 103
  glass-making, III 209–10, 214, 220
  hydraulic engineering, Roman, II 665
  land drainage, III 312
  metallurgy, II 53; III 53
  metal-work, II 417, 454, 464, 467
  military machines, II 722–3; III 349
  milling, II 107
  olives, cultivation, II 122
  photographic arts, 1645, V 736
  printing, III 409
  quarrying, Roman period, II 27–8, 406–7
  roads, II 27–8, 501, 529–31; III 279; IV 544
  sewerage system, 6th century B.C., IV 504
  shipbuilding, Roman period, II 569–70
  timekeepers, III 657
  town-planning, III 276 (map), 279–82, 287–9 (map)
  water-mills, II 602
  wool trade, Roman period, II 192
Romney Marsh, Kent, England, IV 607
Rossel Island, Oceania, I 740
Rosswell, New Mexico, U.S.A., VII 862
Rothamsted, Hertfordshire, England, IV 42; V 254; VI 331
Rothbury, Northumberland, England, V 217
Rothenburg ob der Tauber, Bavaria, III 284
Rother, river, Sussex, England, IV 466
Rotherham, Yorkshire, England, IV 457–8
Rotherhithe, *see* London
Rotterdam, Netherlands, IV 156–7, 556, 639; VII 905
Roubaix, Nord, France, V 576
Rouen, France
  bicycle industry, 1869, V 414
  building methods, V 468
  ceramics, IV 337
  chemical industry, IV 244
  dredging, IV 641
  paddle-boat, 1732, V 142
  refrigeration experiment, V 48
  stained-glass windows, III 241
  textile industry, V 588
  timekeepers, III 652–3
  waterways, IV 548
Roussillon, France, IV 605
Roville, near Nancy, France, IV 41
Roya, river, Belgium and Holland, IV 558
Royaumont, France, II 650
Rudge Coppice, Wiltshire, England, III 504
Rudstone, Yorkshire, England, I 218
Rugby, Warwickshire, England, V 136; VI 158; VII 1251
Ruhestetten, Hohenzollern, I 357
Ruhr, Germany, III 85; IV 105; VI 13, 184; VII 1364, 1396
Rumania, *see* Romania
Rumford Falls, Maine, U.S.A., VI 519
Runcorn, Cheshire, England, IV 241; V 251; VII 889

Rupel, river, Flanders, III 452–3
Rutland, England, III 260
Rye, Sussex, England, III 487; IV 466

Saalfeld, Thuringia, Germany, V 663
Saar, the, Germany, II 68; V 669
Sadu, Japan, I 565
Saffron Walden, Essex, England, V 266
Sag Harbor, U.S.A., IV 58
Sahara, North Africa, III 525, 544
St Acheul, France, I 131, map 2
St Albans, Hertfordshire, England, II 425; III 276: *see also* Verulamium
St Andrews, Fife, Scotland, III 667
St Anne rapids, River Ottawa, Canada, IV 552
St Bernard Pass, Alps, IV 529
St Blaise (lake–village), I 357, map 3
St Catherine's Hill, Winchester, I 313
St-Cloud, Seine-et-Oise, France, IV 337–9
St Denis, Seine, France, IV 239; V 169, 756
Saint-Dié, Vosges Mountains, III 538
Ste-Maxence, near Senlis, France, IV 452–3
Saintes, France, II 288, 293
St Étienne, Loire, France, IV 115; V 556
St Ferréol, Languedoc canal, France, III 468
St Fillans, Perthshire, Scotland, VII 900
St Gall, Switzerland, II 217–18; III 594–5
St Georgenberg, Inn valley, Austria, II 595
St George's Channel, Ireland–Wales, VII 924
Saint-Georges de Bocherville, Normandy, II 490
St Germans, the Fens, England, V 544
Saint-Gobain, France, III 238; IV 368–9, 371
St Gotthard Pass (St Gothard), Switzerland, III 417; IV 529, plate 36B; V 517
St Helens, Lancashire, England
    chemical industry, IV 241; V 240, 802
    copper smelting, IV 128–9
    glass-making, IV 368–9, 373–5, 678; VI 578
St Ives, Huntingdonshire, England, II 679
St John's, Newfoundland, Canada, V 228; VII 1232
St Lawrence, lake, Canada–U.S.A., VII 918
— river, Canada, IV 551–3 (map)
St Lawrence Seaway, Canada–U.S.A., VII 913, 916 (map), 917–20, 921
St Leonard's Forest, Sussex, England, IV 16
St Louis, lake, Canada, VII 918
— Missouri, U.S.A., IV 540; V 62, 564; VII 1322
St Lucia, Istria, II 322
St Malo, France, VI 218
St Michael Island, Azores, III 531
St Moritz, Switzerland, II 664, 666; III 197
St Omer, Pas-de-Calais, France, IV 629
Saintonge, France, III 319
St Osyth, near Clacton, Essex, England, II 440
St Pancras Station, *see* London

St Petersburg (*later* Leningrad), Russia
    chemical industry, VI 518
    industrial unrest, 1905, VI 6
    metals, extraction, IV 144
    roads, IV 529, 545
    sewage system, VII 1384
    ship-design, IV 577
    steam-engine, 1717, IV 166
    telegraphy, IV 655
    water-supply, V 554
— U.S.A., VII 800
St Privé, near Briare, France, III 463
— Rigole de, III 464
St Rémy, Allier, France, II 317
St Rollox, Glasgow, Scotland, IV 241, 245, 248; V 238
St Seurin, Gironde, France, V 57
St Vincent, cape, Portugal, III 506
St-Yrieix-la-Perche, near Limoges, France, IV 340
Salamis, Greece, II 567
Salerno, Italy, II 454
Salford, Lancashire, England, IV 266; V 553, 832; VII 1393
Salindres, Gard, France, V 91
Salisbury, Wiltshire, England, II 428, 431, 433; III 241, 650; IV 518
Salisbury Plain, Wiltshire, England, III 147; IV 610; VII 1232
Salonika, Greece, II 303
Saltash, Cornwall, England, V 514, 534; VII 898
Saltville, Virginia, U.S.A., V 252; VI 507
Salvatori, Lombardy, III 309
Salzburg, Austria, II 51, 689; III 605; V 333
Samaria, Palestine, I 675, 678, map 8
Samarkand, Turkestan, II 307; III 584
Sambre, river, France and Belgium, IV 557–8
Samos, Greece, II 48, 398, 401, 667–8; III 511
Sanchi, India, II 556
San Cugat, Spain, II 641
Sandford-on-Thames, Oxfordshire, England, III 456
Sandhurst, Berkshire, England, VI 154
Sandon Mount, Hertfordshire, England, II 624
Sandwich, Kent, England, IV 467; VII 922
San Francisco, California, U.S.A.
    bridges, VII 887, 888
    building methods, IV 490
    earthquakes, VII 1455
    electronic engineering, VII 1097, 1238
    metals, extraction, IV 135, 142
    radiology, developments, VII 1329
    road vehicles, VII 721
    sea route, New York, VII 916
    telephony, VII 1097, 1230, 1238
Sangiran, Java, I 23, 37 (map)
San Sa inlet, China, VI 217

Santa Fe, New Mexico, U.S.A., VI 272
Santa Giostina, Italy, VI 200
Santarem, on river Amazon, South America, V 774
Saône, river, France, III 459; IV 556; V 142
Sapperton, Gloucestershire, IV 567
Saqqara, Egypt, I map 6, plate 19
  agriculture, I 368, 541, 542–3
  animals, domestication, c.2500 B.C., I 341, 343
  boats, I 732
  buildings, stone, I 475–8, 481–2, 483
  chemical arts, ancient Egypt, I 241, 265
  cosmetic arts, I 291
  culinary arts, I 273
  metals, extraction, I 579, 597 n.
  pottery, I 395
  rotary motion, c.2500 B.C., I 193
  tools and craftsmen, I 689
  transport, c.2400 B.C., I 710, 732
  vase, Egyptian, c.2500 B.C., I 193
  water-supply, I 537
  wood-work, c.2800 B.C., I 696–7
  writing art, I 757
Saragossa, Aragon, Spain, IV 554
Sarata-Monteoru, Rumania, V 111
Sardia, Lydia, I 713
Sardinia, II 44, 45; IV 123; VII 1082
Sardis, Sardes, Asia Minor, II 600, 671
Sarlièves, lake, Puy-de-Dôme, France, III 319
Sargasso Sea, VII 908
Sarine, river, Switzerland, IV 459
Sas van Ghent, Holland, IV 559
Satala, Armenia, II 478
Saulte Ste Marie, Ontario, Canada, VI 464
Savenay, Loire-Inférieure, France, III 96
Savona, Genoa, Italy, III 703
Saxmundham, Suffolk, England, II 440
Saxony, Germany
  agriculture, 18th–19th century, IV 27
  ceramics, IV 335
  coal-mining, III 82
  glass-colouring, III 703
  machinery, 17th century, III 342, 689
  mapping, 1775–1800, IV 610
  metallurgy, III 42, 689, 703; IV 139, 140
  mining, IV 64, 70
  printing, V 694
  roads, IV 527
  spinning-wheels, III 161
  steam-engine, 1729, IV 178
  textile industry, IV 277
  trolleybus, 1901, VII 732
Scandinavia, III 181, 532–3
Scarborough, Yorkshire, England, IV 49
Scarpe, river, France, III 441 n. 1, 453
Schaffhausen, Switzerland, I 443, map 3; III 427–8; IV 108; V 66; VI 663; VII 1293

Scharndorf, Germany, VI 497
Scheldt, river, France–Belgium–Holland, III 452; IV 555–9, 629; VII 905
Schemnitz, Hungary, II 11, 17
Schenectady, New York, U.S.A., VI 253, 255; VII 1331
Scheveningen, Holland, IV 545
Schiedam, Holland, III 442
Schladebach, Germany, V 108
Schleswig, Germany, II 165; V 20
Schleswig-Holstein, IV 35
Schmalkalden, Thüringia, II 76
Schneeberg, Saxony, III 218, 703; IV 120
Schoonhoven, Holland, II 689
Schopfheim, Baden, Switzerland, IV 113
Schussenthal, Ravensburg, I 356, map 3
Schwabach, Bavaria, II 75
Scilly Isles, I 421
Scotswood-on-Tyne, Northumberland, England, VII 767
Scrivia, river, Italy, V 556
Scunthorpe, Lincolnshire, England, VI 414
Seamer, Yorkshire, England, I 303
Searles Lake, California, U.S.A., V 321
Seaton Carew, Co. Durham, England, VII 860
Sedment, Egypt, I 422, 694, 695, map 6
Seefeld, Tyrol, Austria, V 102
Segovia, Spain, II 671; III 152, 341; IV 26
Sehel, Egypt, I 537
Seine, river, France
  aeronautics, V 407
  bridges, III 293; IV 451, 453; VII 895
  canals, III 459–60, 464; IV 555–6
  Eiffel tower foundations, V 476
  merchant-ships, III 477
  steamers, IV 548; V 142, 143, 152
  survey, III 543
  water-pumps, 1670, III 333
  water-supply, Paris, III 333; IV 498
  window-glass industry, III 237
Seine-et-Marne, France, V 16
Seistan, Persia, II 615–16
Seleucia, Mesopotamia, II 316
Selinus, Sicily, II 401, 678; III 273
Sellafield, Cumberland, England, VI 246
Selly Oak, Birmingham, England, V 253
Semna, Nubia, I 538
Sempringham, Lincolnshire, England, II 429
Semur-en-Auxois, France, II 216, 218
Senne, river, Flanders, III 452
Sens, Yonne, France, II 214–15, 217; III 195–6, 198, 202
Serabit el-Khadim, Egypt, I 564
Seraing, near Liège, Belgium, IV 105
Serajevo, Yugoslavia, VI 7
Serbia, Yugoslavia, VI 7, 8
Serchio, river, Italy, III 314

Sergeac, Dordogne, France, I 32
Sète, France, III 465, 467
Severn, river, England
  bridges, IV 102, 105, 455, 457; VII 895, 896
  canals, IV 572
  chemical industry, III 688
  colliery railways, III 84
  iron and steel industry, IV 102, 105
  navigation, III 457
  shipbuilding, V 350
  tunnel, 1873–86, V 516, 518–19
  watermills, IV 199
  wood-cultivation, V 263
—— Canada, IV 552
Seveso, river, Italy, III 444
Seville, Spain, III 703; IV 551
Sèvres, Paris, France, III 679; IV 329, 338–43; V 659, 661, 663
Seyssel, France, IV 539–40
Sezze, Italy, III 311
Shaftesbury, Dorset, England, V 826
Shaheinab, Sudan, I 501, 506, 515–16
Shah-tepe, Turkoman steppe, I 721, table F
Shannon, river, Ireland, IV 572
Shantung, province, China, III 440
Shatt al-'Arab, I 552, map 7
Shechem, Palestine, I 762, map 8
Sheerness, Kent, England, V 439
Sheffield, Alabama, U.S.A., VI 527
— Yorkshire, England
  cutlery manufacturers, IV 108
  diseases, industrial, V 832
  fertilizers, IV 20
  machine-tools, IV 116
  metals, extraction, IV 130
  metals, machine processing, V 610, 612, 614, 624
  railway engineering, V 331, 805; VII 899
  roads, 1875, IV 541
  sewage treatment, VII 1391
  staining technique, microbiology, V 281
  steel industry, V 56, 63, 65, 681; VI 492
  textile finishing, IV 304
  tool-making, III 113, 115
  water-supply, IV 497
  waterways, III 458
Shelton, Staffordshire, England, IV 341, 344–5, 349
Sheppey, Kent, England, IV 448
Shetland Islands, Scotland, II 590; III 186; IV 47, 57; V 452; VI 357
Shinfield, Reading, Berkshire, England, VII 1403
Shippingport, Pennsylvania, U.S.A., VI 254
Shotton, Flintshire, Wales, VI 496
Shrewsbury, Shropshire, England, III 457; IV 455, 531; V 627

Shropshire, England
  bridges, IV 457
  canals, IV 565, 569
  coal-mining, IV 87
  iron industry, III 31; IV 105
  roads, IV 532
Shude Hill, Manchester, England, IV 283
Sialk, Persia, I map 5
  building methods, c.3000 B.C., I 306
  horses, remains, I 721
  pottery, I 394, 396
  textiles, I 432
  tools, agricultural, c.4000 B.C., I 502, 507
Siam, III 159
Siberia, U.S.S.R., IV 529; VI 6, 12, 415, 416
Sicily, Italy, I 770; III 521, 525; V 20; VI 22, 515
Sidcup, Kent, England, VII 1334
Sidon, Phoenicia, II 367, 516
Siegburg, Rhineland, II 287
Siegen, Prussia (Germany), II 74, 76; III 32
Siena, Italy, II 620; III 678; IV 34, 233, 329 n., 544
Siestan, Persia, III 106
Signal Hill oilfield, California, U.S.A., VI 395
Silchester, Hampshire, England
  building methods, Roman, II 414–15, 420
  metallurgy, 4th century, A.D., II 61
  quarrying stone, Roman, II 32
  tools, tradesmen's, III 111
  town-planning, III 277–8
  water-supply, II 671, 678
  weaving and finishing, II 216
Silesia, Germany, I 359; III 151
Siliguri, India, VII 882
Siloam, Pool of, I 764
Silsila, Egypt, I 477, 478, 483, map 6
Silver Creek, Pennsylvania, U.S.A., IV 110
Simms Station, near Dayton, Ohio, U.S.A., V 391
Simplon, Switzerland, V 294 n.
Simplon Pass, Alps, IV 529
Sindh, Indus valley
  potter's-wheel, I 202
  pottery, I 381, 385, 386, 396, 406
  vehicles, 2000 B.C., I 206, 207, 717
  wheels, carts, I 213
Singapore, south-east Asia, V 221, 774; VI 21, 339
Singara, Mesopotamia, I 282
Sinope, Asia Minor, II 5, 361, 701, 760
Sioule, river, France, VII 881
Siphnos, Aegean Sea, II 43, 58
Sipplingen (lake–village), I 355, 356, 357, 359, map 3
Sireuil, near Angoulême, France, V 59
Sirgenstein Cave, Württemberg, I 369

Siris (Sinni), river, south Italy, II 471, 472
Skagen area, Denmark, IV 561
Skagerrak, strait between Jutland and Norway, IV 560
Skåne, Sweden, I 497, 614
Skara Brae, Orkney, I 312, 500, 516
Skegness, Lincolnshire, England, VII 846
Skendleby, Lincolnshire, England, I 313
Skermer, Holland, III 321
Skinningrove, Yorkshire, England, VI 471
Skrydstrup, Denmark, I 442
Slough, Buckinghamshire, IV 657-8
Sluis, Zeeland, Netherlands, IV 558, 629
Smethwick, Staffordshire, England, IV 373; V 88
Smyrna, Asia Minor, II 671, 760; III 159, 704
Snae Fell, Isle of Man, V 345
Snartemo, Norway, II 474
Snowshill, Gloucestershire, England, I 616
Soay, Hebrides, Scotland, I 344, 345
Sodermanland, Sweden, II 551
Sodom, Palestine, I 259
Sofala, south-east coast, Africa, III 521
Soho, Birmingham, England, IV 162, 166, 183, 352, 419, 484
Solani, river, India, IV 555; V 542
Solihull, Warwickshire, England, V 249
Solingen, Prussia (Germany), II 75; III 115
Sologne, north-central France, V 16
Solomon Islands, Pacific Ocean, VI 21
Solothurn, Switzerland, I 443, map 3
Somali coast, East Africa, III 506
Somaliland, I 196
Somerset, England, III 152; V 263, 331
Somerset House, see London
Somme, river, France, IV 555; V 16; VI 9
Sommervillers, near Nancy, France, V 242
Sonderak, Holland, III 304
Sor, river, near Toulouse, France, III 464-5, 467-8
Sorø, Denmark, II 69
Sotiel, Spain, II 676
Souter Point, Durham, England, V 183, 191, 193, 210
Southampton, Hampshire, England
  machine-tools, IV 437, 582
  roads, 1482, II 533
  ships, 16th century, III 473-4, 487
  telegraphy, IV 658
  town-planning, III 283
Southampton Water, England, III 476
South Boston, U.S.A., V 105
South Carolina, U.S.A., III 5
South Flevoland, Netherlands, VII 907
South Foreland, Kent, England, V 182, 183, 192, 210, 228; VII 1232
South Georgia, U.S.A., IV 61; VI 357

South Orkneys, South Atlantic, IV 61
South Pacific, VII 1454; see also Pacific Ocean
South Pole, VII 806
Southport, Lancashire, England, V 563
South Queensferry, West Lothian, Scotland, V 506
South Sea Islands, IV 579
South Shetlands, South Atlantic, IV 61
South Shields, Co. Durham, England, II 455-6; IV 96; V 553
South Wales, VII 889, 1227
Southwold, East Anglia, England, II 436
Spaarndam, Holland, II 688; III 442, 450, 452
Sparrows Point, Maryland, U.S.A., VI 477
Sparta, Greece, II 398, 499, 529, 700; III 273
Speaker's Green, see London
Spelsbury, Oxfordshire, England, IV 212
Speyer (Spires), Germany, II 138, 202, 620, 644
Spiennes, Belgium, I 559, 560, 562 (map)
Spindletop, Texas, U.S.A., VI 376, 381, 397
Spires, see Speyer
Spitalfields, see London
Spithead, English Channel, V 151
Spitzbergen, Arctic Ocean, IV 58, 619
Split Rock, Canada, IV 552
Spree, river, Prussia, III 450, 454
Springfields, Lancashire, England, VI 257-8
Spuyton Duyvil, New York State, U.S.A., V 330
Sri Lanka, see Ceylon
Srinigar, India, VII 874, 884
Sron Mor, Perthshire, Scotland, VII 1072
Stade, near Hamburg, II 150, 159
Staffordshire, England
  ceramics, IV 328-9, 331, 335, 341-56; V 667, 834
  coal-mining, III 84; IV 80 (map), 81
  iron-smelting, IV 105; V 801, 812
  ship-building, V 351
  steam-engines, IV 162, 177, 183
  water-supply, V 553
Stalingrad, U.S.S.R., VI 21; VII 812
Stalybridge, Cheshire, England, V 777
Stane Street, England, III 278
Stanhill, Lancashire, England, IV 278
Stansted, Essex, England, VII 1384
Star Carr, Yorkshire, England, I 499
Stassfurt, near Magdeburg, Germany, V 255; VI 530
Staten Island, New York, U.S.A., VII 891
Stavanger, Norway, III 533
Stecknitz, river, Germany, III 443-4
Steenbergen, Brabant, Holland, III 301, 320
Stenild, Jutland, Denmark, I 514
Stepney, see London
Sterkfontein, South Africa, I 23, 37 (map)
Stjärnsund, Sweden, III 342

Stockholm, Sweden
    canals, III 455
    cartography, III 532 (map), 533; V 447
    chemistry, research, 1683, III 682
    computers, VII 1174, 1188
    glass-engraving, III 224
    steam-turbine, marine, 1892, V 150
    town-planning, VII 976–7
    tunnel excavation, V 520
Stockport, Cheshire, England, IV 196, 300
Stockton-on-Tees, Durham, England, IV 255, 460
Stoke-by-Clare, Suffolk, England, II 624
Stoke-on-Trent, Staffordshire, England, IV 346, 353; V 662; VI 623
Stonar, Kent, England, IV 467
Stonehenge, England, I 81, 195, 315, 490–4, 712, plate 21A
Stonesfield, Oxfordshire, England, II 32, 34
Stonington, Connecticut, U.S.A., IV 58
Stora Kopparberg, Sweden, II 65
Stour, river, Kent, England, IV 467
— — Worcestershire, England, III 458; IV 153
Stourbridge, Worcestershire, England, III 458; IV 131, 237, 371, 374; V 778
Stra, near Padua, Italy, III 447, 451 n.
Stradonitz, Bohemia, II 559
Strasbourg, France
    canals, IV 550–1
    cartography, III 537–8
    farming methods, IV 29
    oil-refining, V 114
    potassium extraction, c.1497, II 354–5
    printing, III 383–4
    roads, IV 539
    telegraphy, IV 646
    timekeepers, 1354, III 655
Stratford, Connecticut, U.S.A., V 404
— Essex, England, III 695
Stratford-le-Bow, see Bow, IV 341, 492
Stratford-on-Avon, Warwickshire, III 458
Straubing, Bavaria, IV 361
Strood, Kent, England, IV 566–7
Stroud, Gloucestershire, England, V 571, 624
Styal, Lancashire, England, V 829
Styria, Austria, II 69, 70, 73, 75–6; III 115
Sudan, I 140; III 521
Sudbrook, Gloucestershire, England, V 518
Sudbury, Ontario, Canada, V 87; VI 413, 418, 420
— Suffolk, England, III 101
Sudeten, Czechoslovakia, VI 18
Suez, Egypt, V 111; VI 25, 250
Suffolk, England, III 101, 152; IV 5, 17, 21
Sukkur, India, VII 886
Sulzbach, Germany, IV 260
Sumatra, I 226
Sumeria, III 417

Sunderland, Co. Durham, England, IV 95, 457–8, 641; V 208, 553
— U.S.A., IV 545
Superior, lake, Canada–U.S.A., IV 553 (map); VII 916 (map)
Superior, lake, N. America
    canals, IV 552
    iron ore transport, V 808; VI 414, 464–5
    metals, extraction, V 79, 84
    mining, opencast methods, VI 411, 463
    St Lawrence Seaway link, VII 917, 920
Suppara, see Ophir
Surbiton, Surrey, IV 494, 503
Surinam, III 6
Surrey, England, III 265; IV 468; V 554
Susa, Persia, I maps 5, 6
    arrow-heads, stone, I 511
    Babylonian mathematics, I 787
    buildings, I 489
    metal-work, I 632, 650
    pottery, I 205, 396, 398
    roads, I 713; II 496 (map)
    textiles, fragments, I 432
    tools, metal, I 602
    water, drinking, II 674
    wheeled-vehicles, c.2500 B.C., I 205, 206, 207, 208–9
Sussex, England
    building methods, 17th century, III 261, 265
    farm implements, IV 4
    glass-making, III 208
    iron smelting, 1796, V 816
    ports, IV 466
    vehicles, 16th century, III 145
    wool industry, 16th century, III 152
Sutlej, river, India, VII 883, 892
Sutton Courtenay, Berkshire, England, II 93 n.
Sutton Hoo, Suffolk, England
    leathercraft, c.660, II 169
    metal-work, I 450, 451, 454–5, 456, 458, 461, 492
    ship excavation, c.655, I 579
Swabia, Germany, IV 527
Swale, river, Kent, England, VII 891
Swanscombe, Kent, England, I 22, 28, 29, 34, 68, table C
Swansea, Glamorganshire, Wales
    metals, extraction, IV 126, 128, 131; V 76–80, 83, 85–8, 96
    metals, working methods, V 607, 610, 629
Sweden, VI 418, 489, 490, 588, 610; IV 922
Sweffling, Suffolk, England, II 173; VII 765
Swindon, Wiltshire, England, V 323, 648; VII 765
Swinton, Yorkshire, IV 355
Sydney, New South Wales, Australia, IV 143, 592; VII 775, 892–4

Syene (Aswan), on river Nile, III 505–6, 510–11
Syracuse, New York, U.S.A., VI 490, 507
— Sicily
  machinery, Roman period, II 637
  military machines, II 604, 658, 699, 702,
    714–15
  quarries, limestone, c.250 B.C., II 26–7
  rut-roads, II 499
Syria, I map 8
  calendar, Jewish, III 574
  cartography, medieval, III 520 (map), 521
  ceramics, 14th century, IV 328
  cotton production, III 159
  dyestuffs, V 259
  glass-making, III 217
  sugar-cane, III 6
  town-planning, 3rd century, III 280
  weaving, figured fabrics, III 193, 196

Tabriz, Azerbaijan, Persia, II 302, 307; III 378,
  380
Taff, river, Wales, III 434
Tagus, river, Portugal, VII 883, 887
Tahiti, Pacific Ocean, IV 598
Takla Makan desert, Sinkiang, I 72
Takoradi, Ghana, VII 925
Tamar, river, Cornwall and Devon, England, IV
  462; VII 898
Tame, river, England, IV 153
Tampa, Florida, U.S.A., VII 800
Tampico, Mexico, V 102
Tanagra, Boeotia, II 277
Tanais (Don), river, Russia, III 519; IV 554
Tangiwai, New Zealand, VII 896
Tapajos, river, Brazil, V 774
Tapiola, Finland, VII 977
Taplow, Buckinghamshire, England, II 482, 483
Tapper, N. Germany, I 209
Taranto, Italy, VII 814
Tarapaca, Chile, V 255
Tarentum, Allegheny river, U.S.A., V 109
— Italy, II 460, 461
Tarkhan, Egypt, I 420, 434, 524, map 6, plates 9,
  11
Tarrant Keynston, Dorset, England, II 37
Tarshish (Tartessus), Spain, II 568 and n.
Tarsus, Asia Minor, II 317
Tartessus, see Tarshish
Tarxdorf, Silesia, II 56
Tasa, Egypt, I 387, map 6, table E
Tasmania, Australia, I 161, 222; V 8, 50
Taungs, Bechuanaland (Botswana), I 20, 22, 37
  (map)
Taupo, New Zealand, VI 220
Tawe, river and valley, Wales, V 76, 78
Taxila, India, II 528

Tay, river, Scotland, V 555
Teddington, Middlesex, England, IV 216; VII
  440, 1194
Tees, river and valley, northern England, IV 460;
  V 524, 810; VII 889
Teesside, England, VII 887
Tegernsee, Bavaria, V 102
Teheran, Persia (now Iran), III 505; VII 788
Tel El-Yahudiya, Egypt, I 237, map 6
Tel Halaf, Syria, II 108
Tel Jefat, see Jotapata
Tell Agrab, Mesopotamia, I 720, 725, map 7
Tell Asmar, see Eshnunna
Tell Beit Mirisim, Palestine, I 381, 383, 409,
  map 8
Tell Defenneh, Egypt, II 231
Tell ed-Duweir, Palestine, see Lachish
Tell Fara, Palestine, I 672, map 8
Tello, Mesopotamia, I 652, 653, map 7: see also
  Lagash
Tema, Ghana, VII 920, 922, 925, 926, 927
Tenda Pass, Alps, IV 529
Tène, La, see La Tène
Tennessee, U.S.A., VII 1333
Tepe Gawra, Mesopotamia, I 205, 460–2, 504,
  512, map 7, table E
Teplitz, Czechoslovakia, VI 495
Tern, river, Shropshire, England, IV 566–7
Terneuzen, Holland, IV 560
Terracina, Italy, II 510, 520; III 311
Tessin, Switzerland, VII 1127
Tessino, Italy, III 218
Tewkesbury, Gloucestershire, England, III 457;
  V 262
Texas, U.S.A.
  agriculture, V 4
  chemical industry, VI 508, 516
  drilling, oil industry, VI 382
  oceanography, Gulf region, VII 1453
  glass manufacture, VI 576
  natural gas, VI 192, 576
  sewage treatment, VII 1391
Thames, river, England
  bridges, IV 454–5, 456; V 506, 512–13, 643
  canals, III 456–7 (map); IV 572
  cartography, IV 618
  cement, use, IV 448, 449
  chemical industry, VI 507
  electric cables, VI 447
  electric-lighting, V 182
  embankment, protection, III 422
  fish preservation, IV 47, 52
  foundations, bridges, III 422, 425
  harbour works, IV 467–8; V 539
  locks, III 441, 456
  navigation, 1585, III 441
  sewage, IV 510–11, 518; VII 1391

Thames, river, England (*cont.*)
  shipbuilding, IV 575, 580, 589; V 351, 352,
    364–5, 371, 372, 810
  steam-propulsion, V 142, 144, 146, 153
  tunnels, IV 427, 449, 463–4; V 361 n., 516;
    VII 901–2, 905
  water-supply, London, IV 491, 493–4; V 554,
    561, 562; VII 1368, 1374
  water-wheels, London Bridge, IV 206–7, 491
Thames Ditton, Surrey, England, V 135
Thanet, Isle of, III 516
Thasos, Aegean Sea, II 43, 130
Thau, Étang de, near Trèbes, France, III 465, 467
Thebes, Egypt, I map 6
  agriculture, I 523, 537, 540–1, 543–4
  animals, domestic, I 329
  calendar, *c.*1500 B.C., 122, 124
  candles, *c.*1300 B.C., I 232
  chemical arts, I 239, 242, 251 (map), 266
  craft skills, I 189, 213, 453, 481, 689–90, 699,
    plate 10; II 150, 162
  food preparation, II 264, 273–5, 277, 278
  glass-ware, I 319
  ivory-work, I 664
  leather-work, II 150, 160, 162–4
  metals, working, I 578, 579
  ointment-compounders, I 292
  paper-making, II 756, 757
  pottery, *c.*1450 B.C., I 389
  rope-making, I 453, 454
  rotary motion, I 189, 213
  stone-work, I 481
  textiles, *c.*1500 B.C., I 437, 439, 440
  tools, iron, I 619; II 228–9
  viticulture, I 290
  water-supply, I 522–3, 537
  weapons, hunting, *c.*1400 B.C., I 156
  wheeled-vehicles, I 213, 726; II 551, 561
  wood-work, I 189, 689, 690, 694, 699
The Fens, *see* Fens, The
The Hague, Holland (Netherlands), II 680; III
  229, 663; IV 545; VI 2, 4
Themesvar, Hungary, IV 550
Thetford, Norfolk, England, II 295; IV 10
Theuben, river Danube, Hungary, IV 550
Thiers, Puy-de-Dôme, France, III 115
Thirlmere, lake, Cumberland, England, IV 496; V
  553
Tholen, Isle of, Zeeland, III 320
Thorny Down, Wiltshire, England, I 316
Thrapsanos, Crete, I 406
Thugga, *see* Dugga
Thule, Arctic, III 506, 509–10
Thurii, southern Italy, III 272
Thurrock, Essex, England, VI 447
Tiber, river, Italy, III 280–1, 288, 312–13
Tiberias, lake, I 259, map 8

Tibet, I 362, 367; V 321
Tibur (Tivoli), Italy, II 27, 406
Ticino, river, Switzerland–Italy, III 444, 446; IV
  555
Tierra de Brea, Trinidad, V 102
Tierra del Fuego, S. America, I 160, 165, 178
Tiflis, Russia, I 588
Tilbury, Essex, England, V 541
Timgad, Algeria, II 530–1; III 279–80, 282
Timonovka, Russia, I 302
Tipton, Staffordshire, England, IV 114; V 472
Tiryns, Greece, I 359, 457, 488, map 4; II 29,
  398
Tisza, tributary of Danube, IV 550
Titel, Yugoslavia, IV 550
Titicaca, lake, Bolivia, I 732, 733
Titusville, Pennsylvania, U.S.A., V 113, 120
Tiverton, Devon, England, V 602
Tivoli, *see* Tibur
Tjaalingermeer, Holland, III 302
Tobolsk, Ural, Russia, IV 529
Todmorden, Yorkshire, England, IV 302
Toeplitz, Austria, V 619, 620
Tokyo, Japan, VII 816
Toledo, Spain, II 57, 680; III 115, 350, 521, 584–5
Tolfa, Italy, II 368–9
Tollund, Denmark, I 355, 359, 370, plate 6B
Tolworth, Surrey, England, VII 905
Tømmerley, Denmark, II 85, 87
Tonbridge, Kent, England, IV 658
Tongataboo, Pacific Ocean, IV 598
Tornea, Finland, III 553
— river, Finland, III 554
Torquay, Devon, England, I 188; V 387, 546, 560
Tortona, Italy, IV 34
Toton, Nottinghamshire, England, VII 786
Toulon, France, IV 582, 633, 635, 644
Toulouse, France, III 329, 430, 459, 464–7
Tours, France, III 463
Towneley, Lancashire, England, IV 490
Transvaal, South Africa, IV 77
Transylvania, eastern Europe, V 321, 322
Traprain Law, Scotland, I 619, 620
Travers, near Neuchâtel, Switzerland, IV 539–40;
  V 118
Trawsfynydd, Merionethshire, Wales, VI 213
Trèbes, on R. Aude, France, III 466
Trebizond, Black Sea, II 760
Treguanda, III 210
Trench, Shropshire, England, IV 569
Trent, river, England, III 84; IV 199, 566, 572
Trentham, Staffordshire, England, IV 352
Trento, Italy, II 66; III 578
Trenton, New Jersey, U.S.A., V 662
Treves, Germany, II 313, 323, 544, 671, 718
Trèves (*later* Trier), Germany, III 279; IV 527
Trezée, river, France, III 460, 462

Trezzo, on R. Adda, Italy, III 446, 448
Trialeti, Georgia, I 209
Trient, Switzerland, II 612
Trier, *see* Trèves
Trieste, Italy, IV 528, 553
Trinidad, West Indies, V 105, 223; VII 1073
Trinil, Java, I 23, 37 (map)
Trinity Bay, Newfoundland, Canada, IV 661
Tripoli, Syria, V 266
Tripolye, Ukraine, I 370
Trois-Frères, Ariége, France, I 151
Troldebjerg, Denmark, I 311
Tromsø, Norway, VI 357
Trowbridge, Wiltshire, England, V 587
Troy (Hissarlik, Ilium), I map 4
    food plants, I 362, 364-5
    fortifications, stone, I 488; II 701-2
    building methods, I 488
    metallurgy, I 584
    mining, copper, I 565
    pottery, I 398, 402-3
    spinning and weaving, c.2500 B.C., I 427, 433,
        435
    tools, non-metallic, I 504
Troyes, Aube, France, II 609; III 371
Trundholm, Denmark, I 213
Truskawiec, Poland, V 116
Tsarskoe Selo, Russia, IV 529
Tuc D'Audoubert, Ariége, France, I 144, 151, 152
Tunis, N. Africa, III 480
Tunisia, North Africa, V 16; VI 529
Tunstall, Stoke-on-Trent, Staffordshire, IV 348
Tura, Egypt, I 453, 714
Turcoing, Nord, France, V 576
Turin, Italy, II 718; III 280, 286-7; V 546
Turkey, III 4, 8, 405, 579; VI 7, 9
Turrach, Styria, Austria, V 57
Turton, Lancashire, England, IV 496
Tuscany, Italy
    boric acid production, IV 320-1
    ceramics, IV 329
    chemical arts, 16th century, III 677
    farming methods, IV 34, 35
    glass-making, III 218
    land-drainage, III 314-5
Tusculum, near Rome, Italy, II 678
Tutbury, Staffordshire, England, II 37
Tuttlinger, W. Germany, VII 1322
Tuz Khurmatli, Mesopotamia, I 262, map 7
Tweed, river, Scotland, IV 459-60
Twickenham, Middlesex, England, IV 242
Tyne, river, England, III 76 (map)
    coal exports, III 74
    industrial change, discussion, V 815, 816
    safety-lamp, IV 94
    shipbuilding, V 121, 371, 381, 810; VII 740
    tunnels, VII 905

Tyneside, Northumberland, England, IV 96, 373;
        V 324; VI 507; VII 1086
Tyre, Lebanon, I, map 8; II 760
    dye, mollusc source, I 247-8; II 367
    fortifications, II 702
    harbour-works, II 516
    military engines, 332, B.C., II 700
    pipes, bronze, II 665
Tyrley, Shropshire, England, IV 566
Tyrol, Austria, III 341; IV 610; V 339
Tyrone, Northern Ireland, III 455
Tyrrhenian Sea, III 311

U.S.A., Great Lakes, IV 553 (map)
Ubaid, El-, *see* El-Ubaid
Udine, Italy, III 286
Ufenti, river, Italy, III 311
Uganda, I 384
Ugarit (*later* Ras Shamra), Syria, I 201, 247, 762-
        3, map 4, plate 30B
Uitgeest, Holland, II 625
Ukraine, U.S.S.R., V 3, 6, 304; VI 10, 507, 538
Ulm, Germany, III 532
Unieux, Loire, France, V 65
Upper Eider lakes, Schleswig-Holstein, IV 560
Upper Great Lakes, Canada, VII 917
Uqair, Mesopotamia, I 462-3, map 7
Ur, Mesopotamia, I 56, map 7, tables E, F
    animals, domestication, I 340
    bitumen, uses, I 253-4
    building methods, 2800-1800 B.C., I 465-9
    buildings, 3000 B.C., I 306
    dyestuffs, I 247
    food preparation, I 278
    glass-ware, II 318, 319
    implements, agricultural, II 83
    lamps, oil-, c.3000 B.C., I 236
    leather-craft, II 163 n.
    metals, extraction, I 582, 584, 591
    metal-work, I 582, 626, 636, 643, 644-5, 649-
        50, 652-7, 659, 661
    potter's wheel, Uruk period, I 199, 200
    soap manufacture, I 261
    town-planning, III 271
    transport, land, I 705, 709
    weapons, Bronze Age, I 614
    wheeled vehicles, I 205, 206, 208-9, 717, 718,
        725; II 163 n.
    wood-work, I 684
Urals, mountains, Europe and Asia, IV 144
Uratu (Ararat), Mesopotamia, I 564, 583, map 7
Urbino, Italy, III 311; IV 329 n.
Urgenj, Turkestan, IV 312 n.
Uruk (Warka), *see* Erech
Usk, river, England, VII 889
Utah, U.S.A., V 3; VI 413
Utica, near Tunis, II 408

Utoquai (lake-village), I 356, 517
Utrecht, Holland (Netherlands), III 197, 441–2;
    IV 15; VI 301

Vaison, south France, II 553
Valencia, Spain, III 210
Valentia, Ireland, IV 661
Valladolid, Spain, IV 554
Valle, Sweden, I 150
Valley of Kings, Egypt, I 440
Valsgärde, Sweden, II 313
Van, lake, Armenia, I 282, maps 4, 7
Vancouver, Canada, VII 788, 898, 917
Vancouver Island, Canada, VII 1082
Väner, lake, Sweden, III 455
Vann Farm, near Chiddingfold, Surrey, England,
    III 207
Vaphio, Greece, I 400, 645, 646, map 4
Västerbotten, Sweden, I 708
Vatican City, III 196
Vatter, lake, Sweden, III 455
Vaux de Cernay, near Paris, II 650
Vebbestrup, Jutland, II 84
Velem St Vid, Hungary, I 568, 588
Velinus, lake, Italy, II 679
Venafro, near Naples, Italy, II 597–8, 602
Vendel, Sweden, II 557, 559
Venezuela, VI 186, 394; VII 1456
Venice, Italy
    agriculture, 18th century, IV 35
    bridges, III 424, 430
    buildings, Renaissance period, III 245, 248–9
    buildings, Romanesque period, II 423, 424
    Byzantine craft, c.1204, II 765–7
    carpenters, c.1500, II 393, 395
    ceramics, 15th century, II 301; IV 329 n.
    chemical industry, III 678, 681, 703, 705; IV 233
    chemistry, pre-scientific, II 356, 368, 372
    dredging, IV 635–6
    dyestuffs, V 259
    glass-making industry, II 326–7, 328; III 209,
        216, 222, 230, 238, 678, 681
    instruments, precision, III 598
    lenses, glass, III 230, 234
    metal-work, II 454
    mirrors, manufacture, IV 362
    navigation methods, 1428, III 526
    printing, III 380, 405–6
    shipbuilding, III 472–4, 476, 479
    soap-making, III 703, 705
    spectacles, manufacture, III 230–1
    spinning processes, II 194–5, 207
    town-planning, III 286
    water-control, III 309, 314
    water-mills, II 608, 648
    waterways, III 309, 441 n. 2; IV 558
    windmills, II 620

Verdun, France, VI 9
Vermont, U.S.A., V 645, 653
Vernassonne, stream, France, III 464
Verona, Italy, III 199, 309; V 318 n. 2
Versailles, France, III 238, 543; IV 154, 363, 500;
    V 448
Verulamium (St Albans), Hertfordshire, England,
    II 169; III 277–8
Verviers, Belgium, V 570, 572
Věstonice, Moravia, I 147, 152, 376
Vesuvius, Italy, III 274
Vettabia di Milano, Italy, III 309
Vettersfelde, Brandenburg, Germany, I 647–8
Vézelay, Yonne, France, II 127, 647
Vicenza, Italy, III 249; IV 337
Vicopisano, on R. Arno, Italy, III 448
Victoria, South Australia, V 8
Victoria Falls, Rhodesia–Zambia, VII 881
Victoria West, South Africa, I 133
Vieille Montagne, Belgium, IV 132
Vienna, Austria
    carpenters, c.1514, II 389
    cartography, IV 607
    ceramics, V 659, 660
    education, technical, V 659, 714
    electric motor, 1873, V 231
    horse equipment, II 559
    instruments, medical, VII 1322
    lamp oils, V 105
    machines, fairground, VII 1129
    metal-work, II 455
    motor-carriages, 1874, V 426
    phosphorus discovery, V 253
    photography, V 722
    printing, V 704, 705, 707, 714
    railway engineering, V 328
    roads, IV 528, 539
    water-supply, V 555, 558
    waterways, IV 553
Vietnam, see Indo-China
Vigevano, Lombardy, Italy, III 286
Villach, Carinthia, Austria, IV 665
Villefranche, near Toulouse, France, III 459, 464
Villeneuve-l'Archevêque, Yonne, France, V 555
Villepinte, near Castelnaudary, France, III 467
Vilvorde, Belgium, IV 528
Vincennes, Paris, France, IV 338–9; V 284
Virginia, U.S.A., III 1–2, 9; IV 536; V 4; VI 363
Virginia City, Nevada, U.S.A., V 559
Vitry-en-Perthois, Marne, France, III 290
Vitry-le-François, Marne, France, III 289–90
Vitznau, Switzerland, V 346
Vix, Châtillon-sur-Seine, France, II 451
Vogtland, Germany, I 508
Volga, river, Russia, IV 554; VI 21
Volterra, Italy, II 52, 60
Vonges, near Dijon, France, V 289

Voorzaan, the, Holland, IV 638
Vosges, mountains, France, III 538; IV 29
Vounous, Cyprus, I 403, 420, 433, 435, map 4
Vreeswijk, Holland, III 442, 444
Vučedol, Yugoslavia, I 311
Vulci, near Rome, II 51
Vyrnwy, river, Wales, IV 497; V 522, 553, 557

Waddenzee, Netherlands, VII 907
Wadi Alaki, Nubia, I 526
Wadi 'Amd (Hadhramaut), Arabia, I 532
Wadi Gerrawi, Egypt, I 528, map 6
Wadi Ghazzeh, Palestine, I 232, 420, map 8
Wadi Halfa, Egypt, I 477, map 6
Wadi Hammamat, Egypt, I 477, map 6
Wadi Maghara, Palestine, I 564, map 8
Wadi Natrum, Egypt, I 260, map 6; II 354
Wakefield, Yorkshire, England, III 458; IV 448;
  V 563
Walcheren, Holland, V 18
Waldburg, Germany, IV 425
Walker-upon-Tyne, England, IV 238
Wallsend, Northumberland, England, II 31; IV
  84; V 152
Waltham Abbey, Essex, England, III 451; V 290
Walthamstow, Essex, England, II 577
Walton-on-Thames, Surrey, England, VII 1374
Waluta (Majorca), II 597
Wandle, river, Surrey, England, IV 207, 499
Wangen, Baden, I 357, 358, 367, map 3
Wapping, see London
Warka (Uruk), see Erech
Warrington, Lancashire, England, IV 241; V 632
Warsaw, Poland, VI 9
Wart, the, Holland, III 321
Warwickshire, England, IV 177
Wash, The, England, III 315, 320; V 545; VII
  846
Washington, District of Columbia, U.S.A.
  economic crisis, 1930s, VI 15
  printing, V 687
  roads, IV 540, 543
  rockets, VII 859
  space programme, NASA, VII 869
  telegraph system, 1842, IV 660
Watervliet, New York, U.S.A., VI 497
Watford, Hertfordshire, England, VII 1140
Wath, Yorkshire, England, VII 785
Watling Street, England, III 277-8
Weald, Kent–Sussex, England, II 10, 61; III
  31-2
Wear, river, Co. Durham, England, IV 94, 457-8;
  V 810; VII 881
Weaver, river, Cheshire, England, III 458; IV 349
Wednesbury, Staffordshire, England, IV 113
Wednesfield, Staffordshire, England, V 253
Weidgreb, the, Holland, III 302

Weimar, Germany, IV 527; VII 800
Welland, river, England, III 316; V 544
Wells, Somerset, England, V 282
Welwyn Garden City, Hertfordshire, England,
  VII 964
Wenlock, Shropshire, England, III 688
Wessel, W. Germany, VII 896
West Africa, III 4; V 2, 23, 95, 828
West Bromwich, Staffordshire, England, V 625
West Drayton, Middlesex, England, IV 657; V
  218
Westergeln, E. Germany, VI 524
Western Islands, Scotland, V 450
West Harptree, Somerset, England, IV 209
West Hartlepool, Co. Durham, England, V 477
West Indies
  ceramics, imported, IV 349
  cotton production, III 159-60
  dyestuffs, III 694; V 265
  food production, III 3, 6-7, 21-3
  navigation competition, 1714, III 557
  rubber source, V 754, 773
  ships, 18th century, III 477
West Middlesex, England, VII 1391
Westminster, see London
Westmorland, England, III 152
Weston Favell, Northampton, England, II 295
Westphalia, Germany, IV 81, 527
Wettingen, Switzerland, III 428
Wetzlar, W. Germany, VII 1288
Wey, river, England, III 456
Weybridge, Surrey, England, III 456-7
Wheeling, West Virginia, U.S.A., IV 459
Whitby, Yorkshire, England, IV 58, 574
Whitechapel, see London
Whitehaven, Cumberland, England, IV 95, 258
Whitehill, Midlothian, Scotland, IV 172
White Horse, Vale of the, Berkshire, England, III
  147
White Leaf Hill, Buckinghamshire, England, I
  313
White Sea, the, U.S.S.R., III 533
Whittlesey Mere, The Fens, England, V 545
Wicklow, Leinster, Ireland, V 245
Widnes, Lancashire, England, IV 241; V 238,
  240, 802, 821, 832; VI 263
Wieringen, island, Netherlands, VII 907, 909
Wietze, Hanover, Germany, IV 539; V 102, 110,
  111
Wigan, Lancashire, England, IV 258; VII 892
Wilderspool, Cheshire, England, II 56
Wiley–Dondero Canal, New York, U.S.A., VII
  918
Wilkowitz, Czechoslovakia, IV 105
Willamette, river, Oregon, U.S.A., VII 890
Willebroeck, Flanders, III 452-3
Willendorf, Lower Austria, I 152, 153

Willey, Shropshire, England, IV 104
Willington, Derbyshire, England, IV 350
Wilmington, Delaware, U.S.A., VI 240, 560, 660
Wilton, Wiltshire, England, IV 241; V 585; VI
    534, 544, 561
Wiltshire, England, III 147, 152, 262; IV 19
Wimborne Minster, Dorset, England, II 237
Wimereux, France, VII 1232
Winchelsea, Sussex, England, II 582; III 284
Winchester, Hampshire, England, III 277; VII
    875-6
Windermere, Westmorland, England, V 835; VI
    212
Windmill Hill, Wiltshire, England, I 510, 515
Windscale, Westmorland, England, VI 246-50,
    255, 263-4, 266, 281
Windsor, Berkshire, England, VII 800, 905
— Vermont, U.S.A., IV 436, 439
Winkel, Holland, V 18
Winnington, Cheshire, England, V 88; VI 558
Winnipeg, Canada, VII 788
Winsford, Cheshire, England, III 458; IV 350
Wisbech, Cambridgeshire, England, V 263, 563
Wisconsin, U.S.A., VII 726, 1295, 1346
Wissembourg, France, IV 29
Witham, river, England, III 316, 516; V 544-5
Witwatersrand, South Africa, VI 420, 424
Wollaton, Nottinghamshire, England, III 82, 84
Wolvercote, Oxfordshire, England, I 26; VI 608
Wolverhampton, Staffordshire, England, V 230;
    VII 732
Wolverton, Buckinghamshire, England, IV 658
Woodford Bridge, Essex, England, IV 207
Woodhenge, Wiltshire, England, I 314-15, 493
Woolwich, London, England
    concrete, 1835, V 487
    education, technical, IV 426, 444-5
    electricity generators, V 182
    glass-making sand, III 702
    guns, machining, III 367
    machine-tools, IV 423, 426, 437
    metal cartridge, development, V 630
    shipbuilding, III 478, 485, 487
    steel, cast, IV 108
Wor Barrow, Dorset, England, I 313
Worcester, England
    ceramics, IV 338, 341, 343-4, 354, plate 18
    pitch manufacture, III 688
    sewage treatment, VII 1391
    silk, figured fabric, III 201
    water-supply, II 671
— Massachusetts, U.S.A., IV 436; V 63; VII
    861
Worcestershire, England, III 264; IV 100, 241
Worms, Germany, II 138
Worplesdon, Surrey, England, IV 16
Wroxeter, Shropshire, England, III 278

Württemberg, Germany, I 366, 369; IV 113, 610;
    V 427, 687
Würzburg, Bavaria, IV 261
Wyandotte, Michigan, U.S.A., V 57; VI 507
Wye, river, England–Wales, III 458; IV 461; VII
    888

Xativa, Spain, III 412

Y, river, see IJ
Yang Shao, China, I 419
Yangtze, river, China, III 439
Yarmouth, Norfolk, England, IV 48, 54, 647, 658
Yattendon, Berkshire, England, I 612
Yawata, Japan, V 70
Yelista, S. Russia, I 209
Yellow River, China, III 439-40
Yenangyuang, oilfield, Upper Burma, VI 398
Yèvre, river, France, III 450
Ynyscedwin, South Wales, IV 111
Yonkers, New York State, U.S.A., V 584
York, Yorkshire, England
    building methods, II 425, 428, 429, 444
    ceramics, II 293
    glass windows, II 425, 429; III 241
    model village, V 830; VII 962
    quarries, Roman, II 31-2
    road-survey, 1635, III 544
    road-vehicles, steam, V 422
    tools, tradesmen's, 1860, III 118
    town-planning, III 277, 278, 284; VII 962
Yorkshire, England
    agriculture, IV 18, 21
    building materials, IV 450
    building methods, III 260, 263, 265-6
    ceramics, IV 352
    education, discussion, V 778-9
    electricity, generation, VII 1070
    forge-hammer, 1820s, IV 104, 105
    industrialization, discussion, V 816, 829
    knitted-fabrics, III 184
    machine-tools, IV 423
    railway engineering, V 333
    spinning machinery, IV 295, 298, 315; V 572,
        577
    steam-power, IV 166, 177, 179
    steam-vehicles, VII 727
    textiles, preparation, IV 304; V 580, 587
    tinplate manufacture, III 689
    waterways, III 458
    water-wheels, IV 166
    windmills, III 101
    woollen industry, III 152; V 570, 572, 574-5
Youngstown, Ohio, U.S.A., V 629
Yperlée, river, Belgium, IV 558
Ypres, Flanders, II 208, 209, 213, 438; III 454;
    IV 629

Yser, river, Flanders, III 454
Ystalyfera, near Swansea, Wales, IV 113
Yucatan, Central America, I 83
Yugoslavia, IV 120; V 93; VI 7, 11

Zaan, river, Holland, III 106; IV 156
Zabad, Aleppo, Syria, I 767
Zagazig, *see* Bubastis
Zaire, VI 256; VII 788
Zalavrouga, Russia, I 709
Zambezi, river, Africa, VII 881
Zambia, IV 77; VI 445; VII 420, 788
Zeeland, Netherlands
    dike-construction, II 684–5; III 301, 318
    dredging, IV 629–30
    farming methods, II 682; IV 15
    land-reclamation, II 681–4; III 302–3
    settlement-mounds, Iron Age, I 322; II 681
Zemzem, Mecca, I 528
Zimbabwe (Rhodesia), VII 781
Zinjirli, Syria, I 675, 768, map 4
Zlokutchene, near Sofia, II 105
Zofingen, Aargau, Switzerland, II 306
Zug, lake, I 361, map 3

Zuid Beveland, Holland, III 301
Zuiderzee, Netherlands, III 302 (map); VII 910 (map)
    agriculture, VII 1474
    canals, IV 560; VII 909
    dredging, IV 640
    fish trawling, IV 52
    land-reclamation, V 545; VII 907–10, 1474
    sea-encroachment, 1300, II 683
    ships, 17th century, III 485
    town-planning, III 293
Zuni, U.S.A., I 406
Zürich, lake, I 355, 356, 360, 366, 421, map 3
Zürich, Switzerland
    computers, VII 1166, 1200
    education, industrial, VI 139
    electric motors, V 232
    gas, petrol enrichment, V 119
    steel industry, V 64
    structural theory, study centre, V 493, 501
    turbines design, Niagara Falls, V 533
    vineyards, II 138
Zwierzyniec, Poland, I 342
Zwijn, inlet of sea, Flanders, IV 558–9

PART 4

INDEX OF SUBJECTS

INDEX OF SUBJECTS

Page numbers in **bold type** refer to subject content of chapters.
For best results consult 'hints for using indexes' in the *Preface*.

Abacus, II 766, 768; III 501; VII 1158–9
Abies cilicica, I 288
Abyssinia, metallurgy II 55, 57
Acacia arabica, I 243
Anilotica, I 244–5, 288; II 151
Académie de l'Architecture, IV 442
Académie des Sciences, Paris, 718
  architectural theory, IV 479
  cartography, IV 596, 599
  chemical industry, IV 234, 239
  cinematography (1888), IV 741
  civil engineering, IV 442, 443
  foundation (1666), IV 234, 442
  'gasometer' instrument, IV 406
  hydrogen illuminant, IV 260
  lubricants discovery (1847), V 119
  nautical tables, III 553
  rubber report (1736 expedition), V 754
  structural theory, V 494
  timekeeping mechanisms, III 345
Academy of Motion Picture Arts and Sciences,
  U.S.A., VII 1313
Accadèmia del Cimento (1657), III 315, 636–7
Accadèmia Fisicomatematica, III 312
Acetic acid, V 309–11
Acetone, VI 542–3
Acetylene, VI 541, 547
Acheulian cultures, I 25–8, 68, 131–2, table D
Acids, II 356–7; III 65, 176, 407, 678, 706; *see
  also named specific acids*
Acorus calamus, I 285, 288
'Acrilan' textile fibre, VI 651
Acts of Parliament
  Alkali (1781, 1863), IV 237; V 236; VII 1205,
    1480
  Arbitration, VI 98
  Adulteration of Food and Drink (1860), VII
    1402
  Adulteration of Food and Drugs (1872), VII
    1402
  Artificial Cream (1929), VII 1406
  Atomic Energy (1946–54), VI 281, 282
  British Reservoirs (Safety Provisions) (1930), V
    523
  Clean Air (1956), VII 976
  Corn Laws (early 19th century), V 7, 10, 27

  Defence Regulations (Sale of Food) (1943), VII
    1407
  Distribution of Industry (1946), VII 968
  Education (1870–1902), V 776, 787, 793, 796
  Education (1899–1944), VI 142, 143, 144, 168
  Electricity (1882–1957), VI 284, 286, 287, 288,
    289; VII 1128
  Emergency Powers (1920), VI 101
  Employers' Liability (1880), V 833
  Explosives (1875), V 290
  Factory (19th century), V 812, 833
  Food (1860–99), V 31, 52
  Food and Drugs (1899–1955), VII 1402. 1403,
    1407
  Locomotives on Highways (1896), VII 705
  London Building (1909), V 477
  London Improvement, 18–19th centuries, IV
    505, 510
  Merchant Shipping (1894), V 378
  Metropolis Water (1852), V 563
  Metropolitan Paving (1817), IV 500, 506
  Navigation, repeal (1849), IV 590, 592
  Public Health (1848), IV 490; VII 962
  Public Health (Drainage of Trade Premises)
    (1937), VII 1396
  Restriction of Ribbon Development (1935), VII
    872
  Rivers (Pollution) (1876–1951), VII 1396
  Road Traffic (1930. 1934, 1949), VII 842, 872
  roads (1766), IV 524, 530
  salt taxation (1781), IV 237
  street cleaning, IV 505–6
  Taxation (Customs & Excise) (1890), V 796
  Technical Instruction (1889), V 28, 779, 794,
    796; VI 139, 142
  Textile Factory (1844), IV 151
  Town and Country Planning (1947), VII 970
  Tramways (1870), VII 706
  Turnpikes (1720–30), IV 530
  Water (1945–8), VII 1365, 1397
  water supply, IV 489–90, 493–5, 497, 503
Admiralty, British
  armoured plate (1861), V 614
  bridge design influence, V 503
  charts, IV 618; V 449–50
  hydrodynamics, V 386, 387, 389

Admiralty, British (*cont.*)
  hydrographic department, V 449, 454
  ironclad ships, IV 588
  lighthouses (List of Lights), V 454
  navigation instruments, III 557; V 455–6, 461
  Notices to Mariners, V 454
  scientific advice, 19th century, IV 674
  ship-building, IV 574–5, 579–80, 584; V 373;
    VII 747
  sounding-machine (1872), V 453–4
  survey expertise, IV 616, 618
  telegraphy, IV 646–7, 651–2
  tide tables, V 454
Advanced Gas-cooled Reactor (A.G.R.), VI 251,
  259
Advertising and town-planning, VII 970, 976
Advisory Committee on uranium, VI 268
Adzes; *see also* Axes
  Assyrian tools, *c.*8th century B.C., II 228
  axes, difference, I 505–6, 601
  boat-building use, 13th century, II 390
  classical period, figures, II 231, 232
  digging implements, form, II 98
  metal, I 504, 601–2, 687, 688
  modern tools, II 36
  post Roman period, II 243, 252, 253
  stone, I 497–8, 502, 507
  tradesmen's tools, 17th–19th century, III 110,
    117
  windmill construction (*c.*1700), III 107–8
Aegilops, I 366–7
Aeolipiles (steam injector pumps), II 633; III
  52–3; IV 168, 672
Aeronautics, V **391–413**, VI 296; *see also* aircraft,
  Military Technology, Transport
  aerodynamics, V 403–4, 407
  aeroplanes, powered, V 391–3, 408–11
  airships, V 400–3; VI 439; VII 796–800
  autogiros, V 405
  balloons, V 395–401
  gliders, V 405–7, 413; VII 806–7
  helicopters, V 404–5; VII 821–4, 834, 987–91
  kites, V 398
  ornithopters, V 403
  parachutes, V 398–9, plate 23A
  rockets/rocket propulsion, V 412
Aes Marianus, II 52
Afghanistan, windmills, II 616–17
'Agilon', textile yarn, VI 660
Agriculture, I **520–57**; III 12–15; IV **1–43**; V
  1–25; VI **298–41**; *see also* Animals, Food,
  Plants
  arboriculture, I 371, 450; III 13–14
  alternate husbandry, IV 15, 18–20, 22–5
  ancient man, I **353–75**
  artificial grasses, III 12
  biological controls, VI 314–15

  Board of, IV 5, 10, 22, 40–1, 621; V 13
  books, III 14
  cereals, I 362–70, 501–2, 522, 550, plate 7
  chemistry, IV 38–9, 41
  collectivization method, Russia, VI 14
  corn-drying kilns, II 99–100
  dairy products, V 32–6
  digging-tools, II 98–9
  drainage, V 11, 14, 16, 24
  enclosures, open fields, IV 1, 17, 30–1, 36, 615
  farm-tools, II 81–102; III 14–15, **134–50**; IV
    1–12, 13, 17, 21, 24–5, 36, 192, plate 2
  fencing, II 99
  fertilizers, *see* Fertilizers
  fruit, I 357–8, 551–4
  genetic studies, VI 304–7, 328
  green vegetables, I 354–6
  harrows and rakes, II 94
  harvesting implements, II 94–7
  hatcheries, VI 315
  implements, *see* farm-tools, Tools
  Institutes, IV 41–2
  irrigation, *see* Irrigation
  Koppelwirtschaft system, V 17, 19 n. 1
  labour-saving techniques, VI 44, 312–14, 344–6
  livestock, *see* Livestock
  lunar calendar, I 794–5
  management systems, VI 298, 303, 312–16,
    326–7
  manure, *see* Fertilizers
  market-gardening, III 13–15; V 6
  marling, IV 19, 41
  mechanization, IV 7–12; V **1–25**; VI 334–8
  mustard and sugar, I 370–1
  Neolithic, I 42–7, 79, 340–1, 349–63, 501–2
  Nile valley, I 539–42, 795
  Norfolk four-course rotation, IV 18–20; V 12
  nuts and oil-seeds, I 358–60
  pest and disease control, VI 303, 315–38
  plant breeding, VI 327–30
  plant nutrition, IV 37–9; VI 330–3
  ploughs, II 81–93
  pulses, I 360–2
  root vegetables, I 356
  scientific investigations, III 679; IV 37–40
  seed sowing, II 100
  Societies, IV 22, 28
  soil-erosion, V 6
  sugar-beet, IV 28–9
  threshing and winnowing, II 97–8, 105–6
  Tigris and Euphrates valleys, 547–54
  tools, *see* farm-tools
  tropical, V 2, 21–4
  vehicles, III 139–48
  viticulture, *see* Viticulture
  Wisconsin Agricultural Experiment Station, V 33
Ahrensburgians, I 160, 161, table E

Air, alchemical element, IV 214
Air conditioning, VII 989
Aircraft, VII **789–836**; *see also* Aeronautics,
    Supersonic aircraft named
Aircraft, civil
  B.A.C. One-Eleven, VII 829
  Boeing 700 series, VII 827, 829
  Concorde, VII 834, 835, 1469
  Convair CV-240-440, VII 828
  De Haviland Comet, VII 701, 827, 828
  Douglas DC series, VII 825, 826, 829
  Ilyushan IL-62, VII 829
  Lockheed, Constellation, VII 825
  Sud Caravelle, VII 829
  Tupulov Tu-100 series, VII 828, 829
  Vickers Viscount, VII 701, 828
Aircraft, military
  Aichi, VII 813
  Albatross, VII 793, 794
  Beaufighter, VII 812
  B.E.2c, VII 793
  Blériot, VII 792
  Boeing, VII 813, 814, 815
  Breguet, VII 795
  Bristol, VII 794
  Caproni, VII 795
  Consolidated, VII 813
  Curtiss, VII 812, 814, 1012
  C.V., VII 795
  De Haviland, VII 794, 795
  Dewoitine, VII 812
  Dornier, VII 811
  Douglas, VII 813, 814
  F.E.2c, VII 794
  Flyer, VII 790, 791
  Focke Wulf, VII 811, 814
  Fokker, VII 794
  Friedrichshafen, VII 795
  Gnome, VII 792
  Gotha, VII 795
  Grumman, VII 813
  Halifax, VII 814
  Handley-Page, VII 795
  Heinkel, VII 811
  Henri Farman, VII 790, 792
  Hurricane, VII 812
  Ilyushin, VII 812
  IM, VII 795
  Junkers, VII 811, 814
  Lancaster, VII 814
  LeO, VII 812
  Lockheed, VII 813
  Messerschmitt, VII 811, 1020
  Meteor, VII 1019
  Mitsubishi, VII 813
  Morane, VII 812
  Morane-Saulnier, VII 794

  Mosquito, VII 814
  Nakajima, VII 813
  Nieuport, VII 794
  Potez, VII 812
  Republic, VII 814
  SE5, VII 794
  Short, VII 796
  Sopwith, VII 794
  SPAD, VII 794
  Spitfire, VII 812, 813
  Stirling, VII 814
  Superstar, VII 1022
  Voisin, VII 791–2, 793, 795
  Vought, VII 813
  Yakolav, VII 812
Airlines, VII 824–9
  B.E.A., VII 697
  B.O.A.C., VII 697
  Imperial Airways, VII 802
  KLM, VII 802
  Lufthansa, VII 802
  Pan-American Airways, VII 803, 825
  SABENA, VII 802
Airships, V 400–3; VII 796–800
  Beardmore R34, VII 799
  DELAG'S Schwaben, VII 798
  Goodyear naval blimps, VII 797
  Goodyear Zeppelins, VII 799
  Graf Zeppelin, VII 799
  Graf Zeppelin II, VII 799
  Hindenburg, VII 799
  Le Jaune, VII 796–7
  Los Angeles, VII 799
  Metalclad ZMC–2, VII 798
  Norge, VII 797
  R100, VII 799
  R101, VII 799
  Schutte-Lanz SL2, VII 798
  Shenandoah, VII 799
  Zeppelins, V 403; VI 439; VII 798–9
Akkadians, I 246, 262, 464–9, 524, 545; *see also*
    Mesopotamia
Alabaster, II 37–8
Albumen, III 29
Alchemy, II **731–52**; III 27, 676, plate 28; IV
    231, 242; *see also* Chemical industries,
    Cosmetic arts, Distillation, Dyes
  alkahest, II 732
  balances, II 744–6, 749
  definition, II 731–2
  elixirs, II 732
  four elements, IV 214
  furnaces, II 351, 741–5, 750, plates 42, 43A
  materials, cost, II 746–50
  sublimation, II 740–1, 750
  theories and equipment, II 733–9
  treatises, II 736–7, 738–9

Alchymia (Libavius), II 748
Alcohols, I 293
  discovery and distillation, II 141–4
  ethyl, V 301–7
  fermentation techniques, I 275–85; II 2,
    139–41; IV 232–3; V 301–7
  methyl, V 308, 311
  military use, II 375 n., 381
  purity testing, 17th century, III 21–2
Algonkin peoples, I 176
Alidade, III 538–41, 589–90, 593, 606–7, 627
Alkalis, IV 217; *see also* Chemical industries
  flax manufacture, IV 293
  Leblanc process, IV 236, 238–9, 241, 253
  natural occurrence, I 260–1
  plant ash, IV 235
  soap manufacture, III 704; IV 231–2
Alloys; *see also* Metallurgy, Steel
  cannons, III 361, 364
  copper, II 48–55; III 37–42, 51, 361; IV 130,
    133
  density studies, 17th century, III 58
  non-ferrous metals, IV 119–20, 124, 133,
    146
  precious metals, III 45–9
  speculum metal, III 237
  tin, III 42–3, 393; IV 124
Almagest, Ptolemy's 'System of Astronomy', III
    508, 566, 584, 586–9, 591, 612, 615
Almanacs, III 535, 552–3; *see also* Calendars
  Connoissance des Temps, III 553, 557
  'Mariner's New Kalendar', III 553
  'Nautical Almanac', III 557
  'Seaman's Kalendar', III 552
Alphabet, *see under* Writing
Alpenquai lake-dwellers, I 355, 360, 369
Altazimuth instruments, IV 603–4
Alum
  chemical composition, II 368–9, 737; IV 250
  copper alloys, III 41
  dye mordant, II 350, 366–9; III 77
  leather preparation, II 149–50, 151, 154–5
  preservative, pre-classical period, I 256
  production, II 355, 368–9, 746; III 77, 706; IV
    232, 235, 250–1
  Roman Egypt, I 262–3
Aluminium
  aeronautical engineering, VI 439–41
  alloys, V 608; VI 434, 436, 439–41, 447
  die-castings, VI 434–6
  electrical transmission cables, VI 447
  manufacturing processes, V 76, 90–4, 248–9; VI
    410, 428, 458, 459
  mining, bauxite source, VI 413
  sheet rolling, VI 458, 459
  world production, V 72–3
Ambergris, IV 57

America; *see also* North America, South America,
    United States of America
  architecture, III 253
  coal, III 86
  cochineal, III 695
  copper ores, IV 126
  metals, IV 64–6, 68–9, 77, 119
  printing, III 390, 401
  rice, III 5
  voyages, III 477
American Association of Railroads (1934), VII 703
American Bureau of the Census, VII 1154, 1179,
    1181, 1200
American Bureau of Standards, VI 623
American Ceramic Society, VI 589, 623
American Dental Laboratory Association, VII
    1324
American Medical Association, VII 1349
American National Aeronautics and Space
    Administration (NASA), VII 857, 858, 869
American Society, Civil Engineers, V 486
American Society, Mechanical Engineers, VI 81
American Society, Testing Materials, V 486
American Urological Association, VII 1359
Amino-acids, VI 301, 310
Ammonia, II 355; IV 253; VI 467
Ammophila arenaria, II 682
Amratians, I 509, table E
Anaesthetics, V 300. 311–15; VII 1334–40
Anchors, II 76
Andamanese people, I 61, 160, 162, 170, 217
Anemourion, II 615
Anglo-Saxon Chronicle, II 580
Animals; *see also* Livestock
  communication among, I 88–9, 91
  decoy-hunting, I 336, 337–8, 350, plate 6A
  domestication, I 43, 44–5, 75, 79, 175, 327–52,
    501, 704
  draught animals, I 719–24
  morphological effects, I 343–50
  origin theories, I 327–8, 331, 332, 337, 350
  pack-animals, I 705–6, 728
  representations, furniture, II 224, 226
  scavengers/social parasites, I 332–7, 351
  symbiosis of species, I 328–32
  taming, development, I 338–40
Annealing, *see under* Metal industries
Anthracite, IV 110–11
Antiaris toxicaria, I 169
Antibiotics, VII 1427–8
'Antikythera' machine, III 617–18
Antimony, II 47; *see also* Stibnite
  alchemic use, 17th century, III 29 *and* n.
  cast iron, II 74; III 34
  copper alloy, I 590; II 12; III 39–42
  cosmetic use, I 588
  density assays (1679–80), III 58

glaze colouring effect, IV 335
smelting, III 51; IV 119–21
ore, II 12
speculum metal, III 237
tin alloys, III 42, 43, 44
tin coinage adulterant, III 46
type-founding metal, III 44, 393
Antiquaries, Society of, III 436
Antiseptics and disinfectants, V 300, 311, 315–16, 319–20, 565–7
Antlers, see Bone and antler
Ants, I 328–9, 333–5, 338–9
Anvils, I 130–1, 609, 618; II 60, 61; III 111, 113, 352
Apes
    hand, development, I 9–10, 11
    intelligence, I 12–13
    tool makers, I 14–15
    tool use, I 13
    visual powers, I 12
Apollo Belvedere of Carrera marble, II 28
Apollo spacecraft, VII 848, 857, 858–9, 865, 868–9
Apparatus, scientific, III 642–7; see also Instruments
Appliances, see Domestic appliances
Apprentices, III 112
Aqua ardens (alcohol), II 375 n.
Aqua regia, II 356–7, 358
Aqueducts; see also Canals, Hydraulic engineering
    Belleville, 1457, II 691
    construction, 19th century, V 558–9
    Languedoc canal, III 466–7
    Martesana canal, III 446
    medieval towns, II 691
    mining use, II 7, 10, 47
    Ninevah (691 B.C.), I 469–70, 531–2, 553, 715, plate 20A
    origin, II 666–9
    Pont du Gard, near Nîmes, II 404
    Rome, II 405, 418–19, 669 (map), 670–4, plate 39; III 289
    water-mills, II 596–7, 602
Arabia, Arabian peoples, II table III; see also Islam, Muslims
    alchemy, II 735–9, 758
    alum recipe, II 368
    astrolabe (1223–4), III 614
    astronomy, III 508, 584
    camel-breeders, II 497
    cartography, III 521–2
    equatorium instrument, III 616
    food plants, I 355
    glassware, II 337, 341–2
    irrigation, I 530, 532; II 680–1
    knitting, III 182
    metal-work, II 453
    navigation, 10th century, III 553

nitric acid description, II 356
numerals, II 766–7, 772; III 501–2
specific gravity assay, III 22
sugar production, II 372
technological treatises, II 351–2
well digging, desert regions, I 525
Archery, II 542, 705–7; III 351–3
Archimedean screw, II 4, 5, 7, 116, 633, 637, 657, 676–7, 689; V 146, 293
Architecture; see also Building headings, Houses, Ships and ship-building
    Associated Architects, IV 443, 454, 475
    engraved drawings, III 409
    French, Academy of, III 256
    Gothic, III 245–7, 252, 257, 265, 267
    Greek and Roman 'orders' of columns, III 246–7, 253, 256, 260
    Greek styles, II 25
    naval, IV 584
    Renaissance, III 245–6, 248–9, 252, 256–7; IV 469–71
    Roman, II 27; III 245–7, 256
    Royal Institute of British Architects, IV 443
    specialization (1670s), IV 442–4
Architecture hydraulique (1737–53) of Belidor, III 324, 422
Ardagh chalice, II 464, 474, plate 32A
Argentina, coal imports, VI 181
Arithmetic, III 501, 542
Arkwrights, II 254
Armament industry, V 8, 19, 822; see also Explosives, Military technology
    artillery, V 294–5
    Colt Armory, Hartford, U.S.A., V 639, 646
    Enfield small-arms factory, Middlesex, England, V 639
    guns, V 820 n. 1
    small-arms, V 294, 295, 630–1, 639, 646, 653
    Springfield small-arms factory, Ill., U.S.A., V 639
Armillaries
    astrolabon, III 591–2, 606, 613
    equatorial/equinoctial, III 587, 589
    meridional, III 588–9
    spheres, III 613, 625
Armour, II 59–60, 472; III 350, 352–4
    cataphracts, II 566, 570–1, 706
    Greek, II 696, 698
    hardening, III 34–5
    medieval, II 720–1, 727
    Roman, II 704–7
    screws, III 629
'Arnel' textile fibre, VI 650
Arrow-heads, I 33, 169, 171, 496, 499, 510–12, 606, 616
Ars vitraria experimentalis (1679) of Johann Kunckel, III 222, 224

Arsenic
    alchemical description, II 737; III 29
    compounds, uses, I 47–8
    copper alloys, III 38, 40–2
    extraction, IV 119, 121–2, 232, 234
    glass-making, III 226, 229, 237
    lead shot casting process, III 45
    metal alloys, I 47, 74; III 45
    oxide, IV 122
    pigment source, II 48, 362; IV 121–2
    tin coinage, counterfeit, III 46
Art, I plate 2
    ancient man, I **144–53**
    artistic impulse, early man, I 27–8, 144–5
    cave drawings and paintings, I 33–4, 235, 300,
        451, 499, 710, 744, 745
    Celtic, I 82
    Cro-Magnon, I 32
    dating methods, I 146–7
    Egyptian, I 209, 240–3
    engraving, Paleolithic, I 149–51
    pigments, Egypt, I 238–40
    plastic forms, Paleolithic, I 151–3
    sculpture, Paleolithic, I 153
Artemisia absinthia, I 282
Artesian wells, I 528
Artificial insemination, VI 305–6
Artillery; *see also* Guns, Military technology
    breech-loading, III 361–2
    cannon, III 31, 34, 54, 360–8, 374–5; IV
        101–2, 104, 108, 183, 451
    cannon balls, II 34, 75, 726–7; IV 121
    field-guns, III 347–50
    missile engines, II 698–701, 707–15
    mortars, II 378; III 368–9
    naval guns, III 350, 363, 473, 478–81, 499
    origin, II 378–9, 726–8, 771
    ranges and range-finding, III 361, 367, 537
    trunnions, III 362, 365
Artisans, II **383–96**; *see also* Glass-workers,
    Smiths
    arkwrights, II 254
    Byzantine, II 761–4
    carpenters, II 388–95, 641, 652; III 111, 113,
        117–18, 487, 491–2, 623
    chairmakers, III 118
    coachbuilders, III 113, 118
    coffer-makers, II 170–1, 255–6
    coopers, III 112–13, 118, **128–32**
    craftsmen, II 603–4, 761–4
    cutlers, III 115
    farriers, III 111, 113
    glass-painters, III 116
    gold-beaters, III 116
    goldsmiths, III 406
    instrument-makers, III 585–6, 609, 613, 621–5,
        629–31, 642

    joiners, III 123, 623
    leather-workers, II 350
    masons, III 111, 435
    millers, II 601
    moneyors, II 490–1
    paviours, II 525–6, 532–3
    pewterers, III 116
    potters, II 261–2, 284–5, 288–91
    printers, III 386–90, 394–5, 402–3, 409
    sawyers, III 112, 487
    shipwrights, II 230, 573; III 117, 120, 476,
        479, 487–8, 492, 494
    shoemakers, II 166–8, 170
    spur- and bit-makers, III 116
    stone-masons, II 384–8, 641, 662
    turners, III 343, 623
    violin-makers, III 116
    wainwrights, III 145
    watch-makers, III 120
    wheelwrights, II 552; III 111–12, 117–18,
        123–8
    wood-cutters, III 404
Arts, Royal Society of, *see* Royal Society of Arts
Arundo donax, I 371
Asphalt, IV 539–41; V 102, 103, 105; *see also*
    Bitumens
Assaying, metals, III **27–71**
Assaying equipment, III 602
Association of Public Schools Science Masters, VI
    154
Assyrians; *see also* Mesopotamia
    architecture, *c.* 7th century B.C., I 469–73
    battering-ram, II 715
    calendar, seven-day week concept, III 565
    canals/irrigation systems, I 532–4
    carpenter's tools, II 229–31
    cotton, I 374
    food plants, I 358, 360, 371
    glass-making, II 313, 328–9
    ivory-work, I 671
    roads, II 494–5
Astragalus, I 244
Astro-navigation, VII 849, 852
Astrolabes, III 603–9
    calculators, trigonometrical, III 614–15
    early descriptions, III 520–1, 528, 536
    height surveying, III 513, 528, 537
    mariner's, III 506, 546, 549, 608–9, 625; IV
        603
    Nuremberg instrument, III 585
    obsolescence (*c.*1650), III 636
    plane, III 603, 605–9
    texts, III 603, 605, 608–9
    triangulation use, III 539–40
Astrology, I 121–4, 125–6, 802; III 565
Astronomers
    Alexandrian, III 583–4, 586, 599

Alfonsine, III 592
Babylonian, III 582
Merton College, III 585, 593, 617
Muslim, III 599, 608–9, 612, 615
Astronomy, I 785–803, plates 34–6; III 508, 517, 535, 581, 615; *see also* Tables
astronomical observations, III 556–7, 582–4, 597, 605, 618
calendar systems, I 123–4, 793–5; III 558–75, 577–81
cosmographic instruments, III 535–6
eclipses, computation, I 800, 801
'harmony of the spheres', I 126
map-making, IV 597, 605, 618
observatory instruments, III 586–94, 625
planets, I 798–801; III 615–18
precision instruments (to 1500), III 582–94, 614–18
primitive observations, I 112–13, 115, 116–20
quadrants and navigation, III 545–6
scientific development, I 802–3
telescope invention, IV 380, 398–400
Athabaskan peoples, I 176
Athanors, II 739, 740, 743; III 53, 680
Athenians, chronology, III 572
Atlases, IV 600, 614, 618–19, 620, 622, 626
Atomic energy, *see* Nuclear power
Atomic bomb, *see* Nuclear weapons
Atomic weights theories, IV 224–5
Atoms, theory of, IV 223–6; VI 223–4
Atriplex hortensis, I 355
Augers; *see also* Drills, Tools
Greek machinery, II 631
illustrations, II 253, 387, 390, 393, 394; III 119, 329
millwright's tool, III 107
pipe-making, III 329
shipwright's (*c*.600 B.C.), II 231
tradesmen's tools (*c*.1500–1850), IV 110, 114, 119
Aurignacians, *see* Cultures
Australia
aborigines, I 35, 138, 157, 330, 730
aeronautics, V 398, 408
agriculture, mechanization, VI 336
cartography, V 441
cinematography, V 750
farming and food, V 2–3, 6–9, 34, 46–50
fire-making, aboriginese, I 220–1
food-gatherers, aborigines, I 42, 155, 158, 184
gold, IV 141, 143; V 249
language, aborigines, I 94, 98–100
merino wool, IV 28
metals, IV 66–7, 76; V 75
oil, V 116
pest and disease control, VI 333–4
pet-keeping, aborigines, I 330

plant breeding, VI 327
population (1843), V 7
roads, IV 536, 541
water-supply, V 557, 558
wool, V 574 *and* n. 2, 577
Australopithecine hominid, I 11
Australopithecus, I 20, 22, table C
Austria
ceramics, V 659–60, 664, 666
chemical industry, V 243, 251, 253
cinematography, V 736–7
engines, V 160
explosives, V 284, 290
farming, IV 27; V 14
fez industry and kermes dye, V 259
geodesy, V 442
glass industry, IV 360
metal-working, V 619, 620
motor vehicles, V 462–7
photography, V 717, 722
printing, V 704, 707, 714
railway engineering, V 326, 328, 333, 342–3
roads, IV 528
steel industry, IV 108; V 57, 65, 67
textile-machinery, IV 292
topographical survey (1620), V 717
water-supply, V 555, 558
Austria–Hungary
lead production, IV 123
railway engineering, V 339, 341
surveys, IV 610
waterways, IV 553
Automata, II 635, 638
Automation, VI 25, 123; VII 1209
Automobile Club of Great Britain and Ireland, V 431–3
Automobiles, *see* Motor-cars
Avena (oats), I 369
Awls, I 188, 416, 516, 688; III 110
Awners, III 139
Axes; *see also* Adzes
adzes, difference, I 505–6
antler and bone, I 516
battle-, I 512–13, 604–5, 616, 618, 619
bronze, I 609–11, 616
carpenter's, II 100, 228, 252, 389–90; III 110, 117–19
copper, I 588, 601–2, 604–5
hand, stone, I 25–31, 68, 131–3, 188
iron, I 618, 619
mason's, II 37
millwright's tool, III 107, 108
pal-staves, metal, I 610–11
shaft-hole, stone, I 602
stone, I 70, 307, 497–8, 502, 506–8, 512–13, 604
tradesmen's tool, III 110, 117–19

Axles, III 326, 328, 330, 332
Ayrshire Dairy Association, V 34

Baboons, I 8, 13, 19–20
Babylonia, I 457, 550–1, 787–91, 798–801, table
    F; see also Mesopotamia
Babylonian Empire, calendar, III 566–8
Babylonian leather-work, II 151
Back-staff, III 551–2, 636
Badarians, I 502, 606
Bag-press, I 290–1
Bailers (mining use), II 8
Bakelite, VI 555
Bakeries, baking, I 272–3; II 118, 120, 127
Balance-spring, III 344
Balances, weighing, II 744–6, 749; III 60–1,
    66–7; IV 403–7
Balanites aegyptica, I 288
Ballistics, III 256, 359–60, 362, 368, 374–5
    range-tables, III 361, 369, 374
Balloons; see also Aeronautics, Airships
    coal-gas, IV 260–1
    hydrogen, IV 216, 255, 674
Bamboo
    fire-kindling, I 219, 220–1, 227
    tools, I 164, 169, 233
Bambus wrayi, I 164
Bandannas, IV 249
Banking and finance, III 30, 713–14; V 805–6,
    809–10, 820
Baptistère of St Louis, II 453, 457, plate 32B
Barges, II 576–7, 686–7
Barium, VI 224
Bark artefacts, I 518
Barley, I 367–8, plate 7; VI 328
Barns, threshing, 19th century, III 137–8
Barometers, III 543, 630, 636; IV 169, 403, 406,
    605
    surveying, IV 605, 612
Barracan, III 178
Barrels and casks, II 133, 136–7, 255, 354, plate
    3B; III 112, 128–33; see also Coopering
Barrows, hand-, III 139–40; see also
    Wheelbarrows
Basalt, stone, I 477, 506, 675; II 29, 406–7
Basilican churches, II 421–2
Basketry, baskets, I 174, 413–55, plates 9, 10
    bituminous mastic, shaping, I 254
    chronological development, I 413–15, 418–24
    coffins, I 419, 420
    colours, I 422–3
    farm-tools, III 135, 138–9, 139–40, 142
    food storage, I 173, 304
    foraging tools, N. America, I 154–5, 173
    ore carrying (c.1556), II 16
    pottery shapes, influence on, I 376–7, 398
    sieves, I 275, 422

traps, wildlife, I 167, 168
travois (draught device), I 175
wine-making, II 115
Basque fishermen, whaling, IV 57
Bath stone, II 32, 34
Batteries, see under Electricity
Battering-rams, II 702, 715–16
Battles; see also Wars named
    Algiers bombarded (1816), IV 585
    Blenheim (1704), III 347
    Britain (1939–), VI 622; VII 812
    Cape St Vincent (1797), IV 576
    El Alamein (1942), VI 22
    Fair Oaks (1862), V 397
    Fleurus (1794), V 397
    Flodden (1513), V 701
    Jutland (1916), VI 9
    Ladysmith (1899), V 397
    Lepanto (1571), III 350, 473
    Mainz (1795), V 397
    Malplaquet (1709), III 350
    Midway Island (1942), VI 21; VII 814
    Quiberon Bay (1759), III 499
    Sinope (1853), IV 588
    Trafalgar (1805), IV 576–7, 583
Bauxite, V 91, 93, 248, 666; VI 413
Bayeux tapestry, II 91, 128–9
Bays (woollen cloth), II 178
Beads, glass, II 320, 322, 336; III 55, 698
Beakers (ancient pottery), I 401–2
Bearings, roller-bearings, III 327, 658, 660; IV 413
Beavers, intelligent behaviour, I 4–5
Bedaux system, VI 106
Bedouin, I 438, 551
Beer, I 277–81; II 105, 106, 123; III 10–11, 80;
    V 27, 47
    adulteration, II 128
    manufacture, II 139–41
    preservatives, VII 1427
Bees, domestication, I 276
Belgium; see also Europe
    bridges, IV 461
    canals, river navigation, III 451–4; IV 556–60
    chemical industry, V 241, 243, 245, 251
    coal, coal-mining, IV 93, 97; VI 180, 184
    farming, IV 14–16; V 14, 17–19
    flax, III 157
    geodesy, V 442
    glass-making, III 206, 208
    industrialization, V 64
    iron industry, IV 105, 115
    lead production, IV 123
    microscopy, VII 1351
    population, V 823
    printing, V 714
    railway engineering, V 327–8, 335–6
    silk, medieval fragments, III 196

steel industry, V 60
surveys, IV 610
textile industry, V 570, 572, 574, 577, 588, 591
vehicles, V 420
water-supply, IV 497; V 554, 566
zinc, IV 132–3
Bellows; *see also* Furnaces
    alchemical equipment, II 738, 746, 749
    blacksmiths', III 112, 352
    blow-pipes, Egypt (*c.*2400 B.C.), I 578, 579
    box type, wooden (1550), III 32
    furnaces, III 50–2, 53, 60, 363, 690
    hydraulic, II 68–9, 73, 348, 612–13, 655; III 690
    Iron Age smith, I 596
    leather construction, Egypt, II 165
    mechanization, II 643; III 82
    metal-smelting, II 66, 165
    mines, ventilation (*c.*1550s), II 19; III 82
    origin, I 233
    piston-bellows, II 770
    primitive types, I 233, 578
    silver mines, Carthage (*c.*125 B.C.), II 5
    tinplate manufacture, III 690
Bells
    bell-founding, II 64; III 39, 40, 54–5, 57
    bell-metals, III 38, 41, 58; IV 120–1, 124
    church bells, III 108
Bentonite, V 666
Benzoic acid, II 741
Benzole, as coking by-product, VI 467, 468
BEPO nuclear reactor, VI 245, 246
Bergwerkbüchlein ('Essay on Mining'), III 27
Bermannus (Agricola), II 13, 15
Berne Convention (1890), VII 703
Beverages, *see* Drinks
Bible references; *see also Luttrell Psalter*
    artos, meaning, II 105
    bitumen references, I 252, 253
    brass, discovery, II 54
    bread, fermentation techniques, I 275
    bread production, II 105
    bronze moulding/casting, I 633
    communication links, II 495, 496
    cosmetics, I 286–7, 289–90, 292
    cropping, Egypt, I 541
    Dead Sea scrolls, I 766
    dye stuffs, early period, I 248, 249
    fire-carriers, I 230
    food plants, I 359, 365
    food preservation, I 264
    gold, refining-methods, I 582
    horses, I 707
    iron, local taboos, II 59
    irrigation techniques, I 554
    ivory-work, I 664, 676–7
    linen from flax, I 448

mallow (jute), I 451
military engines (*c.*300 B.C.), II 700
mining, I 564–5; II 44
Moabite Stone and writing, I 764
natron, I 260
olive cultivation, I 359
paint colours, murals, I 243
potter's wheel, II 262
salt, I 258–9
ships, reed-built, I 733
silver, refining-methods, I 584
slavery, II 591, 605
sling hunting weapon, I 157
soap, I 261
threshing, flail method, II 102, 106
vinegar, I 285
wines, I 277, 282
writing, pictorial origin, I 756
Bicycles, V 414–18, 437, 640, plate 16A; *see also* Motor-cycles, Motor-tricycles
Bill-hooks, III 117
Bimetallic temperature control, IV 395, 412–14
Birds, instinctive behaviour, I 2–4
Birth of Venus (Botticelli), II 363
Bismuth, III 29, 34, 41–4, 51, 57, 62; IV 119–20
Bits, *see under* Harness
Bitumens; *see also* Asphalt, Petroleum oil
    brick joinery (*c.*600 B.C.), I 471
    building materials, I 250, 255–6, 460, 466, 468
    canal construction, I 469–70
    farm-tools, Neolithic, I 502
    lining, food stores, I 460
    Mesopotamia, occurrence, I 250–4; V 102
    military pyrotechnics, II 374, 376
    quffas (boats), waterproofing, I 737
    road pavements (*c.*2500 B.C.), II 494
    sewers, lining (*c.*2500 B.C.), I 466
    sources, V 102–3, 105
Bixa orellana, III 694
Black Death, II 65, 76–7, 128, 142, 651, 691
Blast-lamps, III 55
Bleaching processes
    alkali, 18th century, III 174–6; IV 235, 237
    chlorine, IV 241, 242, 244, 247–8
    fuller's earth, II 218
    hydrogen peroxide process, VI 672
    paper-making, IV 255
    powder, IV 145–7, 239, 242, 248; V 237, 301, 315; VI 672
    sulphur fumes, woollen cloth, II 218
    sulphuric acid, IV 244
    sunlight, II 218; III 175; IV 244
Blow-guns, hunting, I 163–6, 170
Blowpipes; *see also* Bellows
    glass-makers', II 329, 331, 332; III 208
    oxy-hydrogen, IV 118
    smelting-furnaces, III 55

Board of Health, V 553–4
Board of Longitude, V 458
Board of Trade, V 455, 477, 499, 503, 506 782–3
Board of Trustees for Fisheries, Manufactures,
    Improvements in Scotland (1727), IV 224,
    246
Boats, I 730–43; *see also* Canoes, Ships
    British, I 738, 739
    early development, I 731–2, 738–41
    Egyptian, I 734–6
    mastic for caulking, I 255
    Mesopotamian, I 256, 736–8
    sails, I 733
    war galleys, I 741–3
Bodkins, III 394
Boehmeria nivea, I 431, 449
Bogies (wheels pivot), II 548 and n.
Boilers, steam
    Cornish, V 125, 130–1, 137
    Lancashire, V 131, 137
    locomotives, V 325–6, 336, 339, 340–1; VII
        765–7
    marine, V 152–5
    Niclausse, V 155
    stationary engines, V 126–31, 137–8
    steel plates, V 62, 137
    water-tube types, V 153–5
Boiling Water Reactor (B.W.R.), VI 245, 254, 255
Bolas (entangling weapon), I 166, 175
Bolters (sieves), milling, III 16–19
Bolting-cloth, III 18
Bombazine (cloth), III 178
Bombs, explosive, III 38; VI 11, 550
Bone and antler artefacts
    buildings, use, I 516
    cheek-pieces and bits, I 724
    early working techniques, I 140–1, 188–9
    implements, II 147–9
    sculpturing, I 153
    tools and weapons, I 31–3, 62, 141, 172, 432–3,
        435, 497–8, 506, 514–17, 561
Bones, *see under* Fertilizers
Books and Journals; *see also* Printing
    Annual Register, V 694
    *Application of Cast and Wrought Iron to
        Building Purposes* (Fairbairn), V 473
    Applied Mechanics (Rankine), V 491, 492
    Armenian, III 405
    Ars magna lucis et umbrae, V 717
    Art of Glass (Neri), V 672
    Bible, III 385–6, 388, 404, 413; V 698, 699
    binding, II 169, 173–4, 180–1; III 386, 390,
        404; V 713–14
    Book of Trades (Ypres), II 208, 209, 213
    Books of Secrets, III 27, 33, 35, 39, 42
    Bridges (Claxton Fidler), V 495
    British Birds (Morris), V 705

British Medical Journal, V 756
Building News, V 470 n. 1
Century Magazine, V 712
Constance Missal, III 386, 389
Daily Graphic, V 701, 703, 704
Daily Telegraph, V 702
Das System Monier (Koenen), V 489, 495
Der Eisenboten (Mörsch), V 495
Engineer, The, V 554
Family Herald, V 685
Glasgow Star, V 701
Grammar of Ornament (Owen Jones), V 706
Graphic, V 702
Graphic Statics (Cremona), V 492
Harper's Magazine, V 704
History of Photography (Eder), V 704
Illustrated London News, IV 517; V 40–1,
    42 n., 698, 702
Imitatio Christi (Thomas à Kempis), V 706
Justinian (1468), III 393
Koran, III 380
Lancet, V 314
Lloyd's Weekly, V 698
London Pharmacopocia, V 317
Mechanics' Magazine, IV 272
Morning Chronicle, V 702
Nautical Almanac, V 458
New York Daily Graphic, V 704
New York Herald, V 700
New York Tribune, V 687, 688
Norwich Mercury, V 698
Observer, The, V 702
Oculus artificialis, V 717
On the Theory of Lubrication (Reynolds), V 548
Patrie, La, V 698
Pencil of Nature, V 723
Penny Illustrated Paper, V 752
Penny Magazine, V 702
Philadelphia Inquirer, V 700
Philosophical Magazine, V 492
Psalter (Fust-Schoeffer) (1457), III 405
Punch, V 702, 831
Reports of the Juries (1851 Exhibition), V 690
Scots Magazine (1753), IV 647, 649
Sheffield directory of 1852, V 633
Speculum vitae Christi (1487), III 399
Staffordshire Pottery and its History
    (Wedgwood), V 658
Sun, The (1749), IV 646
Technology of Bread-making (Jago), V 28
Times, The, V 684, 686, 690, 691, 696, 698, 700,
    704, 805–6
Traité du Navire (Bouguer), V 353
*Transactions of the Institution of Naval
    Architects*, V 372, 386
Trewe Encountre, V 701
Warrington Guardian, V 685

Boomerangs, *see* Throwing-sticks
Borax, III 62–3; IV 232; V 320–1
  bead test, III 68, 226, 228
Boric acid, V 315, 320
Boring apparatus, II 655–6, 664
Boring-mills, IV 421–2, 424, 427, 493, plate 27A
Bossing-tools, II 469
Bow and arrows
  arrow-heads, I 33, 169, 171, 496, 499, 510–12, 606, 616
  Bushmen cultures, South Africa, I 171–2
  compound bow, I 163, 173, 182
  invention, I 61
  Mesolithic cultures, I 161–3, 499
  military obsolescence, III 351–3
  South American cultures, I 179
Boyle's Law, IV 215, 223
Brace; *see also* Augers, Drills, Tools
  Assyrian tool (possible), II 230
  bits, II 229
  carpenter's tool, 14th century, 387, 393–4
  hand-brace, bevel gearing, IV 381–2
  joiner's tool (1703), II 253
  mechanization, II 653
  tradesmen's tool, III 110–11, 114, 117–18, 120
Brachiation, I 9, 10, 11–12
Brain; *see also* Intelligence
  early man, I 27, 35
  size, I 3, 7, 10, 12, 16, 34–5
Brakes (hand-tool), III 110
Brass (copper-zinc alloys), II 53–5
  bearings, III 41, 103
  composition, II 53–4; V 607–8
  decorative harness, III 149
  early reports, II 54–5, 63
  founding, III 37–9; V 609–10
  guns, III 38–9, 361
  instrument components, IV 400, 404–5, 407, 414, 601
  invention, II 54
  mechanized handling, III 338–40
  metal-working, V 610, 618, 626–7, 631
  Meuse valley, medieval period, II 67
  nails, II 255–6
  paint ingredient, III 49
  production, 18th century, IV 133
  smelting, III 51, 53; IV 130
  speculum-metal, ingredient, III 40
  tools, III 111
Brassica, I 354–5, 360, 370
B. napus, II 125; III 5
B. rapa, II 125
Bread, *see under* Food
Breakwaters, IV 452, 464–6
Brewing, beer, III 10–11, 80; *see also* Beer
Brewing, Institute of, V 28
Brice Commission (1895), VI 142–3

Bricks and tiles; *see also* Building materials
  ancient buildings, I **456–94**
  English, II 438–40
  firing, II 305, 307–8
  flue-tiles, II 420, 421
  furnaces, II 749; III 52–3; IV 101–2; VI 476–7, 484
  Greek, II 398, 402–3
  medieval, II 304–7, 388
  refactory, V 664–5; VI 476, 484, 630–7
  Roman, II 407–9, 415, 418
  Romanesque, II 429
  street-paving, IV 545
  terracotta, II 278
  tiles, London (1212), II 387–8
Bridges, general, I 714–15; III **417–37**; V 499–521; VII 876–99; *see also* Aqueducts, Civil Engineering
  arches, II 771; III 417–18, 420, 424–36
  caissons, III 425–6
  China, 1st century B.C., II 771
  coffer-dams, III 420–6, 433
  construction, IV 163, 442–5, 452–62, 484
  design theory, V 500–3; VII 876–9
  dimensions, III 428 n., 429
  drawbridges (1635), III 291, 292
  foundations, III 420–7; IV 449, 462; V 513–16; VII 891–2
  houses on, III 430, 432
  girders, IV 461–2
  iron, IV 454–62, 484–6
  materials, construction, I 256; V 499–500, 511–12; VII 892–6
  movable, V 512–13, 521; VII 889–91
  piles, III 420–7
  power and machines, IV 452–3; V 500
  railways, IV 460–1
  Roman, II 508, 510–11, 512, 513, plate 34A; III 288, 417–18
  steam-power, IV 452–3
  stone, IV 453–5, 457, 459
  suspension, IV 458–60; VII 887
  tools, III 420
  truss-girder bridges, III 417, 426–7
  types, III 417
  wooden, I 316; III 417, 423
Bridges, named; *see also* Bridges, general
  Africa
  Lower Zambezi, VII 884
  Australia
  Gladesville, Sydney, VII 892, 893
  Hawksbury River Road, VII 891
  Sydney Harbour, VII 884, 885, 887
  Tasman Road, VII 896
  Canada
  Quebec, VII 885, 896, 897
  Second Narrows, Vancouver, VII 898

Bridges, named (*cont.*)

England
Berwick, IV 460; VII 893
Blackfriars, IV 453–4, 510, 544
Britannia, IV 486
Booth Ferry, VII 890
Buildwas, IV 457
Chepstow, IV 461–2; VII 898
Clifton suspension, Bristol, V 507–8, plate 30
Fleet Bridge, III 298
Holborn Bridge, III 298
Hungerford, London, V 507 and n. 1, 643, 644
Ironbridge, Coalbrookdale, IV 455, 457–8, plate 30A
Kew, IV 494
Kingsferry, VII 891
Lee, IV 494
London Bridge, IV 206, 451, 455–6, 458, 484, 491, 510, 546; VII 885, 894
Medway, VII 894
Middlesbrough, VII 889
Newcastle-upon-Tyne, IV 460–1
Old London Bridge, III 419–20, 430, 432
Queen Alexandra, River Wear, VII 881–2
Rochester, IV 462
Saltash, IV 462
Severn, IV 102, 105; VII 875, 888–9, 896
Southwark, IV 456, 458–9
Sunderland, IV 457–8, plate 29
Sutton, VII 846
Tower, London, V 513; VII 889
Waterloo, IV 455–6, 510
Westminster, III 425–7; IV 510

France
Bayonne, V 511
Neuilly, Seine, IV 451, 453
Pont d'Avignon, III 418
Pont de Châtellerault, V 512
Pont Henri IV, Châtellerault, Vienne, III 430–1
Pont Neuf, Paris, III 293
Pont Neuf, Toulouse, III 420, 430
Pont Notre Dame, Paris, III 333, 423, 430, 432
Pont Royal, Paris, III 425
St Maxence, near Senlis, IV 452–3

Germany
Admiral Graf Spey, Godesberg, VII 884
Montgomery, Wesel, VII 896
Rhine, Kehl, V 514

India
Alexandra, VII 883
Attock, VII 883, 885
Hardinge, Ganges, VII 882
Howrah, Hoogly, VII 885
Jhelum Valley Road, VII 884
Khotri, VII 879–80

Lansdowne, Indus, VII 886
Nerbudda River Railway, VII 883
Shri Yuvraj Karan Singh Road, VII 897
Sui Gas, VII 892
Willingdon, VII 882

Ireland
Essex Bridge, Dublin, IV 449

Italy
Ponte Vecchio, Florence, III 418, 428–30
Rialto, Venice, III 424, 430
Santa Trinità, Florence, III 424, 429–30

Luxembourg
Pont Adolphe, V 511–12

Kurdistan
Callender-Hamilton, VII 894

Pakistan
Ayub Arch, VII 886, 887

Portugal
Salazar, VII 887
Vila Franca, Tagus, VII 883

Scotland
Connel Ferry, VII 880–1
Forth, V 477, 500–3, 505–7, 535, plate 28; VI 452; VII 875, 885, 889, 927
Inchannon, VII 890
Jamaica Bridge, Glasgow, IV 454
Kelso, IV 459
Kincardine, Forth, VII 890
Tay, V 503

South Africa
Victoria Falls, VII 881
Lower Zambezi, VII 884

Switzerland
Freiburg, IV 459
Schaffhausen, III 427–8
Wettingen, near Zürich, III 428

Turkey
Bosphorus, VII 889

U.S.A.
Arthur Kill, New York City, VII 891
Astoria, Oregon, VII 884
Bayonne, New York City, VII 884
Blackwell's Island, VII 881
Brooklyn suspension, New York, V 500, 509–10, 623
Golden Gate, San Francisco, VII 887, 888
Grand Trunk, Niagara, V 508–9
Halstead Street, Chicago, V 513
Harlem river, New York, V 513
St Louis, Mississippi, V 510–15
San Francisco-Oakland Bay, VII 891
Sciotoville, Ohio, VII 884
Tacoma Narrows Suspension, VII 899
Thatcher Ferry, VII 913
Verrazona Narrows Suspension, New York, VII 887
Wheeling, West Virginia, IV 459

Wales

Britannia, V 500, 503–5, 534, plate 29
Menai Straits, IV 459–60, 531; V 507
Pont-y-ty-Prydd, III 434–6; V 492
Bristol Brass Wire Company, IV 130
Britain; *see also* England, Europe, Ireland,
    Scotland, Wales
aeronautics, V 397–400, 402, 404–8, 412; VI
    438, 440–3
agricultural vehicles, III 140–8
agriculture, IV 1–12, 14–22, 25, 28, 33, 40,
    254; V 2, 4, 7–8, 10–19, 24, 26, 29–42, 44–6,
    51–2; VI 298–316, 327–30
aircraft, VII 793–6, 799–807, 811–22, 828–9,
    834, 841–53, 1012, 1014, 1022
air transport, VI 37; VII 697–701
architecture and building, IV 373, 443–51, 470–
    7; V 467–78, 482–95; VII 937–47, 949, 954
atomic energy, VI 223–4, 226, 232, 235–6,
    244–51, 257–9, 263–7, 269–71, 281–2
automobile engineering, VI 429–38
basketry and matting, I 421
boats, I 738–9
bridges and tunnels, IV 452–62; V 62, 499–500,
    503–8, 512–14; VII 846, 881–2, 884–5,
    889–96, 899–903
Bronze Age structures, I 312–13, 315–16
building, *see* architecture and building
canals and waterways, IV 452, 563–72
cartography, IV 598, 620–4; V 438–42, 445,
    447–54
ceramics, V 658–65, 667–9; VI 622, 625–35
chemical industry, IV 236–8; V 237–47, 249–
    54, 299, 301, 312, 315, 317–18; VI 499–501,
    504–7, 510–11, 518, 548–50, 564–8
cinematography, V 735–7, 743, 746; VII 1302–9
civil engineering, VII 871–905, 912, 920–9
coal industry, III 72–3, 76–87; IV 79–98, 166;
    VI 14, 173, 178–80, 183–4, 359–62, 366–9,
    371–4
computers, VII 1151, 1153–4, 1159, 1168–71,
    1174–9, 1183–99, 1200–3
distilling, V 305
dyestuffs, IV 321; V 260–3, 265–6, 269–74,
    277, 279, 281
domestic appliances, VII 1126, 1128, 1129–40,
    1142, 1143–9
electrical communication, VI 122–63
electrical engineering, VI 443–7, 449
electricity, V 177–87, 191–206, 208–18, 220–2,
    229–30, 232; VI 248–97; VII 1068–83,
    1086–8
electronic engineering, VII 1101, 1105
engineering, IV 446, 452; VII 983–4, 987,
    989–91
engines, V 125–39, 142–55, 160–1, 163, 170–1,
    175

explosives, V 284, 287–90, 292–3, 296
farming, *see* agriculture
fire-making, I 219, 220, 226
fish industry, IV 44, 46, 48–54; VI 342–4, 346,
    349–57
flint mining, I 558–61
fluid power, VII 1061–2, 1063–7
food plants, I 355, 358, 359, 365–6, 369
food technology, VII 1399–409, 1411, 1413–17,
    1421–2, 1426–36, 1440–1, 1443–5
gas industry, IV 265–74
Geological Survey, IV 491
glass industry, V 673–4, 676–81; VI 576,
    578–87
Gregorian calendar, adoption, III 579–80
hydraulic engineering, V 522–7, 530, 534–41,
    544–50
hydro-electric engineering, VI 202, 206, 212,
    213, 216
hydrography and oceanography, IV 616–19
industrial success, Middle Ages, III 711–12
Iron Age dwellings, I 317–22
iron and steel industries, III 78–9; IV 99–116;
    V 53–6, 58, 60–70; VI 456, 463, 465–8,
    473–6, 482–9, 492–7, 509
machine-tools, IV 418–37, 440–1; V 636–40,
    642–5, 647–56
measurement, I 778–9
medical technology, VII 1317–21, 1334–40,
    1346–8, 1352, 1354
Mesolithic dwellings, I 303–4, 312
metals, IV 64–5; V 53–66, 68–70, 76–81,
    85–91, 93, 96, 605–7, 610–34
mineral resources, IV 65
mining, I 558–61, 563, 589–90; IV 64–5, 80,
    84–5
navigational aids, V 454–62
Neolithic culture, I 80–2, 195, 303–4, 312,
    510–11
non-ferrous metals, IV 118, 119, 122–4, 126–30
oceanography, VII 1451–3
oil and petroleum, V 103–5, 108, 116–19, 121
paint industry, VI 590–1, 601, 604
paper industry, VI 610–19
photography, V 718–19, 721–5, 727–32; VII
    1281–301
plastics industry, VI 552, 554, 558–9, 561
population, IV 489, 506; V 1, 26, 823
pottery, I 377–8, 385–6, 393, 396–7, 400
printing, V 683–714; VII 1268–79
railways, V 322–48; VII 702–3, 770–87
rainfall records, IV 490–1
road transport, VII 704–6, 708–9
road vehicles, VII 716–20, 723, 726–35
roads, I 712–12; IV 523–5, 530–6, 539, 542–6
rubber, V 756–62, 764–5, 767–74
salt, I 257; IV 239

Britain (*cont.*)
  sanitation, IV 505
  sewerage and sewers, IV 506–18
  ships and ship-building, IV 574–94; V 62, 142,
    350–81, 383–90; VII 693–6, 740–7, 750–7
  silk, IV 308, 311, 314–15, 318–19, 321, 323–6
  silver, IV 137–8
  space technology, VII 859, 869
  spinning wheels, III 161
  steel industry, *see* iron and steel
  surveys and surveying, IV 402, 607, 609–10,
    616
  telegraphy, V 218–25, 227–8
  telephones, V 226–7
  textile industries, III 151; IV 246, 248, 282,
    300, 311, 314–15, 318–19, 323, 325–6; V
    569–88, 591–2, 595–604; VI 656–8, 661–3,
    667–8, 671–4
  textiles, VI 639–51
  tin mining, I 563, 589–90; IV 123–4
  tinplate, III 689; IV 104
  town-planning, III 277–9; VII 958–82
  vehicles, V 414–18, 420–36
  waste-disposal, *see* sewerage, water supply
  water-power, IV 151–3, 155, 199–212
  water-supply, IV 489–97, 499–503; V 352–67;
    VII 1363–80, 1382–97
  waterproof fabric, IV 253
  waterways, *see* canals and waterways
  weapons, metal, I 604, 605, 616, 618
  wheels, solid, I 203, 206, 211
  window-glass, III 237
Britannia metal, IV 121
British Association, Advancement of Science
  Bell telephone exhibit (1877), V 226
  Bessemer process (1856), V 54
  education, V 780; VI 147–8, 162
  electricity generation, V 180
  electricity transmission (1881), V 228
  food resources (1952), VII 1473
  foundation (1831), IV 674, 677
  harbours/estuaries (1887), V 550
  jet-pump principle (1853), V 526
  machine-tools (1861), V 636
  science and research committee (1868), V 792
  ships, hydrodynamics (1868), V 386, 387, 389
British Brass Battery and Copper Company
  (1702), IV 130
British Building Research Station, VII 944
British Ceramic Society, VI 623
British Columbia, I 160, 176–9, 184, 312, 318,
  707, 738
British Commonwealth
  food production, V 27
  population, V 26
  printing, V 694
  railways, V 61

British Cotton, Silk and Man-made Fibres
  Research Association, VI 661
British Dairy Farmers, Association, V 32
British Dominions, VI 60–1
British Engineering Standards Association, *see*
  British Standards Institution
British Fire Prevention Committee, V 482
British G.N.P., VI 18
British Refractories Research Association, VI 623
British Science Guild, VI 153
British Ship Research Association, VII 747
British Standards Institution, V 477, 486–7; VII
  877
British Steel Structures Research Committee, VII
  945, 947
Broadcloth, III 178
Brocade, III 192, 197
Brocatelle, III 200
Bromine, V 319
Bronzes (tin-copper and other alloys); *see also*
  Metal headings
  artifacts, I 213, 399; II 327
  braziers, I 231
  building construction, II 403, 417–18, 454
  cannon, III 361, 363–6
  casting, I 607–8; II 476–80; III 38–9; V 609
  cloth preservation effect, I 414
  composition, I 588, 589; II 47, 54, 63; III
    37–9
  copper alloys, I 588–92
  'Corinthian bronze', II 43
  daggers, II 481
  furniture, II 223–4, 227, 235–6
  manufacture, I 590–2; II 53; III 30, 342; IV
    124
  metal-work, I 646
  mirrors, II 449, 460, 463
  miscellaneous-bronzes, V 608–9
  moulds, cannon-balls (*c.*1697), III 34
  origin, I 81–2, 588, 590–2
  patination, II 484
  pillars, Temple of Solomon, I 633
  Roman period, II 53
  screw-casting, III 338
  shields, II 460, 462, 468
  springs, II 712–13
  tools and weapons, I 504–5, 510, 608–16
  water-pipes, II 665
Bronzo antico, II 291
Brotherhood of man, II 605
Bruzz (triangular chisel), III 119, 126
Buckets
  copper, II 689
  leather, II 184
  ox-hide, II 14, 16
  wooden, II 16, 100, 255, 689
Buckwheat, II 125

Building construction, II 384, **397–48**, plates 29–31; III **245–68**; IV **442–88**; VII **930–57**; *see also* Architecture, Building headings, Civil engineering, Houses

ancient man, I **299–326, 456–94**

arches, II 423, 432, 433–5; III 246–7, 250

architectural standards, VII 952–5

books, 16th and 17th centuries, III 245, 250, 253, 257

Bronze Age burial structures, I 312–17

cantilevers, III 252

catenary curve, III 258

ceilings, III 247–50; IV 450–1, 474–5

chimneys, III 252, 256, 261

churches, III 246–9, 254–9, 267

columns, III 247

cornices, III 247, 249, 256

Crete, I 484–7

domed roofs, II 412–13; III 245–6, 248, 253–5, 258–9; IV 480

domestic, II 445–6, 641–2; III 246–7, 249, 252–4, 256–7, 260–5, 267

doors and locks, II 403, 415–16, 425

drainage, roofs, II 404, 444–5

Egypt, I 304–6, 371–2, 473–84

fire-protection, II 413; IV 450, 475, 477, 485; V 470, 480–3, 665; VII 941–4

fireplaces, III 261

floors, IV 450

flying buttresses, II 435–6

fortifications, II 701–3, 717–19, plates 40, 41

Gothic, II 430–46

Greek, I 485; II 397–404

heating systems, II 404, 419–20

high buildings, V 62, 477–9

Iron Age techniques, I 317–24

mansard roofs, III 249, 253, 265

Mesopotamia, I 459–73

Neolithic Europe, I 306–12

new structural forms, VII 947–52

ornamentation, III 245, 252

plaster-work, II 418, 443

religious structures, I 457–8, 460–1, 462–4, 482–4

Roman, II 27, 32, 404–20

Romanesque, II 420–30

roofing, I 472–3, 476, 484; II 401–4, 410–15, 427–9, 433–5, 441–5; III 245–55, 258–9, 265; IV 450–1, 474–5

scaffolding, II 386, 388

spires, III 265, 267

staircases, III 257, 267

stucco, IV 447

tensile tests, III 250–2, 258

testing-machine, Gauthey's, IV 480–1

theory, IV 477–86; V 491–6, 837; VII 944–7

timber, II 397–8, 413–15, 440–3

trussed roofs, II 442–3

vaulted roofs, II 411–12, 427–9, 433–5, 441–2; III 247–50

walls, wall-panelling, III 247, 261, 266–7

water-pipes, II 418–19, 445

weather-boarding, III 265

windows, II 403, 416, 429, 437–8, 443–4; III 248, 253, 255–6, 260

ziggurats, I 467–9

Building materials, I **299–326**; V **466–98**; VII 930–6, 937; *see also* Building construction

bitumen, wooden buildings, I 250, 253, 255–6

brick, IV 446–7

brick and stone, I **456–94**, plates 18–21A

bricks and tiles, II 278, 304–7, 387–8, 402–3, 415, 418, 429, 438–40; III 247–8, 258, 261–2, 265

carpentry, V 466–70

cast-iron, V 470–2, 479

cement, IV 447–51, 474; V 483–7

ceramic, V 665–6, plate 3

concrete, IV 447; V 487–91

early societies, I **299–326**

glass, IV 476–7

iron, IV 447, 450–1, 474–6, 482, 484–5

lath-and-plaster, III 248, 264

masonry, III 247, 255, 258, 260

slats, III 254, 260

stone, II 31–8, 384, 427–8, 436–7; III 247, 249, 254, 256–8, 260–1; IV 446–7

structural steel, V 62–3, 476–7, 479, 837

terracotta, IV 446

timber, I 143, 241, plates 4, 5A; III 248–50, 253–7, 262–6; IV 446, 472–5

wrought-iron, V 472–6, 479

Buildings, named; *see also* Town Planning

Amiens Cathedral, II 435, 436

Arch of Constantine, II 28

Athenaeum Club, England, IV 272

Athens, Tower of the Winds, III 517, plate 26A

Bank of England, IV 475

Banqueting House, Whitehall, III 256, 267

Basilica Nova of Maxentius, Rome, II 412, 421

Bauhaus, Dessau, Germany, VII 954–5

Blue Mosque, Tabriz, II 307

Boots' Factory, Beeston, VII 950

British Museum, London, V 217, 475, plate 26B

Buckingham Palace, London, IV 447

Burlington House, London, IV plate 30B

Canterbury Cathedral, II 429, 431–2, 436; III 516, 596–7; IV 454

Capitol, Washington, IV 540

Carlisle Cathedral, II 437

Carson Pirie Scott Store, U.S.A., VII 930

Chartres Cathedral, II 444–5

Buildings, named (*cont.*)
  Chatsworth House, Derbyshire, IV 476
  Chertsey Abbey, II 305
  Chichester Cathedral, II 436; III 265, 267
  Chiswick Centre, Britain, VI 604–5
  Christ Church, Oxford, III 257
  Clarendon Palace (near Salisbury), II 305
  Clifford's Inn, London, III 266
  Coal Exchange, London, V 471, plate 26A
  College of Surgeons, London, V 487
  Cologne Cathedral, Germany, V 472
  Conservatori Palace, Rome, II 227
  Corfe Castle, Dorset, III 369
  Crown Hall, Illinois, VII 933
  Crystal Palace, London, V 408, 474, 677, 837;
    IV 273, 476, 478; VII 938, 1132–3
  Customs House, Liverpool, IV 469
  Customs House, London, IV 449
  Dalton Arena, North Carolina, VII 953
  Diana, temple of, Ephesus, II 660
  Dover Castle, England, II 651
  Drury Lane Theatre, London, IV 272
  Durham Cathedral, II 436
  East India House, England, IV 272
  Eiffel Tower, Paris, V 62–3, 476
  Eltham Palace, II 442
  Ely Cathedral, II 438, 442
  Empire State Building, VII 949
  Erechtheum, Athens, II 399, 402, 403
  Eton College, II 384, 436
  Farnese Palace, Rome, III 248
  Florence Cathedral, Italy, III 245–6, 248, 258
  Fontainebleau Palace, France, III 256
  Gaiety Theatre, London, V 196, 211
  Gardiner's building, Glasgow, V 470
  General Post Office, London, IV 449
  Geometric temple, Dreros, Crete, II 469–70
  Gloucester Cathedral, England, III 241
  Great Chalfield Manor, Wiltshire, III 262
  Greenwich Palace, England, IV 443
  Grosvenor Gallery, London, V 198
  Halesowen Priory, Birmingham, England, II
    305
  Hampton Court Palace, England, III 245, 253,
    262, 264–5; IV 443
  High Point One, London, VII 939
  Home Insurance Building, Chicago, V 478–9
  Hospital of the Salpêtrière chapel, Paris, III
    254
  Houses of Parliament, London, IV 515; V 217,
    471
  Ingalls Building, Cincinnati, VII 949
  Invalides, second Church of, Paris, III 254–5
  Jupiter, temple of, Baalbek, II 410
  King's College, Cambridge, II 432–3, 436; III
    242
  Kirkaldy's laboratory, London, V 490

  Lateran Palace, Rome, III 289
  Lever Building, U.S.A., VII 942, 943, 948
  Louvre, Paris, III 253, 341; IV 442, 474
  Mansion House, London, IV 272; V 217
  Market Hall, Algeciras, Spain, VII 951, 953
  Masonic Building, Chicago, V 479
  Mechanics' Institution, London, IV 272
  Menier Chocolate Works, Noisiel-sur-Marne, V
    211, 474–5, 837
  Merchants' Exchange Building, Philadelphia,
    IV 540
  Morgan's Road Junior School, Hertford, VII
    950
  Museum of Modern Art, New York, VII 953
  New Canada Junction, Johannesburg, VII 782
  Newark City Hall, U.S.A., IV 540
  Notre-Dame Cathedral, Paris, IV 472
  Nottingham Castle, IV 450
  Old St Paul's, London, III 254
  Padua Town-hall, Italy, III 249
  Palace of Minos, Knossos, Crete, IV 504
  Palace of Westminster, London, IV 436; VII
    1384
  Palazzo Vecchio, Florence, Italy, III 707
  Pantheon, Rome, II 411, 413, 417; III 258
  Parthenon, Athens, II 399–403
  Pazzi Chapel, Florence, Italy, II 249–50
  Penitentiary, London, IV 449
  Peterborough Cathedral, England, II 428
  Pisa Cathedral, Italy, III 662
  Polytechnic Institution, London, IV 273
  Pompeii, 'House of the Faun', Italy, III 274–5
  'Pompey's Pillar', Alexandria, II 410
  Propylaea, Athens, II 399, 400, 403
  Pulitzer Building, New York, V 478
  Queen Anne's Mansions, London, V 478, plate
    27A
  Queen's House, Greenwich, III 256, 267
  Rand-McNally Building, Chicago, V 479
  Reliance Building, U.S.A., VII 930, 932, 933,
    938, 948
  Rheims Cathedral, France, IV 472
  Riccardi Palace, Florence, III 247
  Ritz Hotel, London, V 477
  Romulus, temple of, Rome, II 416
  Rouen Cathedral, France, V 472, plate 27B
  Royal Academy, London, V 217
  Royal Exchange, London, III 298
  Royal Festival Hall, London, VII 945
  Royal Society of Arts, London, IV 473
  Sainte Chapelle, Paris, II 436
  St Denis, near Paris, II 431
  Saint-Eugène church, Paris, V 471
  Ste-Geneviève (later Panthéon Français), Paris,
    IV 475, 480, 482
  St Germain, Château, near Paris, III 254
  St Gervais, Paris, II 438

St John Lateran, Rome, III 289
St John the Divine Cathedral, Jena, Germany, VII 951
St John's church, Leeds, III 257
St Mark's, Venice, II 328, 393, 395, 423, 424, 464, 466–7
St Martin, Tours, II 429
St Pancras Station, London, VII 930
St Paul's Cathedral, III 254–5, 258–9, 296–8; IV 443, 454, 472, 480
St Paul's (Old) Cathedral, London, II 445
St Paul's Church, Covent Garden, London, III 256; IV 470, 472
St Peter's, Rome, II 464, 467; III 246, 248–9, 255, 258, 289; IV 478–80
St Servatius Church, Maastricht, Holland, III 202
St Sophia, Constantinople, II 423, 424, 454
St-Sulpice Church, Paris, IV 475
St Vitale, Ravenna, II 422–3
Salisbury Cathedral, II 428, 431, 433; III 650–1
Santa Maria dei Miracoli, Venice, III 248–9
Santa Maria Maggiore, Rome, II 410; III 249–50, 289
Savoy Theatre, London, V 217
Sens Cathedral, France, III 198
Sheldonian Theatre, Oxford, III 265–6
Sistine Chapel, Rome, III 248
Sorbonne, Church of The, France, III 254–5
SS Paul and Louis, Paris, III 254
Staple Inn, Holborn, London, III 264
Strozzi Palace, Florence, Italy, III 247
Swinsty Hall, Yorkshire, III 260, 263, 265–6
Tamerlane's mausoleum, Samarkand, II 307
Temple of Christ The Saviour, Moscow, IV 554
Temple of Tellus, Rome, III 504
'The Times' printing-office, IV 272
Theatine Church of Ste-Anne-la-Royale, Paris, III 254
Theatre of Marcellus, Rome, III 266
Theodoric's mausoleum, Ravenna, II 413, 424
'Theseum', Athens, II 399
Tilbury Power Station, Britain, VII 901
Tower of London, II 426–7, 432–3, 436, 438; III 298
Trajan's Column, II 407, 705, 707–8, 713 n., 716, 718, 730
Treasury of Atreus, Mycenae, II 413
Tuileries Palace, France, III 425
United Nations Building, New York, VII 942
Val-de-Grâce, abbey church, Paris, III 254–5
Vatican Palace, Rome, III 280
Vicenza Town-hall, Italy, III 249
Visitation, Church of the (later the Calvinist Church), Paris, III 254

Westminster Abbey, London, II 305, 385, 426, 435, 438, 442, 478; III 203; VII 1129
Westminster Hall, London, V 468
Winchester Cathedral, England, II 427–8
Windsor Castle, England, II 34, 35, 38, 443
York Minster, England, II 425, 429; III 241
Buoys, see under Lighthouses
Burden-creels (peat-carrying), III 140
Bureau International des Poids et Mesures, Paris, V 442, 448
Burial mounds/structures; see also Mummification, Pyramids, Tombs
    beehive tombs, Mycenae, I 403, 486, 488
    Britain, I 81, 195, 312–13, 314–15, 490–4
    Europe, I 314
    France, I 510–11
    Minoan tholos tombs, I 484
    Russia, I 324–5
Burin (engraving tool), III 405–7
Burning-glass, IV 218 and n., plate 8A
Burrowing wasp (Ammophila), I 1–2
Business administration, VI 77–92
Business cycle, VI 177
Business education, see under Education
Buttons, IV 130, 133
Butyl alcohol, VI 543
Byzantine Empire
    architecture, III 245
    scholars, III 530
    scientific instruments, III 584
    silk weaving, III 191–2, 195–6, 198
    sketch-plan of Byzantium, III 279
Byzantium, see Places Index, Constantinople

Cables, electric
    manufacture, V 627, 771, 772; VI 445–7
    submarine, IV 660–1; V frontispiece, 224–6, 609
    telephony/telegraphy, VII 1242–3, 1248
'Cadmean letters', I 767–8
Cadmeia (zinc oxide), II 54, 55
Cadmium, IV 119
Caesalpinia, II 367
Calamanco (cloth), III 178
Calamine (zinc carbonate/silicate), II 2, 13, 54; III 30, 37, 690, 692; IV 130, 133
Calcination, II 742
Calculating/computing instruments, III 614–15, 627, 636; see also Computers
    abacus, II 766, 768; III 501; VII 1158–9
    Arithmometer, VII 1153, 1163–5, 1168, 1170
    automatic, VII 1151, 1153, 1172
    Babbage analytical engine, VII 1175–9
    Comptometer, VII 1168–9
    development, VII 1150–83
    difference engines, VII 1173–5
    electric, VII 1171

Calculating/computing instruments (*cont.*)
   Leibnitz machine, VII 1151, 1153, 1160–3
   Millionaire, VII 1166, 1168, 1171, 1188
   Napier's bones, VII 1151–3
Calendars and chronology, I **110–27**; III **558–81**
   Alexandrian, III 564
   Babylonian, III 566–8
   calculators, III 614
   Calippic, III 571
   celestial phenomena, base, III 558–60
   definition, III 558
   Egyptian, III 560–6
   Greek, III 568–72
   Gregorian, III 575, 578–80
   Islamic, III 580–1
   Jewish, III 572–5
   Julian Period, III 564, 571, 576–8, 580–1
   Metonic, III 570–1
   perpetual, feasts of Church, III 654
   Roman, III 564, 575–8
   seasons, cycle, III 559
   seven-day week, III 565–6, 572–3
   'styles', III 580
Calendering cloths, III 177–8
Calico (cloth), III 178
Calico-printing, IV 147
California Institute of Technology, VII 1332
Calipers, II 331; III 110, 120, 512
Camargue cattle, I 344
Cambric (cloth), III 164, 178
Camel-transport, II 497
Camelina, I 354, 359, 370
Cameras; *see also* Cinematography, Photography
   lenses, IV 360; VII 1286–7
   miniature, VII 1287, 1288–90, 1291–2
   press, VII 1290–1
   professional, VII 1293–4
   shutters, VII 1285–6
Camlet (cloth), III 171, 178
Camshafts, II 635, 643
Canada
   agriculture, V 3–7; VI 320–1, 323, 327, 335
   atomic energy research, VI 241–2
   canals, inland waterways, IV 551–3; VII 913, 916–20, 921
   chemical industry, V 243, 255
   civil engineering, VII 913, 917–20
   copper ores, IV 126
   farming and food, V 34, 43
   geodesy, V 442
   hydraulic engineering, V 533–4
   hydro-electric power, VI 195, 218
   metals, IV 66; V 87–9
   oil, V 110
   railways, V 343
   water-supply, V 564

Canals, inland waterways, III **438–70**; IV **548–73**; VII **912–20**; *see also* Canals named
   aqueducts, IV 567
   books, IV 443
   coal transport, Britain, III 85–6
   contoured, IV 566
   culverts, III 446, 449, 452–3, 467
   dredging methods, IV 629
   irrigation systems, I 521, 531–5, 536–7, 546–50, 552–3; III 444, 446
   lateral, III 440, 464
   lifts, IV 548–50, 569–71
   lock-gates, II 657, 686–8, 771; III 286, 439, 442, 445–7, 458, 462
   locks, III 439, 442, 447–9, 452–7, 460–4, 467; IV 548–50, 552–5, 561–3, 565, 568
   mitre-gates, III 286, 447–8, 450–2, 460
   pound-locks, III 446, 450–2
   reservoirs, III 463, 468
   side-ponds, III 454
   slipways, III 439, 441 and n. 2
   stanches, III 440–1, 444, 456, 464
   summit-level, III 440, 443, 448, 454–5, 462
   tide-locks, III 442–4, 451, 453
   towage, IV 549, 557–8
   traffic and craft, III 439; IV 548, 551–2, 554–5, 558–60, 562, 566, 569, 571, plate 39A
   tub-boat canals, IV 565, 569
   tunnels, IV 462, 553, 556, 558, 563, 567
Canals named; *see also* Canals
   Americas
   Chambly, IV 552
   Champlain, IV 553
   Cornwall, IV 552
   Erie (Albany to Buffalo), IV 555
   Lachine, IV 552
   Morris, IV 549
   Nicaragua, IV 562
   Panama, IV 562; VI 4; VII 913–17
   St Ours, IV 552
   Sault Ste Marie, IV 552
   Welland, IV 552; VII 913, 918
   Williamsburg canals, IV 552
   Assyrian empires (*c*.700 B.C.), I 469–70, 531–2
   British Isles, IV 490
   Birmingham, IV 566
   Bude, IV 569
   Caledonian, IV 572, 641
   Duke of Bridgewater's, IV 563, 567
   Ellesmere, IV 567
   Exeter (1567), III 451, 456
   Fens (drainage), III 316–17, 319–20; V 544–5
   Fleet Ditch, London (1671–74), III 298
   Forth and Clyde, IV 196, 571–2; V 143, 350
   Glamorganshire, IV 189
   Glasgow, Paisley, and Ardrossan, IV 571
   Grand Western, IV 570

Huddersfield, IV 567
Manchester Ship, VII 889
Newry to Lough Neagh (1737–45), III 455
Oxford, IV 566
Regent's, London, IV 568, 572
Sankey Brook, IV 563
Shannon to Dublin, IV 572
Shrewsbury, IV 566–8
Shropshire, IV 569
Somerset coal, IV 569
Staffordshire and Worcestershire, IV 566, plate 37
Stratford-upon-Avon, IV 568
Tavistock, IV 569
Thames and Medway, IV 566–7
Thames and Severn, IV 567
Trent and Mersey (Grand Trunk), IV 563, 565
Wey, England, (1651–3), III 457
Canada
St Lawrence Seaway, VII 916, 917–21
China, 3rd century B.C., II 771
Ch'ang-an to Yellow River (133 B.C.), III 439
Grand Canal (A.D. 610 *and* 1280–93), III 439–40
Ling Ch'u in Kuangai (215 B.C.), III 439
Pien Canal in Honan (A.D. 70), III 439
Shanyang Canal in Chiangsu (A.D. 350), III 439
Egypt, V 542–4
Mehmondieh, IV 555
Nile to Red Sea (*c*.510 B.C.), III 439
Suez, IV 561–2, 591–2, 630–1, plate 38B; V 540–1; VII 891
Europe
Amsterdam (*from* 1383), III 294–5
Beaucaire, IV 638
Bereguardo, Lombardy (1452–8), III 446
Briare (1604–42), III 460–5
Bologna (1491–3), III 447
Bruges to Ostend, 17th century, III 453
Brussels (1550–1607), III 450–3, 462
Canal de Bourgogne, IV 556
Canal de Bruges à Ghent, IV 559
Canal de la Sambre à l'Oise, IV 556
Canal de Miribel, IV 550
Canal du Centre, IV 556
Canal du Nord, IV 556
Canal Lateral à l'Oise, IV 555
Castille, IV 554
Charleroi-Brussels, IV 557
Charolais, III 459
de Bourgogne, III 459
Dunkirk to River Aa (1670), III 453
Eider, IV 560–1, plate 38A
Elbe-Trave, IV 560
Finow (1620 and 1744–51), III 454–5
France, III 373, 459–68
Franz, between Bezdán and Bács-Földvár, IV 550

Friedrich-Wilhelm (1669), III 454
Ghent, IV 560
Göta canal (1808–10), III 455; IV 554
Imperial, Aragon, IV 554
Kiel, IV 560–2; VI 6; VII 913
Lake Hjälmar to River Arboga (1629–39), III 455
Languedoc (1666–92), III 459, 464–8; IV 462; V 522
Lieve, IV 559
Ludwig, IV 554
Lys to Scarpe (1962), III 453
Maastricht-Luik (Liège), IV 558
Mariinsk, IV 554
Martesana, Lombardy (1462–70), III 446, 448, 450
Mecklenburg, III 450
Mons-Condé, IV 557
Naviglio Grande, III 444–6, 449; IV 555
Naviglio Interne, III 445, 448
Nega, from Themesvar to Titel, IV 550
Netherlands (Holland), II 685–9; III 303; IV 560
Nieuwe Rijn, Holland (1148), III 441
Orleans (1682–92), III 463–4
Ostend-Bruges, IV 558
Ourcq, IV 498
Paderno, Lombardy, III 448–50
Padua to River Brenta, III 447, 451 n.
Pavia, IV 555
Plaue, Germany (1743–6), III 455
Po valley, Italy (109 B.C.), II 679
Pommeroeul-Antoing, IV 557
Pontine Marshes, Italy, III 311–13
Rhine-Meuse-Scheldt, IV 556
Richelieu (1631–38), III 291
Roman projects, II 679–80
St Quentin, IV 462, 555–6
Sassevaart, IV 559–60
Stecknitz, Germany (1391–8), III 443–4, 454
Stockholm to Göteborg (1610), III 455
Tikhvine, IV 554
Vychnyvolotschok, IV 554
Willebroek (Brussels ship canal), IV 549, 557
Ypres to Boesinghe (1643–6), III 454
Zola, IV 498
India, V 542, 543
Eastern and Western Jumna, V 541
Ganges, IV 555; V 542
Upper Bari Doab, V 542
Mesopotamia, I 548–50
Palestine
Siloam tunnel, I 764
Canarium strictum, I 245
Candles, I 234–5; IV 56–7
candle-power unit, IV 57
CANDU nuclear reactor, VI 241, 242

Cannabis sativa (hemp), I 373; II 197; III 158
Cannon, *see* Armament, Artillery, Guns
Canoes, I 313
  coastal fishing, I 176–7
  dug-outs, I 143, 730–1, 738–40, plate 27
  Egyptian, I 731, 732
  Indonesian, I 170
  whaling, I 176–7
Canopic jars, I 268–70, plate 3
Canvas, II 363
Caoutchouc, *see* Rubber
Capitalism, II 66–7, 78, 640
  mechanization, III 27, 30
Capitulare de villis (Charlemagne), II 608
Capsella bursa-pastoris, I 370
Capsian culture, *see* Cultures
Capstans, III 367, 483
Caput mortuum, III 63 and n.
Carat, meaning, III 46 n.
Carbolic acid (phenol), V 315–16
Carbon, III 33, 63
  cast iron, effect, IV 450, 678; *see also* Iron and
    Steel, Steel
  glass colouring, IV 376
Carbonization, III 34
Carburettors, I.C. engine, V 167–9, 171–2, 175
Cardan joints, IV 398
Cards (for carding of wool), II 193
Carex, I 450
Carmen de ponderibus, III 67
Carpentry and techniques, IV 381; V 466–70; *see*
    also Building Construction, Furniture
  fine wood-work, I **684–703**
  Greek and Roman, II 233–8, 413–15
  medieval, II 241–3, 440–3
  tools, II 228–33, 388–95; III **110–33**
Carpets, *see under* Textile industries
Cars (automobiles), see Motor-cars
Carthamus tinctorius, I 246, 288, 441; II 366; III
  694
Carthusian monks *and* telegraphy, IV 649
Cartography, III **501–57**; IV **596–628**; V **438–65**;
  see also Charts, Maps, Surveys
  contouring, IV 612–13
  hachuring, IV 611–12, 626
  instruments, III 553–5
  meridian, III 506, 508–11, 521–2, 531, 533,
    535–6, 544, 548–9, 553
  projection, III 505, 508–11, 530–1, 535–6; IV
    609, 624–5
  relief delineation, IV 611
  scale-cutting machine, III 555
  stereographic projection, III 603–4
  world map, IV 596–600
Cartridge-cases, V 630–1
Cartridges, III 348, 358
Carts, *see* Vehicles, Wheeled vehicles

Carving in wood, II 245–7; *see also* Art, Wood
Caryota mens, I 372
Casein (milk protein), VI 553
Casks, *see* Barrels and casks
Cassia tora, I 288
Casting, *see under* Metal
Castor oil, VI 598
Catalytic Research Associates, VI 57
Catapults, II 34, 375, 573, 699–701, 707–15,
  723–4, 727
Catenary curve, III 258, 434 *and* n.
Cathode ray tubes, VII 1106–7
Cattle, *see under* Livestock
Caustic soda, IV 227; V 244; VI 519
Cedrus libanotica, I 288
Celluloid, VI 551
Cellulose, VI 649–50
  acetates, VI 553, 567; VII 1283
  nitrate, VI 551–3; VII 1283, 1304
Celtic peoples, *see under* Cultures
Cement; *see also* Concrete
  lime-mortars, IV 447
  manufacture, 19th century, V 483–6
  Portland cement, II 387 n.; IV 448–9, 451,
    464; V 483, 484–6, 512
  quality standards, V 486–7
  Roman period, II 407; IV 448, 464, 497; V 483
  sand/lime mixture, 15th century, II 386–7
Cementation (alloys, steel), I 594; III 27, 33–5,
  37, 47, 54
Census of Electrical Industries, U.S.A. (1902), VI
  284
Census of Production, Britain (1907), VI 679
Central heating, II 404, 419–20, 446; *see also* Heat
Ceramics, general/pottery, I **376–412**; II **259–310**,
  plates 16–23; IV **328–57**, plates 16–20; V
  658–70; VI **622–37**; *see also* Ceramics, named
  places of factories
  Arretine vases, II 272–3
  artisans, IV 356–7
  assayer's crucibles, III 60, 61
  basic processes, I 379–84
  bone-china, IV 337–8, 341–2, 344, 353; VI 625
  bowls, II 267
  bucchero ware, II 270–1
  building materials, V 665–6, plate 3
  cinerary urns, I 324, 378, 401
  clays, treatment, I 380–1; IV 330–2, 337–41,
    345–9; VI 628–30
  colours, IV 331, 333–6, 340, 342, 354
  crouch ware, IV 345
  cups, II 261, 263, 269, 274–5
  decoration, I 241, 383–4, 400–5, 408; II 273;
    IV 332–5, 345–6, 348, 354–6; V 662–3
  delft ware, IV 334, 336–7, 344
  Eastern influence, II 771; IV 328, 336, 338–9
  Etruscan, II 269–71

filters, V 664
firing, I 382–3, 391–7, 410, 411; II 266, 276, 294–300, 307–9; V 660–1
flints, IV 347–9, 352
fluxing agents, IV 338–9
furnace techniques, IV 230–1, 235; *see also* kilns
glazing, II 261–4, 272, 274, 276, 286, 299–302, 357–8; IV 256, 332–4, 336–8, 342–5, 347–51, 353
Greek, II 258–69
Hispano-Moresque ware, IV 329 *and* n., 335
impasto ware, II 269–70
Italian, 16th century, III 678–9
jugs, II 293, 310
kilns, I 392–7; II 294–9, 305; IV 331, 335, 339, 343, 345, 347–9; V 666–8; VI 627–8, 629–30
lead glazing, IV 256; VI 628
lustre ware, II 303–4; IV 335, 356
machinery, V 668–70
magic, influence, I 400–1
majolica (maiolica) ware, II 300, 302, 358; IV 329 *and* n., 334–6
manganese oxides, IV 147
mass-production, IV 354–5
mechanization, IV 328, 330, 348, 351–2, 357
medieval techniques, II 284–99, 302–3
Megarian, II 269, 272
moulds, IV 332, 353–4
origins, I 376–9; II 258
ornaments, plastic, II 292–4
ovens, IV 350
painted ware, II 264–5, 277, 301–2, 303–4; IV 336
porcelain, II 286–7, 307, 771; III 678; IV 336–43, 353; V 662–4; VI 625
pot-lids, V 662, plates 1, 2
pots and pitchers, I 400; II 267, 268, 292
potter's wheel, *see* Potter's wheels
Queen's ware, IV 351–2
pyrometric cones, V 660–1
raw materials, V 666; VI 628–9, 632–5
refractories, V 664–5; VI 630–6
relief-work, II 268–9, 270–1, 274–9
Roman, II 270–4
saggars, IV 343–4
scientific instruments, VI 623
shapes, ancient world, I 397–405
slip, II 302–3
statuary, II 274–9
stoneware, II 286–7; IV 329, 331, 346, 353; V 664
table-ware and art ware, III 20; V 661–3
terra sigillata, II 274, 297
thermal insulation, VI 636–7
transfer-printing, IV 343, 354–5
vases, II 260, 263, 268, 269–70, 272–3

whiteware, VI 625–8
wine-jars, I 283, 406; II 130–3, 135–6, 263
Ceramics, named places of factories
    Battersea, IV 354
    Bow, IV 338, 341–2, 344, 354
    Bristol, IV 336, 340–1, 343–4
    Burslem, IV 344, 348, 351
    Caughley, IV 344, 354
    Chantilly, IV 338
    Chelsea, IV 338, 341–2, 344
    Coalport, IV 355
    Delft, IV 329, 336–7, 345
    Derby, IV 338, 341–2, 344
    Dresden, *see* Meissen
    Etruria, IV 352, 356
    Florence, IV 337
    Hanley, IV 344, 347
    Hot Lane, Cobridge, IV 349, 353
    Lambeth, IV 336, 344
    Lane End, IV 352
    Liverpool, IV 336, 344, 354
    Longport, IV 348
    Longton Hall, IV 338, 341–2, 344
    Lowestoft, IV 341, 344
    Meissen, IV 329, 337–9, 341–2
    New Hall, near Shelton, IV 341, 344
    Plymouth (Coxside), IV 340–1, 344
    Rockingham works, Swinton, IV 355
    Rouen, IV 337
    St-Cloud, IV 337–9
    Sèvres, IV 329, 338–43
    Staffordshire Potteries, IV 344–53
    Stoke-on-Trent, IV 346 n. 2, 353
    Tunstall, IV 348
    Vincennes, IV 338–9
    Worcester, IV 341, 343–4, 354
Ceratonia siliqua, I 289
Cereals, *see under* Agriculture
Cerium, IV 119
Cesspools, IV 507–8; *see also* Sewage
Chain-mail, II 721
Chains, origin, I 656
Chaldeans, *and* calendar, III 564, 566–8
Chalk quarrying, 17th century, II 34–5
Chalybes tribes of smiths, II 54
Chamfering, III 120–1
Champlevé work, II 450 n.
Charcoal
    antimony smelting, III 29
    black-flux ingredient, III 62–3
    black pigments, I 240; II 359; III 683
    brass-making, III 37
    coal replacement, III 78–81, 685
    copper smelting, I 586; II 48–9, 66; IV 127–8
    date palm stones, I 551
    fuel, I 229, 231; II 369
    gas discovery (CO_2), IV 215

Charcoal (*cont.*)
  gunpowder ingredient, II 380; IV 235
  iron-smelting, II 72–4, 77; III 31, 32, 33, 342,
    350, 683; IV 99, 100
  kiln fuel, I 392
  -making (burning), II 368, 369; III 683–5; IV
    261
  metal smelting fuel, II 41, 62, 68; III 53, 55;
    IV 103
  steel-making, I 573, 576, 594–5; III 33, 34–5
  tinplate preparation, IV 125
Chariots; *see also* Wheeled vehicles
  ceremonial type, II 544
  fighting, II 537, 539, 540–4, 696
  harness, II 163–4, 165
  onagers, I 721–2, 724–8
  origin, I 724–8
  transport, Assyrian Empire, I 713
  warfare use (*c.*2500 B.C.), I 209–12, 713, 718,
    724–8
  wood-work, ancient Egypt, I 700–1
Charts; *see also* Cartography
  hydrographic surveying, V 449–54
  marine, IV 575, 612–13, 615–16, 618, 620
  Mercator, III 548, 550–1; IV 625
  soundings, V 452–3
  triangulation, V 450–2
  weather charts, synoptic, IV 621
  wind, IV 620
Chasing tools, II 469
Chatsworth Apollo (bronze), II 476–7
Chelleo-Acheulian culture, *see* Cultures
Chemical industries, chemistry, III **676–708**; IV
    214–29, 230–57; VI **499–569**; *see also* names
    of elements and compounds
  absorption towers, V 236
  acids, II 356–7; *see also* named acids
  agricultural chemicals, VI 500, 507, 510, 521,
    528–30
  alchemy, II **731–52**; IV 214, 220, 228
  aliphatic chemicals, VI 540–3
  alkalis, II 354–5; VI 505–6, 509–12, 554, 567
  ammonia and fertilizers, VI 500–1, 511
  analytical, III 22, 683
  ancient man, I **238–98**; II 350–3
  bitumen production, I 250–6
  books, II 350–2; III 676–7; IV 233
  calcination, IV 218–20
  Castner's sodium processes, V 249, 251, plate
    11
  caustic soda, V 244, 250–1
  chemical analysis, IV 221, 223–4, 228
  Claus kiln, V 239
  cleansing agents, II 355–6
  coal as raw material, IV 252
  combustibles, II 369–71, 374–82
  cosmetics, *see* Cosmetic arts

  crystallization processes, IV 230, 232–3, 241,
    251
  distillation processes, IV 232–3, 235, 252
  drugs, II 371–2, 732
  dyestuffs, V **257–83**; VI 505–6, 509, 554,
    564–8; *see also* Dyes
  electrochemical processes, IV 678; V 248–52;
    VI 518–21
  electrochemical telegraphy, IV 652–3
  elements, IV 215, 221
  explosives, V **284–98**; VI 500, 508, 522, 542–3,
    547–600
  fermentation techniques, I 275–7, 284; IV
    232–3; V 302–8
  fine chemicals, V **299–321**
  gases, IV 215–19
  glass colouration, IV 373–6
  heavy chemicals, IV 237–52; V **235–56**; VI
    514–30
  high-temperature processes, II 347–8
  Lavoisier's experiments, IV 220
  lead-chamber process, V 245
  Leblanc process, *see* Alkalis
  low-temperature processes, II 348–50
  Mond's sulphur-recovery process, V 238
  nitrogen fixation, VI 501, 504–5, 522–8
  nomenclature, IV 221–3, 228
  painting materials, I 240–5, 248
  petrochemicals, VI 510, 533–5, 543, 546–7
  phlogiston theory, III 57; IV 218
  phosphorus, V 252–4
  pigments, I 238–40, 244, 502, 514; II 359–64
  plastics research, VI 503–4
  pneumatic, IV 215–17, 247
  polymer chemistry, VI 501, 534, 551–60,
    567–8
  pre-scientific, II **347–82**
  preservatives, I 256–70
  quantitative and assaying, III 28; IV 217–18,
    223–5, 228
  raw materials, III 705–6; VI 532–3, 535–40
  refining, early processes, I 581–2, 584–7
  revolution in technology, IV 214, 228
  sewage purification, IV 518–19
  smoke abatement, VI 512
  soda, V 235–44, 471–2; VI 507, 509
  sodium cyanide, V 249
  Solvay, ammonia-soda process, V 235, 241–4,
    672
  sublimation purification process, IV 232, 234
  superphosphate, V 254; *see also* Fertilizers
  symbols and formulae, IV 225–6
  synthesis and substitution, VI 501–4, 506, 508
  textile industry, IV 228, 233, 244, 248–9, 255
  water-pollution, VI 512
  Weldon's chlorine process, V 237
  wood based chemicals, III 683–5; V 308–11

Chemical Society, VI 153
Chenopodiacaea (glasswort), I 261
Chenopodium, I 355, 370; II 157
Chileans, I 179
Chimney-wheel, IV 167
Chimneys, chimney-stacks, III 52, 252, 256, 261
China, *see* Ceramics
China (Republic of), I 35; II 768–71
  agriculture, I 368, 373
  alchemy, II 732 n.
  basketry and matting, I 419
  beer, brewing, II 140
  bridges, III 417
  canals, river navigation, III 438–41, plate 18
  ceramics, II 286, 302; IV 328, 332, 335–6, 358–9
  coal, III 73–4; VI 179, 180
  coins, pure zinc, I 650
  cotton, III 159
  cross-bow, magazine type, II 713
  draw-loom, figure-harness, III 190–2
  East *v.* West techniques, II 755–60, 768–71
  egg-drying, V 51
  explosives, II 370, 377–8
  figured fabrics, III 195, 199, 201
  fish preservation, IV 53
  geographical knowledge, III 507, 524
  Gregorian calendar, adoption, III 579
  harness and saddles, II 538, 553, 556, 557
  iron industry, III 74, 79; V 70
  kites, V 398
  lacquered ware, III 696
  language structure, I 88, 94, 97, 772
  lead as sheet, IV 119, 120 n. 1, 123
  measurement, standards, I 111–12, 783–4
  metal-mining, IV 68
  metallurgy, I 590, 628–31; II 770–1; III 691
  nickel, V 86, 608
  oil, V 106; VI 377
  pack transport, I 705
  paddle-wheels, II 607; V 141, 142
  paper money, III 380
  Pekin man, I 23–5
  porcelain, I 78
  pottery, 385, 393, 401–2, 405
  printing, III 378
  roads, I 713–14
  rocket propulsion, V 412
  rotary winnowing-machine, II 98
  saltpetre, IV 232
  seric iron, II 57
  silk industry, II 197–8, 210; III 195, 200–3; IV 312 n.
  stern-post rudder, II 583
  stoneware, II 286–7, 307–8
  tea, III 5; IV 591
  textiles, I 415, 445
  viticulture, II 131
  watermills, II 614
  wheel, ancient uses, I 198, 200, 213
  wheelbarrows, II 642
  wheeled vehicles, I 716–17, 722, 726; II 544, 549
  windmills, II 615, 617; III 106
  zinc production, III 30, 690; IV 119, 130
China-clay, *see* Kaolin
Chisels
  Assyrian, *c.* 8th century B.C., II 229
  bronze, I 611–12
  cannon finishing, III 365–6
  carpenter's, Middle Ages, II 250, 252–3, 390, 391, 392
  copper, I 603, 687, 688; II 384
  Greek mining, II 3, 5
  iron, I 619
  mason's (bolster), II 37, 384, 385, 386, plate 31A
  metal-work, II 453, 469
  Roman, *c.* 1st century A.D., II 230–1
  stone quarrying, Egypt, I 569
  tradesmen's (*c.*1500–1850), III 110–11, 114, 118, 126
  woodcut carvings, medieval, III 404
Chlorine, IV 247–8, 255, 502; V 237, 250–1
Chlorophora tinctoria, III 694
Christian era, III 572, 578
Christianity, II 605–6, 761–2; *see also* Bible
Chromium, IV 119
  alloy steel, V 64–6, 640; VI 410
  aluminothermic production, V 97, 98
  colour pigments, IV 249
  gas-turbine blade alloys, VI 443
  glass, green colouring, IV 375
Chroniques de Hainaut (Jacques de Guise), II 526
Chronology tables, I xlviii–lv; II lv–lix; *see also* Calendars and chronology
  early metallurgy (tables), I 574–5
  East *v.* West techniques, transmission, II 770–1
Chronometers, IV 409–15, 598, 615, 619
Chuckchi of Siberia, I 225
Church architecture, II 420–45; *see also* Buildings named
Cider and perry, II 139, 637
Cinematography, V 734–51; VII 1301–16
  colour, VII 1305–9
  developments since 1890, V 743–7
  early growth, VII 1302–3
  films, V 747–50, plate 44B; VII 1303–4
  meaning and origin of word, V 734
  pioneers, V 735–47
  sound, VII 1263–4, 1309–14
  studios, VII 1304–5, 1314
Cinerary urns, I 324, 378, 401
Cineres clavelati, II 355

Cinnabar (mercury sulphide), IV 133
Cinnamomum malabrathrum, I 288
'Circle of Pythagoras', II 760
Circumferentor or 'Dutch circle', III 627, 636; IV 602–3, 607
Cistercian monks, II 192
Citrullus colocynthis, I 288
Citrus medica, I 288
City and Guilds of London Institute, VI 159
Civil engineering, IV **442–88**; V **499–521**; VII **871–929**
  ancient world, II **492–536**
  bridges, III **417–37**; IV 452–62; VII 876–99
  buildings, IV 469–86
  canals, VII 912–20
  French leadership, IV 525–6
  land reclamation, III **300–23**; VII 907–12
  ports and harbours, IV 464–9; VII 920–9; *see also* Docks, Ports
  roads, construction, IV 525–6; VII 873–6
  tunnels, IV 462–4; VI 200–1; VII 899–905
  waterways, III 373, **438–70**; IV **548–73**
Civil Engineers, Institution of, IV 433, 443, 486; V 492, 557
Civil Engineers, Society of, IV 206
Civilizations, *see* Cultures and civilizations
Clactonian culture, *see* Cultures
Clapboard, II 244–5
Clay, IV 447–8; *see also* Bricks and tiles, Ceramics
  cannon, casting mould, III 363–5
  potter's, I 379–80, 405–6, 407
'Clay of wisdom', II 734
Cleopatra's Needle (obelisk), I 569
Clepsydra, *see* Water-clocks
Clinkers, III 226, 228
Clipsham stone, II 35–6
Clocks, clock-making and watches, III **648–75**, plate 32
  alarms, III 652
  astronomical, III 653–5, 658
  chronometers, development, IV 409–15
  early mechanical, III 648–53
  escapements, III 664–73; IV 410, 412–15
  machinery and tools, IV 382, 407, 409
  mercury, III 648–9
  metals, thermal expansion, IV 395
  nautical, III 552–3, 557, 672–3
  pendulum, III 552, 636, 662–7, 671–2
  precision mechanics, origin, IV 380
  repeating mechanism, III 669–70
  sand-clocks, I 126
  shadow-clocks, I 112, 113, 797; III 595–6; *see also* Sun-dials
  spring-drive, III 656–60
  toothed wheels, cutting tools, IV 391
  water-clocks, I 113, 122, 123, 126, 797; III 568, 601–4

Clodding-beetles, III 134–5
Cloisonné work, II 450 n.
Cloth types, III 178–9; *see also* Textile industries, Wool
Clothes, I plates 16, 17
  boots, shoes, *see* Footwear
  cleaning, II 355
  fur, I 413
  hemp, I 373
  knitted garments, III 181–2, 184, 186, plate 9
  leather, II 148, 150, 156, 158, 160–5, 175–8
  linen, II 195
  matting sandals, I 422
  wool, I 442
Clothing manufacture, *see* Textile industries
Coal, coal-mining, II 369–70; III **72–88**; IV **64–98**; VI **359–75**; *see also* Mines
  accident prevention societies, IV 95–6
  blasting, IV 80
  bord-and-pillar working, mines, IV 86–7
  briquettes, III 80, 679
  chemical raw material, IV 252
  China, III 74
  choke-damp, IV 89–90
  coalfields, Europe (1851), IV 80
  coke products, *see* Coke
  conveyors, IV 85–6
  cutters, mining, IV 82–3; IV 364–9
  domestic, III 76
  drainage, mines, III 74, 81–3; VI 372
  explosions, mine, IV 83
  fire-damp, IV 90–1, 93, 95–7
  'firemen' term, IV 91–2
  -gas, *see* Gas industry (coal-gas)
  Great Britain (1700), IV 166
  hand tools, IV 80–1, 84, 87–8
  hauling and hoisting, IV 82, 84–6, 90; VI 369–72
  health and safety, VI 374
  industrial, 17th century, III 76–8, 686
  lighting, mines, IV 86, 91–7; VI 373
  long wall working, IV 87
  machinery, mines, III 82–4; IV 79–83, 87, 161–2; VI 359–69, 371–2
  medieval Europe, III 74
  Middle Ages, II 68, 347–8, 355
  nationalization, Great Britain, VI 184
  ponies, mine haulage, IV 84
  production (*c.*1900–50), VI 177–80
  Roman Empire, II 10, 62; III 72–3
  safety lamp, mining, IV 71, 91, 95–7
  transport, III 84–5
  ventilation, mines, III 75; IV 86, 91–7; VI 373
  winding and hoisting, IV 87–9; VI 371–2
  wood substitute, III 78–81, 220–1; IV 234, 371
  world wars, 20th century, VI 178–84, 359–66; VII 759

Coal-tar, V 299–300; VI 467, 535–40
Cobalt; see also Zaffre
    colouring, ceramics/glass, III 218, 678, 698,
        702–3; IV 335, 374; V 79
    extraction, III 703; IV 119, 145–6
    paint-driers, VI 595
    production, Swansea (1874), V 86
    refining, electrolytic, V 97
Coccid insects, II 362, 366
Cochineal, I 245–6; II 366–7; III 695; V 258–9
Cocos nucifera, I 372
Codex Amiatinus, II 243
Codice Atlantico, Leonardo da Vinci, III 161, 172,
    175
Coffee, see Drinks
Coffer-makers, II 170–1, 255–6
Coffins, basketry and matting, I 419, 420
Cog-wheels, II 635, 648, 654, 675–6
Cogware (cloth), III 178
Coins, coinage; see also Medallions, Mints
    alloys, Europe, 15–16th centuries, III 45–7
    assaying, III 60
    blanching, III 46–7, 56
    brass, II 55
    Carthage (c.814 B.C.), II 568
    copper, II 53; III 46–7
    design and manufacture, II 491–2
    die-cutting and stamping, II 485–92
    dies, III 338–9, 340
    early reports, I 584, 585
    gold, see Gold
    mechanization, III 338–43
    medieval period, II 10–11
    metal-work source, I 637 n.
    silver, I 585; II 13 n., 46, 585; III 46–7; V 94;
        VI 411
    techniques, tools, III 337–43
    zinc, Ming dynasty, I 650
Coke
    coking-ovens, IV 101, 259–60; VI 467–8
    iron smelting, III 29, 80, 689; IV 99–101, 235;
        VI 465–7, 537–8
    production, IV 105; VI 465–8, 535–8
    tinplate manufacture, IV 125
Colby's compensation bars, IV 601, 608, 611
Collective bargaining, VI 100, 102, 108–10
Colleges named, see Universities and colleges
Collieries, see Mines
Collotype, V 707
Colossus of Rhodes, II 470
Colour Index, textiles, VI 671–2
Colour photography, VII 1295–8; see also
    Photography
Colours, III 692–8, 702–3; see also Dyes
    assaying, III 64, 68
    colour-knitting, III 182, 185
    oxidation, III 49

speculum-metal, III 40
steel, III 36–7
Columbium, see Niobium
Combs, I 515, 516, 665; II 193, 194, 196
    weaving-combs, II 212, 213–14
Combustion, discoveries, IV 218–21, 223, 228
Combustion engine, see Internal combustion
    engine
Commission des Poids et Mesures, France, IV,
    397–8, 405
'Committee of Ten' (1893), U.S.A., VI 145
Committee on the Neglect of Science (1916), VI
    151, 154, 157
Committee on the Utilization of Rural Land
    (1942), VII 968
Commonwealth of Nations, see British
    Commonwealth
Communicating techniques, VI 50; see also
    Electrical communication, Radio, Telegraphy,
    Telephony
Comparators, IV 395–8
Compasses, magnetic
    bearing variations, III 548–9, 555
    cartographic use, III 517, 533–4; IV 615–16
    direction dials (c.1500), III 600–1
    instrument designs, III 626–8, 630, 637–8
    lodestone 'touched' (c.1180), III 523–4
    meridian type, III 548–9
    surveying instruments, III 538–40; IV 600,
        602–3
    wind-rose 'compass', 13th century, III 524–6
Compasses, mensuration
    alchemical laboratory (1606–7), II 749
    carpenter's, I 195; II 391
    drawing, III 110, 124, 488, 512, 627–8
    instrument-tools, IV 388–90
    mason's, II 386, plate 31A
Compensation devices
    inertia, IV 395
    marine chronometers, IV 409–15
    temperature variation, IV 395, 410, 412–15,
        601–2
    upthrust changes (in balances), IV 407
'Compleat Chymical Dispensatory, The' (1669),
    III 29
Compositiones ad tingenda musiva (Compositiones
    variae), II 63, 329, 351
Compressed-air, IV 71, 83
Computers, VII 1150–203; see also Calculating . . .
    astronomical, IV 382
    binary stored programs, VII 1191–9
    computer industry, VII 1200–3, 1153–7
    digital, VII 1154–5
    electronic, VII 1103, 1154–8, 1171, 1188–202
    languages, VII 1157, 1192, 1196, 1199–200
    programming techniques, VII 1199–200
    software, VII 1157–8, 1198–9

Concert of Europe, VI 2, 4, 7
Concorde (aircraft), VII 834, 835, 1469
Concrete, V 487–91; *see also* Cement
  bridge foundations, III 424; IV 449; V 487
  bridge structures, V 496, 500, 502, 512; VII
    879, 892–4
  building purposes, 20th century, VII 934–41,
    947–53
  fire resistance systems, V 480; VII 942–3
  prestressed, VII 879, 892–4, 935–7, 939
  reinforced, IV 451; V 487–91, 495–6, 502; VII
    892–4, 934–5, 937–44, 945
  roads, IV 539; VII 875–6
  Roman period, II 405, 407, 410–13
  Romanesque period, II 127–8
  tunnels, IV 464; V 502
Condensers
  gas (coal-) manufacture, IV 266–70
  metallurgical furnaces, III 51
  radio invention (capacitors), VII 1233
  steam engines, IV 182, 184–6, 189, 192, 421
Conduit, built by Hezekiah, King of Judah, III
  510–11
Conservatism of craftsmen, II 350, 558, 649
Consultative Committee on Secondary Education
  (1938), VI 167
Contractors, IV 454–5, 631, 642
Chenopodium, I 355, 370; II 157
Chileans, I 179
Convolvulus scammonia, I 249
Coopering, III 128–33, plate 7; *see also* Barrels
Copley medal of the Royal Society, IV 204
Copper, IV 225; *see also* Brass, Bronze, Metal
  headings
  alloys, I 588–92; II 53–5; III 37–42, 46, 49,
    58, 361; IV 130, 133, 146; V 609–10; VI
    428–30, 443–7, 449
  assaying of ores, III 62–3
  'black' copper, II 50; IV 129
  brazing materials, I 650–2
  bulls, temple at El-Ubaid, I 465, 638,
    639–40
  by-products, V 77, 96
  carbonate, III 48
  casting techniques, I 578, 587–8, 601, 606–8,
    627–8, 634–5; V 609–10
  coins, II 53; III 46–7; IV 119
  compounds, IV 122
  cooking pots, I 399; V 605
  early times, I 64, 66, 574–5, 584–8
  enamelled metal-work, II 462–3, 466; III 48
  engraving, copper plate, III 405, 408–10
  extraction, *see* refining
  gilded metal-work, II 462, 466, 467
  glass-making tools, IV 371–2
  glazes, ingredient, II 301; IV 333–5
  gold refining process, II 42–3

metal-working, V 607–9
mining, I 563–6, 567, 574; II 7, 48; IV 64–5,
  74, 77; VI 410, 413–4, 418–19, 424
oxide, III 46–7
pigments source, I 293; II 362; IV 122
plate (sheet) manufacture, V 617, plate 38
precision instruments, IV 601
printing blocks, woodcuts, III 405
production, II 51–3; V 73, 76, 605
refining and smelting, I 585–8; II 11, 48–51;
  III 27, 29, 33; IV 118–19, 126–30; V 72,
  76–88, VI 424
sheet copper-work, I 641
'Sheffield plate', IV 130
ships, copper sheathing, IV 579–80, 591–2; V
  84, 605, 608
silver refining, IV 136–7, 139–40, 142, 146; V
  95–6
smelting, *see* refining
solders, I 651–3; II 352; III 48
tools and weapons, I 46, 59–60, 73, 207–8, 509,
  513, 577, 601–7, 687–9
tube manufacture, V 625–8
tyres, primitive wheels, I 208
uses, IV 130; V 84, 605
verdigris manufacture, II 362
Copperas (ferrous sulphate), III 680, 694, 706; IV
  250–1
Coppersmiths, II 53
Corchorus, I 450
Cordials, *see* Drinks
Cordwain, II 150–1, 155, 177
'Corinthian bronze', II 43
Cork stoppers, III 10
Corn-mills, IV 199–201, 206–7, 209–12
Correcting devices, tools, IV 395
Corrosion, VI 427, 434, 450–3; VII 985
Cosmetics; *see also* Incense
  alabaster cosmetic jar, I plate 3
  ambergris, IV 57
  ancient man, I 238–98
  eye-paints, I 292–4
  face-paints, I 294
  gold cosmetic case, Ur, I 654, 656
  oils and fats used, I 287–9
  ointments, I 291–2
  perfumes, preparation, I 289–91; V 317
Cosmography, III 535–6; *see also* Cartography
Cost-book system, mining, IV 73–4
Cotton, II 199–200; *see also* Spinning, Textile
  headings, Weaving
  bleaching process, III 178–9
  cloth, list of types, III 178–9
  crop production, VI 328–9, 338
  early history, I 373–4, 432, 436, 449
  fulling process, cloth, II 214–16; III 169–72
  Gossypium species, I 449; III 159

machinery, cotton-mills, IV 277–91, 315; V 598–9
Manchester cotton, III 179
-mills, IV 265–6, 268–9, 274
paper-making, III 411
processing after picking, III 160, 163–4
supply sources, III 159
Cow's milk, Greeks' abhorrence, II 119
Craft, change to science, IV **663–81**
Craftsmanship, craftsmen; *see also* Artisans, Tradesmen
ancient man, I **128–43**
human skill, I **1–37**
Cranes, II 602, 607, 638, 656–62, plate 38B; 420, 422–3; VII 938; *see also* Hoists, Pulleys
Cranks, II 393–4, 635, 651–5; III 423
Crease resistance, VI 672–3
Creep (metal deformation), VII 984–5
Crepon (cloth), III 178
Crete
building techniques, I 484–7
chryselephantine work, I 661
dyes, I 246–7
fresco paintings, I 242–3
ivory work, I 672–3
Minoan culture, I 199–200, 202, 204, 212, map 4, table F
Minos Palace, I 232, 235, 486–7, 672, 714, 715
pottery, I 397–9, 401 n. 1, 403–4, 406–7
roads, I 714
windmills, II 618–19
writing, I 758–9
'Crimplene' yarn, VI 660
Cro-Magnons, I 32, 35
Crocus sativus, I 246
Crops, *see* Agriculture
Cross-bow (*gastraphetes*), II 647, 699, 707–9, 721–4; III 353
Cross-shave (smoother, inside pails), III 112
Cross-staff, or Jacob's staff, III 527–8, 537, 546–7, 591, 625
Crow-bars, II 252
Croze (cooper's tool), III 119
Cruciferae, I 360
Crusades, II 756, 764–6, table III
Cubit measures, I 777–9
Cudbear (dye), IV 249 *and* n., 253
Cuir bouilli, II 171, 173, 180
Culinary arts, ancient man, I **238–98**
Cultures and civilizations, I **38–57, 58–84**; *see also* Societies
Abbevillian, *see* Chellean
Acheulian, I 25–8, 68, 131–2, table D
American (Aztec, Inca, Maya), I 82–4, 124–6
Aurignacian, I 33, 190, 497, table D
Capsian, I 501, 509, 511, 606
Celtic, II 537–40, 542–3, 550–1, 560–1

Chellean, I 27, 131, table D
Chelleo-Acheulian, I 25–9, 131–2
Clactonian, I 28–30, 188, table D
Cortaillod, I 373
Danubian, I 309–11, 354–5, 357, 359, 361, 362, 364–6, 368, 372–3, 387, 502, 507
Egyptian, I 46, 49–55, 80, 102–3, 111, 139, 161; *see also* Egypt
Eskimo, I 180–3; *see also* Eskimo people
Greek, I 104–5; *see also* Greece
Hallstat, I 362, 367, 369–70, 399–400, 443, 451, 518, 565–7, 613, 617–18, table E
Inca, *see* American
Indus valley, I 46, 50, 80, 194, 202, table F
La Tène, I 514, 567, 617, 618, table E
Levalloisian, I 29–30, 31, 35, 132–4, table D
Magdalenian, I 32, 134, 136, 157–8, 160–1, 188, 407, table D
Maglemosian, I 175, 190, 498–9, table E
Mayan, *see* American
Mesolithic, I 34, 100, 156–7, 161–2, 495–9, tables D, E
Minoan, I 199–200, 202, 204; *see also* Crete
Mousterian, I 30–2; map 1A, table D
Natufian, I 496, 501–3, 511, 513, 520, 542, table E
Neolithic, I 42–5, 79–84, 192, 304, 306–12, 349–51, 353, 376–7, 418, 499–503, 561; *see also* Agriculture
Palaeolithic, I 22, 27, 30, 31, 32, 68
Phoenician, *see* Phoenicians
Roman, I 82; *see also* Romans
Solutrean, I 32, table D
Sumerian, I 46, 48–9, 50, 52, 54–7, 80, 101–2, 194, 199–200
Upper (Late) Palaeolithic, I 32–4, 101, 134, 136–9, 300, 413, 495, 559, 745, table D
Cupels and cupellation, III 60, 64–5
Curcuma, I 247; III 694
Cutlery, II 125; III 20, 115; IV 108; V 834
Cycloid theory, III 345, 663 *and* n. 3
Cylinders; *see also* Internal combustion engines, Tubes
boring machines, III 367–8; IV 421–2
cannon costs, III 363
coin-making, III 341
diameters, steam-engine, IV 177, 179, 183, 189; V 127
escapements, timekeepers, III 671, 673–4
'McNaught' type, 194–5
paper-making machines, III 412, 414–15
steam-engines, IV 176–7, 179, 181–2, 184–6, 188–9, 190–2, 197, 431; V 124, 126, 130, 134
suction pumps, III 330–4
Cymbopogon schoenanthus, I 288
Cyperus esculentus, I 288
C. papyrus, I 285, 754

C. rotundus, I 288
Cyprus, map 4
  alum production, I 263
  copper mines, I 565
  iron industry, I 598
  ivory-work, I 676, 677
  metal-work, I 623–4
  pottery, I 403
  script, I 761
Czechoslovakia, coal, VI 180

'Dacron' fibres, VI 651
Daggers, *see also* Knives, Swords, Weapons
  bronze I 510, 613–14, 661; II 481
  copper, I 603–4, 605
  flint, I 509–10; II 159, 164
  gold sheath, Ur, I 654, 656
  iron, I 618, 619
  wood, I 517
Dairy products, V 32–6; *see also* Milk industry
Damascene (Damask-steel), II 57 n.
Damascus pigment (cobalt blue), II 301
Damask (cloth), III 165, 192, 199–203, 205
Dams and reservoirs, IV 490, 495–8, 501–2; *see also* Dikes; Water-supply
  design, V 502, 556–8; VII 1366
  earth, VI 197; VII 1366–7
  flood control use, III 310–11
  gravity, VI 199; VII 1368
  land reclamation, III 301–3, 308
  massive buttress, VI 198–9; VII 1368
  multiple arch, VI 199; VII 1368
  rockfill, VI 197, 199; VII 1367, 1369
  single arch, VI 199–200; VII 1368
Dams and reservoirs, named
  Australia
  Lithgow No. 2, V 557–8
  Egypt
  Aswan, V 543–4
  Asyut, V 544
  England
  Blackton reservoir, V 524
  Crown reservoir, V 523
  Dale Dyke dam, V 523, 556
  Haweswater, VII 1368, 1370
  Holmfirth, V 523, 556
  Howden, Derwent Valley, VII 1369
  Keswick pumped storage scheme, VI 212
  Ladybower Dam, Derwent Valley, VII 1367
  Lynton and Lynmouth scheme, VI 212
  Thirlmere Dam, V 558
  Windermere storage scheme, VI 212
  France
  Bouzey Dam, V 523
  Furens Dam, V 522, 556–7
  Lampy Dam, V 522
  Zola Dam, V 522, 557–8

  Hong Kong
  Shing Mung Dam, VII 1367, 1369
  Italy
  Santa Giustina Dam, VI 200
  Vajont Dam, VI 200
  Netherlands (Holland)
  Amsterdam city dams, III 293–5
  Scotland
  Galloway hydro-electric scheme, VI 200
  Loch Sloy Dam, VI 201
  Spain
  Alicante Dam, V 522, 556–7
  Uganda
  Owen Falls Dam, VI 198
  U.S.A.
  Bear Valley Dam, V 557–8
  Boulder Dam, VI 199
  Salt Springs Dam, California, VI 199
  Wales, V 553
  Claerwen, VII 1368, 1369
  Caban Coch Dam, V 557–8
  Dolgarrog Dam, V 523
  Vyrnwy, V 522, 553, 557–8, plate 33B
Danubian culture, *see* Cultures
Date-palms, I 371, 450, 523, 551–2
Daucus, I 356
'David' of Michelangelo, II 28
De Aquis (Frontinus), II 405, 419
De coloribus et artibus Romanorum (Heraclius), II 63, 351; III 207
De materia medica (Dioscorides), II 371
De natura fossilium (Agricola), II 34–5
De re aedificatoria (1485) of L. B. Alberti, III 250, 446
De re metallica of Agricola, II 13–24, 290; III 27–8, 82, 206, 212–14, 324, 677
De rebus bellicis (*c.*370), II 605, 607, 719
Decimal system, III 61, 542
Decoration; *see also* Ceramics
  building, IV 447, 474
  ornamental turning, IV 419–20, 425
  techniques, IV 436
Delhi Durbar (film 1911), VII 1306
Demarcation disputes, VI 98–9
Denmark; *see also* Europe
  agriculture, I 355; IV 16, 35; V 2, 15, 19–20, 32
  boring tools, I 191–2
  building methods, I 302–3, 311–12, 314
  canals and inland waterways, IV 560–1
  ceramics, V 661, 666, 667–8, 670
  chemical research, V 303
  dredging, IV 640
  electricity, V 178, 183–4; VI 216
  fishing, I 168; VI 348
  food-gatherers, I 495–6
  Iron Age man of Tollund, I 355, plate 6B

motor vehicles, V 429
navigational aids, V 456
ploughs, II 83–5, 87, 91
surveys, IV 610
telegraphy, IV 654
textiles, I 442–3, plate 17A
water-supply, IV 497
wheel, I 206, 207, 209, 210, 213
wind power, electricity, VI 216
Density, assay by, III 66–8
Dentistry, IV 145; VII 1323–6
Department of Science and Art, Britain, V 782–4,
    790, 794
Department of Scientific and Industrial Research,
    Britain, V 550
Design methods
    fabrics, III **187–205**
    stained glass, III 240–2
Detergents, I 249, 260, 261; VI 547
Detonators, VI 549; see also Explosives
Development areas, VII 968–73
Diagrams, engraving, III 405, 409
Dials; see also Instruments
    clocks, IV 414
    micrometer screw, IV 396
    telegraphic equipment, IV 409
Diamonds, IV 66, 68; see also Jewellery
    cutting, IV 387
    glass-cutting, III 233
    glass-engraving, III 229
    industrial, IV 68
Diaper (linen fabric), III 178
Dibbling (sowing method), III 135–6
Dictionaries; see also Encyclopaedias
    chemical substances, IV 223
    technology, 17th century, III 682
Dictionnaire des Sciences (1765), see *Encyclopédie*
Die-stamping, III 338–9
Die-stocks, IV 419, plate 28
Diesel engines, see under Internal combustion
    engine
Diffusion, ideas/information, I **58–84**; see also
    Chronology, Invention, Trade
    Chinese technology, II 768–71
    fire-drills, I 226
    glass manufacture (c.750 B.C.), II 321
    gunpowder uses (c.1300), II 379–80
    metal-working skills, I 573, 576
    microliths, I 496
    Old World and the New, I 82–4, 166
    primitive writing, I 761
    wheel, I 203–4, 207, 209–11, 716
Digging-sticks, I 59, 502, 510
Dikes and polders, II 680–9; III 293–4, 301–3,
    308, 310, 318, 320, 441–2, 452, 468; VII
    907–12; see also Canals, Dams, Drainage
Dilatometers, IV 395–7

Dimity (cotton fabric), III 178
Dioptra instrument, III 591, 609–12, 641
Dioscorea, III 2
Dipsacus fullonum, II 193
Disarmament, VI 282–3
Disc sound recording, VII 1262–3
Discovery, I **58–84**; see also Diffusion, Invention
Diseases; see also Medical technology
    cholera epidemics, IV 510, 624
    geography of, IV 624
    industrial, IV 347, 349, 356; V 237, 253, 515,
        520, 636–7, 831–2
    plague (Black Death), London (1349), IV 505
    occupational, IV 517
    water-borne, V 561–2, 567
Distemper, VI 602; see also Paints
Distillation; see also Chemical subjects
    alcohol manufacture, II 141–4, 375 n.; III 2,
        11–12, 678; V 303–8
    ancient techniques, II 348–9
    apparatus, III 60
    chemical processes, III 679, 686–7; IV 230
    furnace design, III 54
    medieval alchemy, II 739–45
    mercury, III 51
    'pelicans', II 749, 750
    wood products, V 308–11
Diversarum artium schedula (Theophilus), II 63–4,
    351
Dividing-engines, III 641–3; IV 380, 388–95, 409
Divining-rods, II 24
Division of labour, VI 78; see also Management
Docks and harbours; see also Civil engineering,
        Hydraulics, Ports
    Bristol, England, IV 468
    building work, 19th century, IV 452, 464–8; V
        539–41
    Cherbourg, France, IV 465–6
    dredging, II 656–7
    East India Docks, London, IV 467
    Glasgow, Scotland, IV 468
    Greek, II 517–18
    Howland Great Dock, London, IV 467
    Liverpool, England, IV 468
    London Dock Company, IV 492, 494
    London Docks, Wapping, England, IV 467
    machinery, IV 426–7, 467–8, 581–2
    medieval, II 536, 607, 625
    Plymouth, England, IV 466
    Portsmouth, England, IV 424, 580–1
    Ramsgate, England, IV 467
    Roman, II 516–17, 518–21
    Victoria Docks, London, IV 449
    West India Docks, London, IV 467
Dolomite, V 664
Domesday Book, II 35, 91, 93, 611
Domesday Survey (1086), corn-mills, IV 199, 209

Domestic appliances, VII **1126–49**
  electric, VII 1132–5, 1140–1, 1144–8
  gas, VII 1135–6
  heating, VII 1132–5
  hot water, VII 1142–4
  kitchen design, VII 1135, 1137–8
  manual, VII 1129–31
  solid fuel, VII 1138–9
  vacuum cleaners, VII 1129–31
Door-sockets, I 193–4, 306, 472
Dorsal-car sledge, III 140–1
Dowlas (linen cloth), III 178
Dowsing, *see* Divining-rods
Drainage; *see also* Irrigation, Land reclamation
  administrative organizations, III 308, 319, 322
  dikes, II 685–9; III 301–3, 308, 310, 318, 320
  drainage-mills, III 303–7, 317, 321–2
  drying, inland meers, III 302–6, 308, 320
  fen drainage, III 315–22; IV 32
  flood-gates and sluices, III 302–3, 305–6, 312, 320
  land, II 680–5; III **300–23**; IV 22, 32
  London, England, IV 449
  marshes, III 311–13, 318–19
  Paris drains, IV 505, 512–13
  polders, III 301, 304, 306, 310, 320; VII 907–12
  pumps, III 303–5, 307–8, 321–2
  river regulation, III 308–16, 318–20
  scoop-wheel, III 303–6, 321–2
  sea-defences, III 300–2, 308, 316
  steam-power, IV 452
Draught-animals, II 90–2, 94, 514–15, 546–7, 592–3; *see also* Harness, Horses
Draw-plates, III 47
Dredgers, dredging, III 420, 425; IV **629–43**
Drills, drilling; *see also* Augers
  arrow-headed, I 613
  beam-drill, III 114
  bore-holes, mines, IV 67–8
  bow-drills, I 182, 187, 189–90, 224–5, 612–13, 679–80, 688, 689; II 230–1, 631
  coal mining, IV 80
  dentistry, VII 1326
  fire-drill (includes fire-making), I 198, 220, 222–4
  hand-drill, I 189
  Iron Age tool, I 619
  machine-tools, IV 385, 441, plate 22A
  oil industry, V 107–12, 185; VI 377–92
  Mohole project (1958), VII 1456–7
  pneumatic, V 527
  pump-drills, I 190, 191, 225–6; II 231
  rock-boring, IV 71–2
  scoop-drills, II 230–1
  seed-sowing, III 135–6
  strap drills, I 197; IV 381
  thong-drills, I 220, 221, 224–5
  tradesmen's tool (*c*.1500–1850), III 110
  tubular drills, I 191–2
Drinks, II **103–46**; III **1–26**; *see also* Alcohol, Beer, Viticulture, Wine
  alcoholic, III 10–12, 21–2
  ale, III 23
  barley-water, II 119
  beer, II 105, 106, 123, 128, 139–41; III 10–11, 80
  brandy, II 142–3, 380
  chocolate and cocoa, III 5–6
  cider and perry, II 139, 637; III 10
  coffee, III 1, 5–6, 23
  liqueurs, II 142
  milk, II 119, 120, 123; III 23–4
  spirits, III 11, 23
  tea, III 1, 5, 23
  wine, II 106, 112–13, 116, 119, 123, 128–31, 134, 137; III 11
  wine-manufacture, II 131–3, 135–6, 137–8
Drugget (cloth), III 179
Drugs (medicines), II 371–2; III 1; V 300, 317–19; VII 1326–7
Duchy of Cornwall, Stratford corn-mill, IV 209
Dudley Observatory, New York, VII 1174
Durio, I 169
Dust, protection from, III 351
Dutch metal (brass), IV 133
'Duty' of engines, IV 164, 175, 179, 193–4
Dyaks of Borneo, I 117
Dyes, dyestuffs and industry, V **257–83**; *see also* Pigments
  alizarin, V 261, 276–8
  ancient craft, I 248–50, 422–3, 441, 677–8
  aniline dyes, IV 321; V 267–74
  annatto, V 267
  archil (lichen extract), II 366
  basil-wood trees, V 263–6
  chemical industry, II 359, 364–9; III 681, 692–5; IV 231, 233–4, 246, 248–50; VI 505–6, 509, 563–8
  classification, IV 248; V 280
  cochineal and kermes, II 366–7; III 679; V 258–9
  cutch (catechu), V 266
  disperse class, VI 554
  fastness, VI 566–8
  fustic, V 264–5
  indigo, II 364–5; III 679; V 261–3, 279
  indigotin, II 364–5; V 261–2, 278–9
  insect sources, II 366; III 695; V 258–9
  leather dyeing, I 156, 168
  logwood trees, V 263–6
  madder, II 366; V 260–1, 276–7
  manganese sulphate, IV 147
  manufacture, 20th century, VI 505–6, 644–5

mollusc source, II 367
mordants, II 350, 364–9; III 77; IV 231, 233–4
organic sources, I 245–8
phthalocyanine pigments, VI 568
plant sources, II 364–7; III 692–4; V 260–7
purple, II 367
safflower, V 266–7
saffron crocus, V 266
silk dyeing, IV 320–1
'sulphur' dyes, V 279
sulphuric acid use, IV 244–5
tannin sources, II 151 n.
uses, miscellaneous, V 281–2
weighting process, silk, IV 320–1
weld (dyer's rocket), II 365–6; V 265–6
woad, I 247, 249; II 349, 361, 365; V 261–3
Dynamics, III 344–5
Dynamos, IV 679; *see also* Electricity

Earth, alchemical element, IV 214
Earth technology
  calendar, *see* Calendars
  Mohole project (1958), VII 1456–7
  ocean depths, VII **1446–66**
  shape and size, III 504–6, 518–22, 544, 553–4, 613; IV 596, 600, 607
  space exploration, VII 857–9, 861, 866
Earthquake effects, II 423, 424, 470, 522
East India Company, IV 309, 328, 336, 576, 578–9, 590; V 449, 454
Easter, dating, III 578–9
Ecclesiasticus, quoted, IV 666
Echo sounders, VI 344, 350; VII 844, 1449–53, 1458
École Polytechnique, Paris, IV 409, 444, 482, 526
Echinochloa, I 369
Economic depression, VI 15, 102–3, 178; VII 694, 750–1, 777, 872, 967, 976
Economic influences, 20th century, VI **48–76**
Education, V **776–98**; *see also* Universities
  business, VI 83–6, 89
  ceramic industry, V 659
  industrialized societies, VI **138–71**
  instruction methods, IV 664–5
  schools, V 776, 778, 784–5, 790–4, 821, 835; VI 155–7
  sciences, 19th century, V 779–97
  secondary (age 14–18), VI 142–6, 167–9
  technical, V 778–97; VI 140–2, 158–61
  trade unions, VI 105
  universities, V 780–2, 785–92, 796–7, 821; VI 146–52, 153, 160–2, 170
Egypt, I map 6, tables E, F
  alphabet, I 762
  alum production, I 262
  animals, domestication, I 329–30, 340–1, 350, 352

aqueducts, II 670–1
basketry and matting, I 418–19, 420, 422–4, plates 9, 10
boats/ships, I 731, 732–6; II 564
bridges, III 417
building construction, I 304–6, 371–2, 473–84, plate 19
calendar systems, I 121–4, 793–8; III 560
canals/waterways, III 438–9; IV 555
capacity, standards, I 781–2
carpenter's tools, II 229–31
ceramics, I 377, 387, 388, 389; IV 332, 334–5
chariots, I 725–8; II 541
chronology, III 564–5
coffee, III 5
coinage, II 488
copper production, I 564–5
cosmetic arts, I 285–95
dams, I 529
dials, III 595–7
draught animals, I 722–3
drills, drilling, I 190, 192, 193
dyes, I 246–50, 441
early civilization, I 49–55, 139, 161, 168, 496, 509
fabrics, figured, III 194, 196–7
fire-making, I 224–5
fish preservation, IV 45
flax cultivation, II 195–6
food (culinary arts), I 271–85
food plants, I 355–6, 359–62, 365–8, 371, 501–2, 514, 539–42; V 2
food preservation, I 263–5
furniture, II 221–2, 235
gardens, I 523, 543–4
geodetic measurements, V 442
glass-making, II 312–18, 329, 335–6, 340
glassware, II 318–20, 322, 324–5, 327
gold-mining, I 579–81; II 41–2
graphic art, I 209, 238, 240–5
hieroglyphics, I 80, 192, 222–3, 237, 263, 281, 439–40, 747, 752–7, 776
hydraulic engineering, V 542–4
iron production, I 594, 596–7
irrigation, I 46, 526, 528–9, 535–42, 795, plate 20B
ivory work, I 663–4, 666–70, 674, 679
kilns (pottery), I 394, 395
knitting, III 181–3
language, I 102–3, 287, plate 32
lead production, I 583–4
leather-work, II 147–51, 156, 162–4, 172
lighting methods, I 234, 237
mathematics, I 791–3
measurement, I 111, 775–7, 781–2
metal-work, I 627–8, 641–2, 649, 655–60
metallurgy, I 578–9

Egypt (*cont.*)
  milling sieves, II 104
  mummification, I 266–70, 277, 756
  pharos construction, Alexandria, II 521–2
  pigments, I 238–40; II 359
  ploughs, II 85
  potter's wheel, I 199–200
  pottery, *see* ceramics
  printing, III 380
  religion, I 123–4
  roads, II 494
  rope-making, I 452–4
  Roman quarries, II 29
  salt production, I 257–60
  shadow-clocks, III 595–6
  ships, *see* boats
  silver production, I 583–4
  stone quarrying, I 477–9, 526, 568–70
  sugar production, II 372; III 6–7
  surveying, III 502, 510
  syllabic signs, I 747–50
  tapestries, III 195
  textiles, I 414, 427, 431–2, 434–5, 436–41, 449,
    plates II, 12A, 13–16
  tools/weapons, metal, I 504, 604, 606
  town-planning, III 269–70
  transport, I 706, 710, 714, 722–3, 725–8
  viticulture, I 282–4
  water-clocks, III 601
  water-lifting devices, I 523–4, 542; II 671, 677
  water-supply, I 526, 528
  weaving, II 212; III 193
  weights, measurement, I 779–81, 783, plate 31B
  wheeled vehicles, I 209, 211–12
  wood-work, I 685–7, 688–702
  writing, I 102–3
Elasticity
  'Hooke's Law', III 257
  mathematical, III 58–9
  measurement, IV 163
  rubber, V 753
Electrical communications, IV **644–62**; VII
  1220–67
Electrical Engineers, Institution of, London, V
  206
Electricity, VI **284–97**
  A.C. and D.C., IV 83; V 200–2, 228–32
  appliances, VI 285–95; VII 1128
  armatures, V 188–93
  batteries, IV 145, 226–8, 653, 678; V 177–8,
    181, 183, 187, 203–6, 210
  cables and transmission lines, VI 447; VII
    1073–8
  commutators, V 182–3, 185, 189, 191, 193
  consumption, VI 291–7, 314–15, 360, 371, 373,
    386
  distribution, V **208–34**

  dynamos, V 90, 131, 134, 138–9, 183–96, 198,
    200–3, 210–11, 679
  electric motors, IV 409; V 230–3, 301, 641;
    VII 1086–90
  'electrical machines', III 638, 644
  frictional machine, IV 648–9
  galvanometers, IV 655–6
  generation, V **177–207**; VII **1068–90**
  hydro-electric power, VI 196–212; VII 1071–3
  insulators, V 220–3, 663
  lighting, IV 274; V 178, 181–3, 195–7, 208–22;
    VI 286, 292; VII 1082–6
  luminous sphere, evacuated, III 638; IV 648
  measurement, VII 1208
  motive-power, IV 79, 81, 83, 409, 548, 571,
    679
  National Grid, Britain, VI 287, 447; VII 771
  power stations, V 133–4, 139, 197–203; VII
    1068–71
  public ownership, Britain, VI 289
  standardization of systems, Britain, VI 288–9
  steam turbines, VII 1069–71
  storage schemes, VI 212–13
  supply industry, VI 284–8
  switchgear, VI 448; VII 1078–80
  transformers, V 179, 198–200, 230; VII 1081–2
  utilization, V **208–34**
Electrochemical industry, IV 227, 678–9; V 608,
  633; VI 297, 444
Electrochemical telegraphy, IV 652–4
Electrolytic corrosion effect, lead, III 45
Electrolytic effect, copper, IV 580, 592
Electromagnetism, IV 679
Electronic engineering, VII **1091–125**; *see also*
    Calculating, Computers, Radio
  cathode ray tubes, VII 1106–7
  circuit and system theory, VII 1091, 1103–5,
    1111–15
  electronic components, VII 991
  microwave devices, VII 1107–8
  modulation, VII 1109–11, 1238–40
  passive components, VII 1120–3
  photoelectric devices, VII 1105–8
  solid state devices, VII 1116–20
  thermionic valves, VII 1092–105, 1237–8
  transistors, VII 1156
  X-rays, VII 1092
Electroplating, V 608, 633
Electrum, I 623; II 485, 486; IV 146
Elevators, *see* Lifts
Elixir of life, II 732
Embalming, *see* Mummification
Embroidery, III 182
Emery, II 29
Empire Cotton Growing Corporation, VII 838
Employers/employees in engineering, IV 418
Emulsion paints, *see under* Paints

Enamels, enamelling; *see also* Lacquers, Paints,
    Varnishes
  bassetaille, II 461–3, 465, 468, plate 33
  champlevé, II 450 *and* n., 458, 460–1, 462–8
  cloisonné, II 450 *and* n., 458–61, 462, 464–7,
    765, 767
  encrusted, II 462
  glass, II 328, 341–2; III 215–16, 699
  jeweller's, III 698
  metallurgy, III 55, 692, 695
  painted, II 462
  preparation, III 697–8
Enclosures, open fields, *see* Agriculture
Encyclopaedias, IV 674
Encyclopédie (1751–76) (Diderot's), III 204, 226,
    360, 367; IV 674–5
Encyclopédie Méthodique, IV 304
Energy; *see also* Fuels, Natural/Nuclear power
  industrial revolution, IV 148–67
  output comparisons, IV 154–9
  research, IV 677
  thermodynamics, laws of, IV 163–4, 679–81
Engineering; *see also* Civil/Electronic/Production
    engineering
  analysis of problems, scientific, VII 984
  atomic power, VII 991
  creep, metals, VII 984–5
  fatigue, metals, VII 984–5
  fluid mechanics, VII 987, 991
  fortification, III 370–3
  hydraulic power transmission, VII 988
  jet propulsion, VII 990
  land transport revolution, VII 986–7
  light, III 325, 334, 340, 345
  machines, II 629–62; III 324–46; VII 986
  military, IV 442–4
  plastic materials, VII 991
  power sources, II 589–628
  precision, III 629, 641–2, 647
  scientific basis, III 344–5; VII 983–96
  space flight, VII 990
  stress analysis, VII 984, 985, 992
  temperature control, VII 988–91
  timekeepers, mechanical, III 648–75
  training, IV 442–6; VII 983–4
  vibration, VII 987
  welding, VII 986
Engineers, Society of ('The Smeatonian'), IV 443
Engines; *see also* Internal combustion engines,
    Steam-engines, Turbines
  Boulton and Watt, IV 77–9
  Corliss compound, IV 164
  'Cornish', IV 79, 164, 192, 197, 499
  Diesel, V 163–4
  gas-engines, V 157–61, 164, 301
  gas turbines, VII 760, 988
  Kinneil, IV 162

  Maudslay's table-engine, IV 196–7
  Newcomen, IV 78–9, 153, 173–8, 180–1, 406,
    421, 672–3
  oil-engines, V 160–3
  petrol, V 164–76
  rotary, III 84
  Savery's, IV 79, 171–5, 672–3
  Smeaton's, IV 179–81
  Trevithick's, IV 188–90, 192–4, 197, plate 10
  Watt, IV 153, 161, 163, 181–8, 192, 194–5, 197,
    380, 421–2, 452, 673
England; *see also* Britain
  agriculture, III 14–15, 134–49; IV 2–42
  alcoholic spirits, III 11
  aqueducts, II 671, 691
  archery, III 348, 351–2
  architecture and building, IV 443–51, 470–7,
    480–6
  astrolabes, III 609
  brass, III 37; IV 133
  bridges, III 418–20, 425–7, 430; IV 452–62,
    484
  building, general, III 245, 247, 252–3,
    256–67
  building, Norman, II 426–30
  building, Roman, II 408–9, 414–15, 419–20
  building, Saxon, II 425–6
  canals and waterways, III 438, 441, 451–2,
    456–8; IV 462, 563–72
  cartography, III 522; IV 611–12, 614–15,
    620–4, 626
  ceramics, II 289, 293, 295–7, 300, 305–6; IV
    328–9, 331, 336–56; VI 622–8
  chemical industry, IV 242–5, 250
  chronometers, IV 410–15
  clocks, III 553, 663–4, 674
  coal, II 68; III 73–4, 685–6, 688–9; IV 79–87,
    91
  cobalt, IV 146
  coinage, III 46
  copper, IV 126, 129
  corn-mills, IV 199–202, 209–12
  cotton industry, II 199; III 159–60
  cropping-machines, III 174
  Crystal Palace, *see under* Exhibitions
  date, New Year's Day, III 580
  diet, II 123–8; III 1, 3, 5–6, 8, 24
  drainage systems, II 679–80; III 300, 308–9,
    315–22; IV 506–18
  dredging, IV 631, 640–1
  dyeing industry, III 694–5; IV 247–8, 320
  engineering, IV 443–5
  fish industry, IV 48–54
  flax, III 158
  fulling cloth, III 171
  gas industry, IV 263–74
  gig-mills, III 172

England (*cont.*)

glass manufacture, II 323, 326, 443–4; III 86, 206–9, 216–23, 225–34, 237–42, 700, 702–3; IV 360–2, 367–9, 371–6; VI 588

Great Exhibition (1851), *see under* Exhibitions

industrial revolution, IV 148, 328

instrument-making, III 585–6, 621; IV 382, 387–8, 390, 393–9, 402, 404–6

iron industry, II 61, 70, 76–7; III 31; IV 100–6, 113–14

japanning and lacquering, III 696–7

kilns, pottery, II 295–7, 300

knitting, III 184–6

land reclamation, III 300, 308–9, 315–22

lead production, IV 123

leather-work, II 155, 157, 169, 182–4

lighthouses, II 523; *see also* Lighthouses

machine-tools, IV 418–37, 440–1

marine works, IV 466–9

metal-work, II 450, 451–2, 454–5

millstones, IV 212

mining, II 9–10, 12–13

navigation, III 550–7

nickel, IV 146

paper-making, III 416

ploughs and ploughing, II 86–7, 89–90, 92–3

pottery, *see* ceramics

printing, III 392, 399, 409–10

quarrying, II 31–8

railways, III 85

roads and streets, II 503, 505–6, 525, 532–3; IV 524–5, 527–8, 530–2, 535–6, 538–46

science, impact on industry, III 679–83

sewerage, IV 506–18

ships, II 576–9; III 475, 477–88, 490–2, 494, 496–9; IV 574–94

silk industry, III 196; IV 310–11, 315, 320

soap, III 703

spinning-wheels, III 161–2

steam, III 687

steam-engines, IV 172–97

sun-dials, III 596–7

surveys, III 516, 541–3; IV 402, 443, 600–4, 607–8, 610

telegraphy, IV 645–9, 651–2, 654–60

textile industry, III 163, 173; IV 279, 282, 284, 286–7, 291–9, 304–5, 310–11, 315

tin-mines, IV 174

tinplate, III 689; IV 104

tobacco, III 8–10

town-planning, III 277–9, 284, 295–8

tunnels, IV 462–4

velvet, III 205

viticulture, II 138

wall-linings and panels, II 244

wars, III 349, 369

water-supply, IV 491–3, 499–503

watermills, II 610–11; IV 151–3, 156, 199–212

weapons and armour, II 721–2

weaving looms, III 164–5, 167–70

whaling, IV 58

wind-power, IV 156–9

windmills, III 89, 91, 93–107

wood uses, III 685–9

wool, II 192, 193; III 151–3

zinc, IV 130–2

Engraving

copper-plate, III 586; IV 355, 625–6

craft, III 586, 623, 625

dry-point method, III 406–8

glass, IV 377

instrument-making, III 623–5

intaglio, IV 625

printing process, III 405–10

steel-plate, IV 626

tools, III 407–8, 624

Ephemerides, III 503, 535, 547, 552

Epinetron, II 201

Equatorial telescopes, IV 361, 399

Equatorium (planetary device), III 616–18

Ertebølle people, Denmark, I 303, 495–6, 499, table E

Erysimum cheiranthoides, I 370

Escalators, VII 941

Eskimo people

engravings, I 41, 180, 336

fire-making, I 219, 222–6, 229

food-gatherers, I 41–2, 180–5, 336

lighting/heating, I 229, 235–6

pelts, treatment, II 148–9

reindeer, domestication, I 336–7

skin-boats, I 736; II 582

tools/weapons, I 62–3, 69, 156–63, 166, 180, 182, 190

Esparto-grass (halfa), I 450; III 415; V 713; VI 607, 609

Etamine (cloth), III 179

Etching, *see* Engraving, Printing

Ethanol, VI 541–2

Ethylene, VI 56, 546

Etruscans, II table I lv

alphabet, evidence, I 769–70

architecture, IV 404, 408, 410, 413

chariots, II 542–3

furniture, II 223–4, 235, 237

gold bowl, I 657, 658

lathe (*c.*700 B.C.), I 518

leather craft, II 167

metal-work, II 470

pottery, I 518; II 269–71

sewers design, II 530–1

statuary, II 275–6

Euphorbia, I 172

Europe; *see also* individual countries
  aeronautics, V 392–3, 400, 406, 411–12
  building, V 466, 477, 485, 491, 493
  cartography and surveying, V 440, 442
  ceramics, V 659
  chemicals, V 290, 292, 299
  cinematography, V 734, 743, 744
  Coal and Steel Community, VI 25
  Common Market, VI 25
  farming and food, V 2–4, 8, 10, 19–21, 24, 51
  metals, V 60, 606, 608, 625
  photography, V 721–2
  population, V 1, 26
  printing, V 688, 694
  railways, V 328, 330, 335, 342, 344
  rubber, V 754–5, 765
  textiles, V 569, 572, 588, 592
  wars, their influence, VI 8–26
  water-supply, V 554–5, 559, 564–6
Exhibitions
  Aeronautical (1868), Crystal Palace, V 408
  Berlin (1879), V 346
  Caxton Celebration (1877), V 686
  Centennial, Philadelphia (1876), V 131
  Columbian, Chicago (1893), V 64; VII 1133
  Earls Court, London, VII 1129
  Frankfurt (1891), V 232
  Ghent, Flower (1859), V 116
  Glasgow (1888), V 623
  Great Exhibition (1851), Hyde Park, V 817–18;
    see also entries below
  Amalgamated Soc. of Engineers, V 800
  autoclaves, V 39
  brickwork, iron reinforced, IV 451
  ceramic machines, V 669
  ceramic-ware, V 662
  combing-machine, IV 297
  Crystal Palace, IV 476–7; V 474
  education, V 778, 780–1, 782–7, 789
  food management, V 27, 39
  gold extraction method, IV 141–2
  housing (1850s), V 829
  hydraulic engineering, V 545
  machine-tools, IV 431, 440; V 637
  matches, V 253
  photography, V 727
  Portland cement, V 483
  printing, V 683, 690, 698, 709
  reaping machines, IV 7–8; V 11
  Great Exhibition (1862), VI 552
  International Exhibition (1862), IV 432; V 637,
    643–5, 650–2, 710
  Milan (1959), VI 663
  Paris series (1854), V 669; (1855), V 472, 488;
    (1867), IV 631; V 108, 131, 789, 791; (1881),
    V 192; (1889), V 476, 609; (1900), V 640, 663
  Vienna (1873), V 610, 641

Explosives, II 378; III 374; IV 246; V **284–98**;
  see also Pyrotechnics
  blasting, V 292–4
  cartridges, V 293
  commercial, VI 549
  detonators, V 296–7; VI 549
  dynamite, V 292
  gun-cotton and nitro-cotton, V 284, 290–2,
    294–6
  gunpowder, II 377–82, 771; V 284, 285, 292;
    see also Gunpowder
  incendiaries, VII 816
  mercury fulminate, V 296–7
  military, VI 547–50
  nitroglycerine, V 285–90, 292–315, 296, plate
    11C
  Oxyliquit, V 294 n. 1
  picric acid, V 296; VI 548
  priming, 17th century, III 355 *and* n. 1
  propellants, V 294–6; VII 859–66
  saltpetre, II, 377, 379–82; IV 235
Exposure-meters, VII 1284–5
Eyes, artificial, III 699

Fabrics, figured, III **187–205**
Fabric finishes, VI 671–4
Factory Inspectorate, VI 592
Factory-ships, VI 357
Factory system, III 151–2
Fagopyrum esculentum (buckwheat), II 125; III 5
Faience, meaning, II 311 n.
Falcon clan, rulers of Egypt, I 49–50
Fan, winnowing, II 97–8
Farm-tools, II **81–102**; III **134–50**; *see also*
  Agriculture
Fatigue, metal, VII 985
Fats, ancient cosmetics, I 289, 290
'Feed-back' furnace mechanism, III 680
Felt fabric, III 186
Fermentation, I 275–7, 284; V 302–8; *see also*
  Brewing, Chemical subjects, Distilling
Ferrous sulphate in sewage, IV 519
Fertilizers, agricultural, IV 37; VI 323, 325–7,
  528–30
  ammonium compounds, VI 500, 511, 528–9
  basic slag, V 16, 60
  bones, IV 254; V 10, 14; VI 529
  chalk, II 34–5, 37
  Chilean nitrate, IV 20; V 10, 14, 16, 254; VI
    521–2
  farming practices, V 17
  guano, IV 20, 254; V 4, 10, 14, 16; VI 529
  industrial waste, IV 20; V 15
  legislation, V 16
  lime, V 4, 9
  manures (animal), IV 13–14, 15–16, 18, 31,
    36–7

Fertilizers, agricultural (*cont.*)
  manures (plant), IV 15, 16, 20; V 17
  marling, III 15; IV 19–20, 37
  mixed, V 254; VI 500, 530
  Netherlands, 16th century, III 13
  nitrogen compounds, V 255; VI 325–7, 521–2,
    528–9
  phosphatic, IV 20, 246; V 16, 254–5; VI 500,
    529–30; *see also* superphosphate
  plant-nutrition, IV 37–9; VI 330–3
  potassium salts, V 255; VI 530
  production, 20th century, VI 528–30
  refuse/sewage, IV 506, 517–18, 519
  sodium nitrate, V 255
  superphosphate, IV 20, 253–4; V 8, 14, 16, 17,
    19, 254–5; VI 529–30
Feuerwekbuch (1954), II 379–80, 727; III 34
Fezzan sheep, I 345
Fibre plants, I 355–6; *see also* Cotton, Flax,
    Hemp
  cultivation, I 372–4, 522
  rope-making, I 453–4
  spinning, preparation, I 424
  textiles, I 447–51
Fibre Research Institute, VI 657
Fibreglass, VI 589; VII 937
Fibres, animal, VI 299
Fibres, man-made, VI 299, 501–2, 560–3, 648–51,
    659; *see also* Textile industries
Ficus sycomorus, I 371
Figure-harness, III 187–95, 197–8, 201–3
Files, I 613, 620; II 60, 230, 395; III 110, 343,
    352, 365
Filters, II 674; V 301, 561–7, 668–9
Fine ivory-/metal-/wood-work, *see* Ivory-work,
    Metal-work, Wood-work
Finland, water-supply, IV 497
Fire, I **216–37**
  alchemical element, IV 214
  bow-drill, fire-maker, I 224–5
  discovery, I 25, 28, 58, 216–18
  fire-drill, fire-maker, I 198, 220, 222–4, 226
  fire-fighting equipment,, II 184–5; IV 638
  fire-piston, fire-maker, I 226–8
  fire-plough, fire-maker, I 221–2
  fire-saw, fire-maker, I 220–2
  firing clays, *see* Ceramics
  glass-making, II 312
  metallurgic uses, I 65, 576–9, 594, 624–5, 627,
    633–4
  military pyrotechnics, II **374–82**
  mining uses, I 565–6, 567; II 8
  percussion fire-making, I 218–20
  pottery-making, I 376, 382–3, 391–7
  prevention, VI 673; VII 746, 941, 943
  protection, *see* Building construction
  pump-drill, I 225–6

    thong-drill, I 220, 221, 224–5
    transfer and transport, I 229–30
    weapon of war, II 374–82
    wood-friction, fire-making, I 229–30
Fire-engines, IV 638
Fire-grates, III 20
Firearms, guns; *see also* Artillery
  air-guns, III 359
  caliver, III 354
  cannon, III 360–9
  casting, III 38–9
  field, III 347–50
  firelock, III 354
  flint-lock, III 347, 354, 356
  gun-barrels, III 356–9; IV 271
  gun-locks, III 355–6, 360
  hand-guns, III 347–8, 351–60, 362 n.
  harquebus, III 352, 354, 367
  invention, II 378, 772
  matchlock, III 354–6, plate 17
  Middle Ages, II 75, 726–8
  military strategy (1500–1700), III 347–50
  muskets, IV 437–8
  muzzle-loading guns, III 347, 355
  muzzle-velocity, III 356, 360
  naval guns, III 350, 362–3; IV 575–6, 578,
    583–4, 587–9
  pistols, III 348–9, 355, 358–9, 362 n.
  revolvers (Colt), IV 439
  rifles, III 358
  Royal Small Arms Commission (1853), IV 440
  sporting guns, III 354, 358, 360
  wheel-lock, III 355, 358, plate 17
Fireworks, II 377–9; *see also* Explosives
Fish, fishing, VI **342–58**
  ancient man, I **154–86**
  blow-gun, I 164
  cod, IV 49–50
  conservation, VI 354–6
  echo-sounders, VI 344
  flares, I 119
  fleeting, IV 53
  fresh fish, IV 51–2
  harpoons, I 158, 160, 176–7
  herrings, IV 45–8
  hooks, I 166–8, 176–7, 498–9, 515–16, 732
  medieval fish supply, IV 44
  navigational aids, VI 349–50
  nets, I 166, 176–7, 180, 265; VI 345–9
  poison method, I 173
  preservation, IV **44–54**; VI 350–7
  refrigeration, IV 53–4; VI 350–4
  steam propulsion, VI 343–5
  smoked-, IV 48–9, 54
  stockfish, IV 48–50
  technology application, VI 343, 348–9
  traps, I 168, 177; II 596

trawling, IV 52; VI 343–8
whaling, *see* Whales, whaling
Flails, I 70; II 97, 98, 106, 164; III 134, 137
Flame resistance, VI 673
Flanders
   agriculture, IV 13–19, 29, 31–6
   building in brickwork, IV 446
   canals and waterways, III 441 n. 2, 452–4; IV 558
   diet changes, III 6
   dyeing industry, III 694–5
   Gregorian calendar, adoption, III 579
   stained-glass windows, III 241
   surveys, IV 606
   weaving, III 159
   windmills, III 93, 95, 97
Flannel (cloth), III 179
Flax/linseed; *see also* Linen
   cultivation, I 540; II 195–7; III 156
   early history, I 372–3
   fibres, preparation, II 196; III 157–8
   linseed oil, I 288, 358; III 381, 697; VI 595, 597–8
   retting process, III 156–7
   spinning machinery, III 160, 163–4
   textiles, I 448–9; II 195–7; III 156–8; IV 291–4, 315; V 580
Flint; *see also* Stone, tools/weapons
   book-binder's, II 180
   building stone, II 35
   ceramics, from 17th century, IV 347–9
   fire-making tools, I 218–20
   mines, I 558–60
   tools and weapons, I 561, 562
Flour and bread, *see under* Food, Milling
Fluid power, VII 1060–7; *see also* Hydraulics
Fluxes
   glass-making, III 206–7, 209–10, 217, 226–8
   metallurgy, III 39, 41, 48, 62–3
   welding, V 631
Flyers, spinning-wheel, origin, II 203–5
Flying clubs, VII 804, 806
Fodder (animal feed), III 5, 12–13; VI 307–12
Food, food technology, II 103–46; III 1–26; VII 1399–445; *see also* Agriculture, Fish, Food-gatherers/production
   additives, VII 1409, 1415
   adulteration, II 128; III 20–2; VII 1401–3
   agriculture, beginnings, I 42–4, 46–7, 79, 340–1, 349–50, 353–4, 362–3, 374
   analysis, III 20–2, 24–5
   ancient man, I 154–86
   beverages, III 5–7; V 3, 21–2
   biscuits, V 36–7; VII 1433, 1438–9; *see also* flour
   bread, II 103, 105, 118, 120, 126–7, 140; III 23; V 28–9, 31; VII 1433–6; *see also* flour

buckwheat, III 5
butter, II 127; III 14, 23–4
canning, V 28–9, 38–45, plates 4B, C; VII 1403, 1405, 1418–21
cheese, III 23–4
cooking, I 31, 41, 231, 270–5, 401; III 19–20
dairy products, V 13, 33–6, 49
dehydration processes, V 29, 51; VII 1406, 1413–17
diets, Greek, II 128–21
diets, medieval, II 123–8
diets, Renaissance changes, III 1, 7, 19, 22–5
diets, Roman, II 128–21
digestion discoveries, III 24–5
eating habits, III 19–20
eggs, III 23; V 49, 51
energy values, VI 309–10
fish, II 119, 120, 123–4; III 22–4; V 43, 49, 51
fish preservation, IV 44–54; VI 350–7
flour, II 103–4, 118, 120–1, 594, 599–601, 605; III 15–19; VII 1406, 1429–36; *see also under* Milling
fruit, III 23–4; V 49–51
groats, II 118–19
honey, II 127, 138, 139; III 6
jams, V 32, 37–8
maize, III 1, 3–4, 13
meat, II 119, 120, 123–4; III 12, 23–4; V 31–2, 47–51
milk, *see* Milk industry
milling, II 104, 107–12; III 15–19; IV 200–1; V 29–31; VII 1406, 1429–33
nutrition, 20th century, VII 1401–9
pasteurizing processes, V 33; VII 1417–22
potatoes, III 1–3, 12–13, 23
pounding, II 106–7
preservation, I 173, 175, 184–5, 263–4; III 22, 679; V 26–52; VII 1411–28
pressing processes, II 112–18
primitive societies, I 40–4, 64
refrigeration, V 8, 28–9, 32–3, 45–51; VII 1400–1, 1422–5
research, VII 1403–9
rice, III 1, 4–5
salting and smoking, I 258, 259, 263–6; V 31–2
sterilization, VII 1417–22, 1428
storage, early civilizations, I 42, 43, 47, 50, 52, 263, 537
sugar, III 1, 6–8, 23
sweetening agents, II 127, 139, 372
turkeys, III 1
vegetables, II 120, 124; III 19, 23–4; V 49–51
vitamins, VI 300–1, 310
Food and Agriculture Organization of United Nations (FAO), VI 318, 342

Food-gatherers, I **154–86**; *see also* Agriculture,
 Food headings
 Arctic regions, I 179–83
 cultivation, beginnings, I 353–4, table E
 economies/social organization, I 184–6
 foraging, I 154–5
 forest lands, northern, I 175–9
 grasslands, high latitude, I 174–5
 honey-collectors, I 275
 hunting and fishing, I 155–68, 336–8, 350,
  498–9, 510–12, 515–16, 518
 Mediterranean regions, I 173–4
 primitive economies, I 40–2, 154, 495–6
 scrub and desert, continental, I 172–3
 techniques, diversity, I 183–4
 tropical regions, I 168–72
Food-production, V **1–25**; *see also* Agriculture,
 Food-gatherers
 animals, 20th century, VI **298–318**
 cereals, cultivation, I 362–70, 522, 550
 crops, 20th century, VI 318–22, 327–34
 dairy products, V 2, 7–9, 13–14, 18–21, 32–6
 fish, 20th century, VI 342–3, 354–6
 fruit, cultivation, I 357–8, 551–4
 industry, 20th century, VII 1473–5
 market influences, VI 338–40
 manufacture, VII 1401, 1406, 1411, 1429–45
 margarine, V 13 and n. 1
 meat, V 2, 4, 6–9, 14, 18, 27
 milk, *see* Milk industry
 mustard, cultivation, I 370–1
 nuts and oil-seeds, I 358–60
 oils, vegetable, I 358–60; V 20, 23–4
 pulses, cultivation, I 360–2
 sugar, I 370–1; III 7–8; IV 230, 232; V 3, 9,
  17, 20–1, 23
 sugar-beet, IV 28–9, 40
 vegetables, cultivation, I 353–6
Footwear, I 706
 boots, II 167, 175–7
 buskins, II 167, 175
 coloured, II 168, 177
 monodermon, II 167
 sandals, I 422; II 150–1, 160, 162–3, 166–8
 shoes, II 166–8, 170, 175–7
Forces, structural-theory, III 432–3
Forel (parchment), II 180
Forges; *see also* Furnaces
 armourer's workshop, III 352
 assaying, III 52–3
 cannon-balls, casting, III 34
 charcoal fuel, III 684
 coal fuel, III 78–9
 hammer-, III 32
 smith's, II 59–60; III 51–2, 62–3, 350
Forks
 agricultural, II 98, 99; III 134, 137

glass-making, II 331
 table-ware, II 126, 127; III 20
Fortifications; *see also* Military technology
 Amsterdam (from 1204), III 293–5
 castles, III 369–70
 cities, from 4th century B.C., II 701–3, plate
  40A
 French towns, *c.* 15th century, III 290–1
 Hadrian's Wall, II plate 40B
 literature, medieval, III 256, 370
 medieval towns, III 282–4, 369–73
 Roman period, II 717–19, plate 41; III 276–9,
  286–7
Fosse Dyke, II 680
Fossil fuels, VI **172–94**; *see also* Coal, Petroleum
 balance of utilization, VI 172–4
 energy provision, VI 172–7; VII 1481–2
 requirements (1950s), VI 176–7
Fossilia, III 28 n. 1
France; *see also* Europe
 Academicians, III 544
 aeronautics, V 392–3, 395–402, 404, 407–10
 agriculture, III 12–13; IV 13, 16, 27–34, 36,
  40–1; V 14, 16–17
 aircraft, VII 791–6, 800–3, 807, 816, 821–4,
  841, 1011–12
 architecture, IV 442–4, 446, 449–51, 470, 472,
  474–5, 479–80, 482; VII 935–6
 asphalt, discovery, IV 539
 bridges, III 418, 420, 423, 425, 430–1; IV
  452–3; V 502, 511–12
 building, III 247, 252–7, 262; V 468–9, 471–2,
  474–6, 483–4, 487–96
 canals and waterways, III 86, 439, 441, 448,
  450, 459–68; IV 452, 462, 548, 550, 555–7
 cartography, III 544, 554; IV 596–600, 611–14,
  621, 625–6; V 440–2, 448
 ceramics, III 678–9; IV 329, 331, 337–40,
  342–3; V 659–66, 669, 670
 chemical industry, IV 238–9; V 241–3, 245,
  248–9, 252–3, 302, 312, 315
 chemistry, discoveries, IV 218–21, 222–3
 chronometers, IV 410, 414–15
 cinematography, V 734, 736–43, 745–6, 748–50
 civil engineering, VII 879
 clocks, III 651, 674
 coal, II 68; III 72, 86
 coinage, II 489–90; III 341
 colour photography, VII 1296
 computers, VII 1151, 1159, 1162–3
 copper, IV 129
 diet changes, III 2, 5–6
 drainage systems, II 681; IV 505, 512–13
 dredging, IV 631, 633–5, 638, 641
 dye industry, III 695; IV 248–9, 251, 321
 dyestuffs, V 260, 272, 282
 electrical communications, VII 1238–9

electricity, V 179–83, 188–93, 196, 204, 206, 210–13, 232

engines, steam-, V 125, 137, 140, 142–3, 149, 152, 155, 157–8, 162, 169, 171–2, 174

explosives, V 284, 289, 296

fish industry, IV 48–50

flax, III 156

food technology, V 31, 38–9, 45, 48, 304; VII 1399–1400, 1426

fortifications, III 372–3

fustian industry, III 159

gas-lighting, IV 258, 259–64

glass industry, III 86, 206–8, 214–15, 237–41, 677; IV 360, 364 n., 367–9, 371, 375; V 674, 678

Gregorian calendar, adoption, III 579

hydraulic engineering, V 522, 526, 528–9, 531–2, 538–9, 546

hydrography and oceanography, IV 615, 618–19

instruments, scientific, III 585, 621; IV 384, 388, 391–4, 396–7, 399, 404–9

iron industry, II 70, 76; III 31 n. 1; IV 101–2, 105, 111–12, 115

knitting, III 183, 185–6

lacquering and japanning, III 696–7

land reclamation, III 300, 309, 318–19, 322; see also drainage systems

lead production, IV 123

machine-tools, IV 437–8; V 638, 641

marine works, IV 465

metal-work, II 474–5

metals, II 62–3; V 57, 59–60, 62–3, 65–6, 70, 82, 91–3, 608–9, 622, 627

millstones, IV 212

natural gas, VI 407

navigation and aids, III 553, 557; V 458 n. 1

oil and petroleum, V 106–8, 113–14, 116–19

paper-making, III 415–16

photography, V 718–20, 722, 724–5; VII 1296

platinum metals, IV 144

population (1800–1950), V 1, 26 n. 1, 823–4

pottery, see ceramics

printing, III 384, 392, 397–8; V 685–6, 689, 691, 697–9, 701, 706–8

railways, V 61, 322, 325–6, 330–2, 336, 340, 341, 344, 345; VII 763, 765, 771–3, 776

road-transport, VII 716–20, 722, 733

roads and streets, II 524–6, 532; IV 521–8, 535–6, 538–40, 542, 544

rubber, V 754–6, 765, 773

sewerage, IV 505, 512–13

ships and ship-building, III 478, 483, 485, 496; IV 576–9, 586, 588, 593; V 142, 189, 353, 386, 387

silk, medieval fragments, III 196, 202

silk industry, III 165; IV 309–11, 319–21, 324

soap, III 703

spinning-wheels, III 161

steam-navigation, IV 548

surveying, III 540, 543, 554; IV 402, 442, 600–7, 609–10, 613, 625; V 440–2, 448

telegraphy, IV 644–7, 650–2, 655, 660; V 225

textile industry, III 170–5, 178; IV 247–8, 279, 282, 292, 295, 297–300, 304–5, 311, 324

textiles, V 571, 574, 576–7, 585, 587–8, 591, 602, 604

tinplate, III 689; IV 104

tobacco, III 8, 10

town-planning, III 283, 290–3

tunnels, IV 462; V 502, 511–12

vehicles, V 414, 418–19, 423–5, 429–30, 433–6; VII 715, 718, 719

water-power, IV 154–5

water-supply, II 691; IV 495, 500

water-wheels, III 328–9

watermills, II 598–9, 608–10, 639

weapons and armour, II 720

weaving-looms, III 164–9

windmills, III 89–93, 95–7, 99, 101–5

wine production, II 137–8

wool, III 152, 177

Friction; see also Lubricants
  -band, spinning-wheel, III 161
  early machines, I 198; II 652–3
  electrical machine, III 638
  hydrodynamic effects, V 386–90
  laws of, IV 163; V 118

Frit, II 317–18

Froe (rending-axe), III 119

Fruit, see under Agriculture

Fuegians, South America, I 184
  canoes, I 731
  glass, I 563
  dwellings, I 300
  fire-making, I 219
  fire transport method, I 230
  weapons, I 158, 159, 178

Fuels; see also Coal, Energy, Petroleum
  atomic energy, VI 248–56
  early man, I 228–30, 277, 382, 391–2, 576
  economy, III 686

Fuller's earth, III 169–71

Fulling, II 193, 214–17, 355

Fumifugium (1661), John Evelyn, III 296

Fundamental Research Centres, see Research

Furnaces, III frontispiece, plate 3; see also Bellows, Forges, Kilns
  alchemical, II 739, 740, 741–5, 750
  automatic (athanors), II 739, 740, 743; III 53, 680
  automatic design, III 680
  'beehive', III 685–6
  Bessemer converter, V 53–7, 606; VI 480–4

Furnaces (*cont.*)
  blast-furnaces, II 62, 66, 70, 655; III 30, 31–4, 50, 52, 79–81, 363; IV 109–14
  bloomery-fire type, II 45, 50–1, 56, 71
  blowing-apparatus, IV 103–4, 111
  catalan type, II 71–3
  ceramics, IV 337–8, 339, 344
  chemical, II 347–8; III 66–7; IV 230–1, 236–7, 240–1; V 242, 246, 249, 253–4
  coal-burning, III 78–9, 86, 221
  Corsican type, II 71, 72
  crucible types, II 45–6, 64; V 609; III 37–8, 55; IV 609
  electric, VI 489–91, 492, 577
  furnace-gas, utilization, IV 113
  gas-making, IV 262, 263–7
  glass-making, II 329–30, 332, 358; III 78, 86, 207–14, 217, 220–1, 226, 238–9, 699–700; IV 364, 370; V 672, 680–1
  hot-blast invention, IV 109–10, 253
  iron and steel, II 56–7, 72–5; IV 99–106, 111–15, 117, 183, 253; V 53–60, 66–70
  linings, refactory, III 50–1; V 664–5; VI 476–7, 484, 630–7
  metallurgy, medieval, II 66–9
  metals, non-ferrous, IV 121–9, 131–2, 133–6, 139, 141; V 78–83, 89, 93–4, 609, 615
  muffle type, III 600, 698
  open-hearth, V 57–60, 606; VI 462, 485–8
  open-shaft type, III 49
  osmund type, II 70, 72
  reducing, III 49–51; VI 489
  reduction, IV 131; VI 489
  refractory materials, *see* Refractory . . .
  'resistance' type, VI 489
  reverberatory, III 27, 39, 49, 54–5, 57, 80, 210; IV 106, 122, 126
  Roman times, II 45–6, 49–51, 72
  Saigerhütten (smelting-works), II 78
  'Scottish' type, IV 111
  self-stoking, III 53, 55
  shaft-furnace, II 50, 56, 73
  Siemens-Martin type, IV 678
  Snyder 'coffee pot lid', VI 491
  solar-, III 55
  Stückofen, II 72–5
  Swedish Lancashire hearth, IV 115
  Swindell-Drysler, VI 491
  tank type, VI 575
  thermostatic controls, III 679–80
  Tysland-Hole type, VI 490
  wind-furnaces, III 51–2
Furniture, I plates 23–5; II 221–58, plates 13–15; *see also* Carpentry, Joinery
  ambries, II 241, 248
  arks, II 254
  armchairs, II 223

beds, I 689, 698–9, 700; II 221, 235, 248
bookcases, II 243
caskets, I 686, 687, 690–4, 697
chairs, I 689, 698–700; II 178–9, 221–2, 248–9, 252–4, 256–7
chests, I 696–7; II 170–1, 180, 222, 237, 241, 247, 253
coffers, II 170–1, 255–6
coffins, I 686
couches, II 221, 223
cupboards, II 225, 241, 246
drawers, II 240–1
fine wood-work, I **684–703**
Gothic, II 240–1, 245
Greek and Roman, II 221–6
ivory-work, I 670, 673, 676–8, 679
leather, II 171, 178–9
medieval, II 241–3, plate 11
metal, II 235–6, plate 10
painted, II 247–9, 363
panelling, II 179–80, 243–5, 247–8, 363
post-Roman, II **240–58**
Roman period, II **221–39**
side-boards, II 224–5, 249
stick-furniture, II 251–3
stools, II 221, 249, 251–2
tables, II 224, 249, 250–1
theatre-seats, II 234
toilet caskets, I 686, 687
thrones, II 222–3, 226, 235
veneers, II 227
wicker-work, II 237, 253–4
woods, kinds used, II 226–7, 228
Fusee, cutting machine, IV 384, 420, 425, plate 27B
Fustian (cloth), II 199; III 159, 171, 179

Galena (lead sulphide)
  alchemical classification, II 737
  eye paint, ancient Egypt, I 293, 583
  glaze ingredient, IV 332–3, 344, 346
  Greek mining, II 1–2
  lead source, I 583, 584; II 1
  smelting, I 584; II 43–4, 50; III 122–3
Galeopsis, I 370
Gardens
  Babylonian, I 550–4
  Egyptian, I 523, 543–4
  Kew, Botanic Gardens, V 774; VI 338
  Paris, Botanic Gardens, III 9
Garments, *see* Clothing
Gas industry (coal-gas), IV 258–76; VI 181–2; *see also* Natural gas
  'barrel', IV 271
  burners, IV 259, 262–3, 266, 268, 271–3
  by-products, IV 261–2, 271
  cookers, IV 264, 273–4; VII 1136–8

Cupar Gas Company, IV 271
domestic appliances, VII 1127, 1128
engines, *see* Engines
fires, heating, IV 264, 272
furnaces, V 58–9
Gas Light and Coke Company (*once* National
    Light and Heat Company), IV 252, 268–70
gas-making plant, IV 261–2, 266, 268–9, 272,
    plates 12, 13
gas-rings, IV 273
geysers (hot water), IV 273–4; VII 1142–4
Glasgow Gas Works, IV 253
hydraulic seal, IV 266–7
lighting, IV 246, 252–3, 258–74, 674; V 209
mains distribution, IV 267–8, 271; VI 182
mantle (incandescent), IV 274
meters, IV 270–1
origin of word, II 745
pipes, IV 259–63, 265, 267, 270–1
purification, IV 261, 266–71
refrigerators, VII 1146
retorts, IV 259, 261, 263, 265–72, plates 13B,
    14A
Gas-turbine engines, VI 438, 443; VII 760, 998,
    1016–24
Gases; *see also* Airships, Chemistry
    alchemical concept, II 745; IV 215
    blast-furnaces, V 58
    early experiments, IV 215–21
    fire-damp, mines, IV 90–1, 93, 95–7
    furnace techniques, IV 230
    hydrogen, IV 216, 221, 224, 227, 259–60
    laws, IV 163, 216–17
    methane, IV 215
    oxygen, IV 218, 220, 223–4, 226–7
    physical properties, IV 223
    sewers, occurrence, IV 516
Gasoline ('gas'), *see* Petroleum oil
'Gasometer' balances, IV 406–8
Gauges, carpenter's, III 112–13
Gay-Lussac tower, V 245
Gears
    bevel gearing, IV 381–2, 428, 429
    clock mechanisms, III 345, **648–75**; IV 391,
        412–15
    cutting-tools, IV 391; V 655–6; VII 1039
    gear-boxes, V 640–1; VI 431–2; VII 719–20
    gear-wheels, IV 420–2, 425, 429
    Greek machines, 4th century B.C., II 632, 635
    lens grinding tool (1671), III 235
    machines, 16th century, III 325, 327, 328, 331,
        359
    mill-mechanisms, II 648, 653–5; IV 207, 208
    motor-vehicles, VI 431–2
    rack-and-pinion, IV 207
    ratchet devices, II 646–7
    screw reduction-gearing, II 647

    sun-and-planet, IV 185, 187
    worm-gears, III 331; IV 429
Gems, *see* Jewelry
Genetic studies, VI 304–7, 328
Genista tinctoria, III 694
Geodesy, instruments, IV 400–3; V 441–4
Geographike syntaxis, Ptolemy, III 502, 509, 530
Geography, III 502–3, 505, 507–9, 517
Geology; *see also* Coal/metal mining, Petroleum
    Geological Society (1807), IV 491
    maps, IV 621–2
    sea-floor, VII 1453–66
    Survey of Great Britain (1835), IV 491, 622
    Survey of India, V 112
    Survey of U.S.A., V 44
Geometry; *see also* Mathematics
    cartography/surveying, III 507–13, 520–2, 537
    dividing engines, design, IV 388–90
    Egyptian, II 791–2
    Greek, I 126; II 632–3, 675; III 502, 520
    navigation, III 526–9
    optics, III 633
Geoponica, I 405–6, 412 (59)
Geothermal power, VI 219–21
Gerard's *Herball* (1597), III 1–3
German Chemical Society, V 270
Germany; *see also* Europe
    aeronautics, V 397, 399, 402–6, 412
    agriculture, III 12; IV 13, 16, 19, 22–9, 31,
        33–4, 36–7, 40–1; V 14, 16, 304
    aircraft, VI 17; VII 17, 793–803, 806–7,
        811–21, 828–9, 840–9, 853, 1019–20
    architecture, VII 934, 954–5
    asphalt, IV 539–40
    astronomical instruments, IV 399, 401
    atomic energy research, VI 226, 268–9
    beer production, II 140–1
    brass-making, III 37
    bridges, V 502, 514
    building, V 472, 482, 484, 489, 491, 493,
        495–6; VII 934, 954–5
    canals/waterways, III 441, 443, 449–50, 454–5;
        IV 548, 550–1, 554–6, 560–2
    cartography, III 534–9; IV 614, 620, 622, 624,
        626; V 449
    ceramics, II 295, 300; IV 329, 331, 335, 338–9,
        342; V 662–7, 669; VI 631–6
    chemical industry, III 682; V 242–3, 247–8,
        251–2, 255, 302, 310; VI 499–510, 514, 517,
        522–8, 530, 540–1
    civil engineering, VII 872, 913
    coal, III 82; VI 362, 367
    cobalt, IV 145
    coke, blast-furnaces, IV 235
    coinage, II 491–2
    copper, IV 126, 129
    diet changes, III 2–3, 5–6

Germany (*cont.*)
   distilleries, alcohol, V 305–6
   dredging, IV 633, 637, 640
   drilling technology, V 107, 108
   dyes, dyestuffs, III 694; IV 249, 320–1; V 261,
     269–71, 273, 275–9, 281–2; VI 644–5
   electricity, V 184–5, 187–9, 205, 210, 217, 218,
     221, 232–3; VI 290; VII 1069
   engineering, VI 439; VII 983, 1037
   engines, V 140, 143, 158–9, 164–9
   explosives, V 284
   fire-arms, III 358
   flax, III 156
   food, V 14, 16, 304
   fulling-stocks, III 170
   furnaces, IV 122, 124–5, 129, 235
   glass industry, IV 360–1, 367, 374; V 673, 674
   glass-making, II 325; III 206, 220–2, 224–5,
     229, 237–8, 240–1, 700
   Gregorian calendar, adoption, III 579
   hemp, III 158
   houses, half-timbered, III 257
   ink, III 381
   iron and steel, III 31 n. 1, 689; IV 113,
     115–16; VI 468, 482, 490, 497
   japanning, III 696
   knitting, III 186
   lead production, IV 123
   machine-tools, V 656
   maps, III 533–5
   medical technology, VII 1326–7
   mercury, IV 135–6
   metallurgy, II 65–8, 76–7; V 57, 60–1, 63–4,
     67–9, 87, 96, 615, 625, 629, 634
   metals, non-ferrous, IV 119–20
   millstones, IV 212
   mining, II 11, 13–24, 65; III 678
   navigational aids, V 456
   nickel, IV 145
   oil and petroleum, V 102, 103, 107, 108, 110,
     113, 115
   paints, VI 602–3
   photography, V 717–18; VII 1281–1301
   plastics, VI 553, 556–60
   platinum metals, IV 144
   playing-cards, III 380
   population (1871–1914), V 1
   pottery, *see* Ceramics
   printing, III 377, 382–5, 388–90, 404, 406–7; V
     683, 694, 704, 706–8, 713–14
   railways, III 84–5; V 61, 326, 328–30, 332, 333,
     335, 336, 338, 340, 345–6
   river boat services, IV 548
   roads and streets, II 524–5, 526, 531–2; IV
     520, 527, 539, 542; VII 707–8, 718–19, 726,
     730, 732, 767–8
   rubber, V 764

scientific books, III 677
scientific instruments, III 585, 609, 613,
   621–2
sewage disposal, IV 518–19
shipping, IV 558, 594
silk industry, III 196; IV 319
steel industry, IV 107–8
sulphuric acid, IV 244
telegraphy, IV 654–6; V 227, 228
telephones, V 226
textiles, II 199–200; IV 248, 299; V 574, 577,
   587, 589–90, 598, 599, 602, 604
Thirty Years War (1618–48), III 349, 622
tin industry, IV 104, 124–5
town-planning, III 284–5
vehicles, V 427–8, 434, 436; VII 718, 726, 730,
   732, 734
water-supply, IV 498; V 554–5, 564, 566
watermills, II 611–12
windmills, III 97, 99, 106
wine production, II 137, 138
zinc, III 691–2
Gerzeans, I 509, 512, 516, table E
Gesso, II 249, 363
Gesture systems, supplementing speech, I 94–5
Gig-mills for raising cloth, III 172–3
Gimlets, II 394
Girders, IV 460–2, 477–8, 486
Glaskonst ('Art of Glass'), Peder Månsson, III
   209–10, 212
Glass, glass-making, II 64, 300, 301, 311–46,
   plates 24–8; III **206–44**, 698–702, plates 16,
   30; IV 358–78; V 671–82; VI **570–89**
   ancient ware, II 314–15, 318–22
   annealing chambers, IV 373
   batches, VI 570–1, 573–4
   blowing, II 322, 337
   Bohemian, III 206, 220–3, 209
   books, III 206–7, 209–15, 217–19, 222, 226,
    230–1, 698
   bottle-glass for windows, VI 570
   bottle-making, I 286; IV 376; V 674–7, plate
    4A; VI 580–5
   cane-form, II 322, 335; III 216
   casting-tables, IV 362, 369, 371–2
   chemical apparatus, II 733–5, 739–50; VI 570
   cold-cutting methods, II 322, 336
   colours, III 209–10, 214–15, 218–19, 224, 229,
    240–2
   containers, VI 580–5
   core-winding, II 336–7
   craftsmen, Roman period, II 323–5
   craftsmen, post-Roman, II 327
   crown, III 240; IV 358, 360–1, 365–7
   crystal, III 209, 215, 221–2, 702; V 672
   cut-glass, IV 376–7
   cutting techniques, III 225–6, 233

decoration, II 323–4, 328, 338–43; III 214–16,
    222–5, 229
electrical uses, VI 586–8
engraving-wheels, IV 377
flat glass, V 677–80; VI 578–80
flint, III 86, 678, 702; IV 362–3, 368–73
float glass, VI 570, 579–80
Florentine glass-blowers, III 636
fluxes, III 206–7, 209–10, 217, 226–8
food containers, V 42–3
frits, II 317–18
furnaces, III 78–9, 86, 207–14, 217, 220–1, 226,
    238–9, 699–700; IV 359, 361, 364, 367,
    370–1, 373; V 680–1; VI 575, 577
glass-houses, III 206–7, 214–15, 224–5, 226–8
glass-maker's 'chair', III 219–20
glassworks, IV 237, 343, 361, 368, 372–5, plate
    21B
gold-sandwich type, III 229
household ware, V 677
industrial relations, VI 585–6
inlay work, II 320
lead-glass, III 219, 225; V 672
marvering operation, IV 364 and n., 365–6
masks, protective, IV 366
materials, II 328–30, 332, 354, 358–9
mechanization, V 671; VI 588
medieval, II 325–8
mirrors, IV 362–4
mould-pressing, II 322, 335–6
nature/appearance, II 311–14, 333–5
opaque, II 333; III 215–16, 219, 224, 229
optical, III 229–37, 699; IV 358–62; V 673
origin, II 311–18
obsidian, I 563, 642
plate, III 237–40; IV 362–3, 368–73, plate
    21A
recipes, I 260, 261, 262
Roman period, II 322–5
ruby, III 224–5, 229; IV 375
sand-core technique, II 321, 322, 336–7
sheet, III 237–40; IV 365–8
still-bodies, II 348–9
stirrers, IV 359–60
tableware, III 20; VI 570
techniques, III 206–29
textile fibres, VI 589
tools, II 331, 332–3; III 208–12, 217, 219–20,
    222–9, 231–2, 234–6; IV 363–4, 366, 368,
    371–2, 377
tubing and rods, IV 363–5, 601, 607
Venetian, III 209, 216, 222, 230, 231, 678
vessels, III 20, 208, 214, 222–3, 228, 698
weathering effects, II 333–5
window-, III 209, 237–42, 699
windows, II 358–9, 416, 425, 429, 443–4
Glauber's salt (sal mirabile), II 745–6

Glazes, glazing, I 471, 485; II 311–46; III 261,
    295; see also Glass, Ceramics
Gleep, low-energy atomic reactors, VI 243
Globes, celestial, III 612–14
Glockendon Bible (1524), III 160
Glover tower, denitrating (1859), V 245
Gloves, early use, II 163, 177–8
Glues, I 695, 696; VII 937
Glycerol (glycerine), II 356 n.; IV 253
Glycine max (soya bean), I 361
Gnomon calendar stick, III 560, 594–6
Gold
    alchemical outlook, II 731–2, 747
    alloys, II 63–4; III 45–9
    aqua regia solvent, II 356–7
    assaying, III 59, 63–8
    book-cover, Monza, II 451
    coinage, III 46–8, 342; VI 411
    cross of St Cuthbert, II 451–2, 457
    cross of St Peter's, Rome, II 464–5, 467
    crown of Charlemagne, II 455, 459
    cups, II 461–2, 465, 469
    discovery, I 64, 66
    Dutch gold, IV 375
    extraction, I 575, 579–82; II 41–3, 63; III 29,
        33, 327–8; IV 119, 140–2; V 72, 74, 76, 79,
        94–7, 249–50; VI 424
    filigree, I 654–8
    fulminates, V 285
    furnaces, III 53
    gliding, II 342–3; III 48–9; IV 355–6
    glass pigment, III 224–5
    granulation effect, I 654–8
    inlay work, II 481
    jewelry, I 656, 659
    leaf, I 677–8, 698–9, 700, 701; III 48–9; IV
        355
    leather, -tooling, II 173–4, 180
    machines, III 327–8, 342
    medieval period, II 63–5
    mining, I 526, 563, 572, 574, 579–82, 589, 594;
        II 10, 12, 65–6, 77, plate 1; VI 413–17, 421,
        422
    objects, fine art, I 641–8, 654, 657–8, 660,
        frontispiece, plate 22
    patination, II 483–4
    pigment source, II 362
    plating, II 227; V 632–4
    powder, paint/ink use, III 49
    purse-mount, Sutton Hoo, II 451, 456
    refining techniques, II 12, 42–3, 63–4, 65; III
        47–9; IV 142–4; V 96–7
    Renaissance writers, III 27, 29
    repoussé work, II 468, 470–4
    solders, I 650–1, 652–3; III 48
    staining effect, I 243
    Standard, VI 2, 13, 411

Gold (*cont.*)
  Sword of Childeric, II 451, 453
  Tara brooch, II 473–4
  thread, III 48, 192, 205
  value-ratio with silver, I 585
  wire, I 654–6; II 481–3
  weight by balance, I 779
Goods vehicles, VII 722–8
Gossypium, I 373, 432, 449; III 159; *see also* Cotton
Gouges, carpenter's, I 612, 618; IV 404
Government, 20th century, VI 113–37
Granite, I 477, 478, 569–70, 574; II 30, 406; IV 543–5
Graphic art, I 144–53; *see also* Painting, Printing
Graphometers, III 627, 636; IV 403, 602–3
Gravettians, I 190
Greece, Greeks, I map 4, table F; *see also* Europe
  agriculture, I 355, 356; II 103, 121–3
  alchemy, II 733–5, 758 n.
  architecture, III 246
  arts, liberal *v.* mechanical, II 603–4, 638
  astronomy, I 802–3; III 520, 566, 612, 615, 620
  bauxite, V 93
  beer, Greeks' aversion, II 140
  building techniques, I 485, 487–8; II 397–404, 445–6
  calculations, numerical, III 501–2
  calendar, I 126; III 568–72, 579
  carpenters, II 232–3
  cartography, III 504–7
  ceramics, I 386, 390, 393, 395, 401, 403; II 259–69
  chemical techniques, II 350–1, 355, 359, 361–2
  coinage, II 485–7
  copper production, II 51–2
  dyestuffs, V 266
  electricity, meaning, IV 648
  food and diet, I 355, 356; II 103, 105, 118–23
  furniture, II 221–3
  geographers, III 613
  geometry, I 126; II 632–3, 675; III 502, 520
  Gregorian calendar, adoption, III 579
  harbour installations, II 517–18
  horse-riding, II 555–8
  iron production, I 598; II 58–9
  irrigation and drainage, II 678
  ivory work, I 674–5
  language, I 104–5; *see also* writing
  lead production, IV 123
  leather, use, II 150, 166–8
  lighting torches, medieval, I 234
  machinery, II 630–6
  mathematics, I 802–3; III 501–2, 505, 520, 620
  measurement, I 777
  metal-work, I 645–6, 658, 660–2, 681; II 449–50, 469–70
  metallurgy, II 41, 44–5, 476–8
  military pyrotechnics ('Greek-fire'), II 375–7, 700, 726, 771
  milling techniques, II 108–10, 112
  mining techniques, II 1–6
  navigation, III 517
  olive-oil production, II 112–13, 121–3
  ploughs, II 82–5
  pottery, *see* ceramics
  presses, olives and grapes, II 112–13
  quarries, II 24–7
  reckoning-apparatus, III 501–2
  roads and streets, II 498–9, 528–9
  science, III 584, 615, 618
  scientific instruments, III 618
  sewerage, classical period, IV 504, 518
  ships, II 563–9
  slave-labour, II 591
  statuary, terracotta, II 275–6
  surveys and surveying, III 502, 510–11
  textiles, I 443–5; II 191–7, 200–1, 211–12
  time-reckoning, I 121, 126–7
  town-planning, III 271–5
  viticulture, II 128–33
  water-engineering, I 531; II 663–4, 667–9
  watermills, II 593–5, 599–600
  weapons and armour, II 695–703
  wheeled-vehicles, II 539–40, 541–5, 549–50, 554
  writing, I 767–8, 769–71
Greek Anthology, II 228
Gresham College, London, III 257, 681
Grimani Breviary, II 192
Grindstones, III 110, 115; *see also* Machine-tools
Grit-stone, II 31–3
Groats (crushed grain), II 118–19
Grogram (cloth), III 179
Groma (surveying instrument), III 511–12
Gross National Product (G.N.P.), VI 25, 116–17
Growth, business corporations, VI 86–7, 88–9
Gruit in medieval beer, II 141
Guaiacum (lignum vitae), III 672
Guano, IV 254 *and* n.
Gulf Stream, VII 1447
Gunnery techniques, III 627–8
Gunpowder; *see also* Explosives
  alcoholic spirits, assay, III 21
  blasting explosive, II 1; IV 71, 75, 80; V 284, 292
  Chinese invention, II 771
  fireworks, II 377–9
  manufacture, II 379–82; III 684; IV 232; V 285
  military use, II 64, 377–82, 726–8; III 347–8, 350, 355, 359–62, 366, 368; VI 548
  power source, IV 170–1, 672
  rocket fuel, VII 859

Guns, *see* Artillery, Firearms
Gunter's chain, land-measurement, IV 600
Gutta-percha, V 221–3, 764, 771; VII 1242
Gypsophila struthium, II 215
Gypsum, I 469, 482 n.; II 737; *see also* Plaster of
    Paris
Gyroscopes, IV 408–9

Hacks, clod-breaking, III 134–5
Hadendoa people, I 427
Hadow Committee (1926), VI 167
Hadrian's Wall, II 31, 719, plate 40B
Hague Court, VI 2, 8
Halberds, I 614
Hallstatt culture, *see* Cultures
Hammers, hammering
    Bronze Age, I 608–9, 610
    carpenter's, II 231, 243
    claw, II 252, 394–5
    Egyptian, I 481, 688–9
    firearms, III 358
    forge-, III 352; IV 104
    glazing, paper-making, III 414
    gold-beating, III 49
    hammerstones, I 129–35, 608–9
    helve-, IV 104–5
    hydraulic, II 69, 73, 75, 396, 645–6
    Iron Age, I 618
    machines, III 342–3
    mason's, I 481; II 36, 385
    metallurgical, II 66, 643; III 49, 54, 82, 83
    miner's, I 567; II 8–9, 12, 15
    pile-drivers, III 422–3
    power-driven, III 341, 350, 690
    printing operation (c.1450), III 387
    steam-, IV 116, 430–1, 441, 461, plate 24A, B;
        V 610–11
    stone-working, I 129–35
    tradesmen's (c.1500–1850), III 110, 115, 119
    tilt-, III 32, 54, 115, 689; IV 418, 431
Hammurabi, Code of, I 256, 280, 548–9
'Handmaid to the Arts, The' (1758), R. Dossie,
    III 226, 228
Hands
    apes, I 9–10
    communications by gesture, I 18
    finger-tip painting, I 146, 148
    left hand/right hand, I 19, 24–5, 31
    man, I 10–12
Hanseatic League, fish industry, IV 45, 48–9
Hapsburg monarchy, VI 1, 9
Harbours, *see* Docks, Civil engineering, Ports
Harness, II 537–62
    ancient, I 719–24; II 552–3
    bridle-bits, I 722–4; II 558–61
    cart-horses, 19th century, III 148–9
    chariots, II 163–4

harness-leather, II 157, 165, 172–3, 183
horse-collars, II 90, 91, 527, 538, 554, 592, 639,
    770
rein and stirrup, II 516, 538, 556–7, 559
Roman, II 515, 592
saddles, II 182, 538, 554–7
shafts, II 547, 553–4
traces, II 554
Harp, Egyptian, II 149
Harpoons, I 158–61, 176–7, 178, 181–2, 498, 516
Harrows, II 91, 94; III 134; IV 4–5
Hatchets, II 243, 253; *see also* Adzes, Axes
Healds, figured-fabrics, III 187 n., 188, 191–3,
    195, 197–9, 201–3; *see also* Heddles
Health and safety, coal-mines, VI 374
Heat, heating; *see also* Central heating, Lighting
    development, indoor fires, I 231–3
    hearth accessories, I 232–3, plate 5B
    primitive man, I 230–1
    Roman period, I 231; II 4, 404, 419–20, 446
    tapers and candles, I 234–5
    thermodynamics, IV 163–4
Heat engines, VII 988–90; *see also* Internal
    combustion engines
Heavy Clothing Working Party Delegation from
    Britain to U.S.A., VI 677
Heddles, III 164, 166, 187 *and* n.
Helena, Empress, sarcophagus, II 29
'Helenca' textile yarns, VI 660
Helianthus tuberosus, III 2
Helical springs, IV 381, 383, 385, 391–3, 415
Heliometers, IV 361
Heliotropaeum europaeum, I 288
Helwanians of Egypt, I 496
Hemp, I 373; II 197, 363; III 158–9, 174, 176;
    VI 613
Henna dye, I 246, 289, 294, 441
Herbals, III 1–4, 8
Herminiera elaphroxylon, I 732
Hevea brasiliensis (Para rubber), V 752–3, 773–5
Hibiscus cannabinus (Deccan hemp), I 373
Highways, *see* Civil engineering, Roads
Hinges, II 235–6, 242 n., 429, 443
Hippocratic Collection, II 119–20
Hispano-Moresque ware, II 304
Hittites, II 151, 163, 167, 494
Hodometers, III 512–13, 628; IV 600, 616, plate
    23B
Hoes, I 502, 507, 539–40, 620; II 5, 16, 98
Hoists, II 388, 602–3, 637, 641–2; III 30, 32, 105
Holland (Netherlands); *see also* Europe, Low
    Countries
    agriculture, III 13–14; IV 14–16, 26; V 2, 13,
        18–19, 31
    alcoholic spirits, III 11
    brick-work, III 262
    building, IV 446–7

Holland (Netherlands) (*cont.*)
  canals/waterways, III 439, 441–2, 443–4, 450, 452; IV 452, 556, 558–60
  cannon, III 367
  cartography, IV 612
  ceramics, IV 329, 334, 336–7; V 666, 668
  diet changes, III 5, 23
  drainage systems, II 680–9; III 300–10, 315, 318–22
  dredging, IV 629–32, 635–40, 642
  dye industry, III 693, 695; V 260
  electricity, V 177
  electronic engineering, VII 1102
  fish industry, IV 45, 47–8, 53
  foods, V 2, 13, 18–19, 31
  geodesy, V 441
  glass, III 216, 220, 229, 700
  Gregorian calendar, adoption, III 579
  hydraulic engineering, V 545
  ink, III 381
  japanning and lacquering, III 696
  knitting, III 186
  land reclamation, III 300–10, 315, 318–22; VII 907–12
  lens-making, III 231–2, 234, 236
  longitude, determination, III 662
  military technology, III 349
  navigational aids, V 459
  paper-making, III 412, 414–15
  population (1800–1950), V 823
  printing, III 383–5, 592–3, 400–1, 406–7, 409–10; V 699
  roads and streets, IV 529, 543, 545
  ships, III 482–3, 485, 496–8; IV 576, 578, 586
  silk, medieval fragment, III 196, 202
  spinning-wheels, III 161
  States-General, III 662; IV 630, 632, 640
  sulphuric acid, IV 243–4
  surveys, IV 606, 610
  textile industry, III 151, 156–9, 170, 174–5
  town-planning, III 293–5
  wars with English and French, III 349
  water-supply, IV 498; V 554
  weaving-loom, III 164, 167–8
  windmills, II 619–20, 625–6; III 89, 93, 95–102, 106–8; IV 156–60
  zinc, IV 130
Hominidae, I 12–13, 18, 20, 22, 34, 36
Homo neanderthalensis, I 22; *see also* Man, Neanderthal
H. sapiens, I 22–5, 27–30, 32, table C
Hong Kong, land reclamation, VII 912
Hooke's law, IV 482–3
Hooks, agricultural, III 136–7
Hoppers, III 53
Hops, cultivation, II 140–1; III 14
Hordeum (barley), I 367–8, plate 7

Hornblende-granite, quarrying, II 29
Horologium (1658), Huygens, III 344, 663
Horologium oscillatorium (1673), Huygens, III 663 *and* n.
Horse-power, III 340; IV 164
Horses, VI 421; *see also* Harness
  agricultural use, III 142–50; VI 298–9
  ancient empires, II 495, 555–6
  bridle-bits, II 558–61
  cavalry, II 697–8, 705–6, 720
  draught-horses, III 592–3; VI 298–9, 421
  horse-cloth, II 556
  horse-whim, II 18
  horse-shoes, II 76, 77, 515–16
  messenger-service, II 495–6, 508, 524, 527
  Nesaean horse of Media, II 706
  pack, III 139, 141–2, 417
  power-source, III 45, 82–4, 162, 170, 172, 178, 317, 322, 342; VI 308
  rein and stirrup, II 516, 538, 556–7, 559
  saddles, II 182, 538, 554–7
  spurs, II 557–8, 559
Horticulture, III 13
Hose-pipes, II 184–5
Hosiery, *see under* Textile industries
Hosiery and Allied Trades Research Association, VI 667
Hospitals named; *see also* Medical technology
  Guys, London, VII 1347
  Jefferson, Philadelphia, VII 1331
  Peter Brent Brigham, Boston, Mass., VII 1358
  Philadelphia General, VII 1330
  Presbyterian, New York City, VII 1330, 1339
  St Bartholomew's, London, VII 1335
  University of California, VII 1332
Houses; *see also* Building, Lake Villages, Town-planning
  bitumen use, I 250, 253, 255–6, 460, 466
  Bronze Age, I 315–17
  coastal villages, British Columbia, I 177–8
  Cretan, I 484–5
  door-sockets, I 193–4, 306, 472
  drainage systems, I 466
  Egyptian, I 304–6, 473–5
  environment effect, I 299, 307, 456–7, 459
  Iron Age, I 317–24
  Mesopotamian, I 306, 459–60, 469, 472
  Neolithic, I 306–12
  Palaeolithic and Mesolithic, I 299–304
Hovercraft, VII 744–5
Huguenots, *and* English glass, IV 368
Human skill, I 1–37
Hundred Years War (1338–1453), II 76, 651
Hungary; *see also* Austria-Hungary, Europe
  asphalt, IV 539
  canals and waterways, IV 550, 553
  ceramics, V 662, 668

chemical industry, V 243
electricity, V 187
explosives, V 284
flour-milling, V 30
gold, IV 141
horse-riding, II 557, 559
mining, IV 70
oil, V 115
platinum, IV 144
silver, IV 139
tools/weapons, I 602, 614, 616, 625
vehicles, II 547-8
Hunting, ancient man, I **154-86**
Hydraulic engineering, II **663-94**; V **522-51**; *see also* Pumps, Water-power
dams and reservoirs, V 522-4
drainage, V 541-5
fluid power, VII **1060-7**
Greek machinery, II 633-5
harbours, V 539-41
Hydraulic Research Station, Britain, V 550
irrigation, V 541-5
land reclamation, III **300-23**; V 541-5
lift, agricultural tractors, VII 337
power, V 513, 514, 527-35
presses, V 610, 627, 628
pumps, III 325-6, 330-4; V 524-7, 545; VII 1060-3
rams, V 504, 505, 507, 520, 628
science, V 386-90, 545-50
spades, V 501, 535
turbines, V 528-34, 536
valves and meters, V 536-9
Hydro-electric power
dams, VI 197-200
development, V 231; VI 196; VII 1071-3
generators, VI 211-12
Niagara Falls, V 201-2
pressure penstocks, VI 201-2
tunnelling, VI 200-1
valves, VI 202-3
water turbines, VI 203-10
Hydrochloric acid, II 357; III 12; IV 124, 133, 142, 233, 241-2; V 115
Hydrodynamics, III 303, 319
Hydrogen, *see* Gases
Hydrographical Office (1795), IV 618
Hydrography, IV 615-20; V 449-54
Hydrometers, III 12, 22, 68
Hydrostatics, III 303, 305
Hygrometers, IV 403
Hyphaene thebaica, I 271, 423, 450
Hypocausts, II 419-20
Hypsometer and hypsometry, IV 605, 614

Iatrochemistry, II 745
Ice, fish preservation, IV 51-4

Iceland, VI 220, 256-7
Ignition, I.C. engines, V 157, 167-70, 175, 428-9, 435, 436
Implements, metal, I **600-22**
Incense, I 55, 260, 289, 294-5
Incubators, III 679-80; VI 314
India; *see also* Indus valley
architecture, III 249
borax, V 321
bridges, V 514; VII 879-80, 882-6, 892, 897
canals, waterways, IV 555
cartography, V 440, 442
city-streets, II 527-8
cotton, III 159-60
dyestuffs, V 258, 261, 266, 267
fire-making, I 220, 224, 229
food production, V 2, 21, 23
geodesy, V 443-4
Geological Survey, V 112
hydraulic engineering, V 523, 541-2
iron industry, I 597; V 70
lacquering, III 696
metal working, V 608, 617
oil, V 112
population growth, 19th century, V 1
potter's wheel, I 198, 202
railways, V 61
rubber, V 752, 773-4
saltpetre, IV 232; V 32
scripts, I 771-2
silk production, III 193; IV 309, 326
steel (*wootz*), II 57; III 35; V 64
sugar cultivation, I 370-1
surveys, IV 599, 610-11
textiles, I 432, 449
town-planning, III 270
voyages to, III 476-7, 506-7
water-engineering, II 666
water-supply, V 563
western technology, effect, II 772
zinc, III 30
Indians, North American, I 174, 176; II 148
Indigo, I 247, 249, 441; II 350, 361, 364-5
Indigofera tinctoria, I 247, 249; II 364-5; V 261-3
Indonesia, oil, VI 185
Indus valley civilization; *see also* Index of Place Names: Harappa, Mohenjo-Daro
agriculture, I 46, 50, 522
door stone-sockets, I 194
Harappa citadel, I 52
irrigation, I 521-2
kermes dyestuff, 5th century B.C., I 246
maps, I map 5, 251
script, I 80, 760
wheeled vehicles, I 717, 719
Industrial factors; *see also* Education areas, concentration, VI 50, 53-4, 57-60

Industrial factors (*cont.*)
  Eastern superiority, ancient, II **753–76**
  economics, VI **48–76**
  government role, VI 113–37
  growth, VI 61, 64, 119–20
  innovation, VI **27–47**
  management, VI **77–92**
  organization, V **799–813**; VI 53–5, 60–1, 86
  psychology, VI 35, 84–6
  social effects, V **814–41**; VII 1057–8
  trade unions, VI **93–112**
  unrest, VI 6, 99, 101
  Western advances, III **708–21**
Industrial Rayon Corporation, U.S.A., VI 649
'Industrial Revolution', IV 148–51; VII 958,
  962–3, 983
Information, *see* Computers, Diffusion, Discovery,
  Invention
Inks, I 245, 247; II 359–60; *see also* Pigments,
  Printing
Innovation, VI **27–47**
  business management, VI 86–7
  financing, VI 50–1, 57, 65–6
  government influence, VI 42–5, 66, 125–32
  industrial structure, VI 36–42
  invention, time factor, VI 51–3, 65–7
  social distrust, VI 94–5
  sources, VI 34–6, 50, 62–4
Inscriptions, I plates 28–34; *see also* Writing
  Arabic (A.D. 512), I 767
  Assyrian (*c.*650 B.C.), I 713
  boustrophedon, lines arrangement, I 758, 761,
    767–8, 770, 772
  Byblos, I 761, 762, 763, plate 31A
  Chinese oracle bones, I 772
  Egyptian, I 453, 537–8, 775–6
  Etruscan, I 769–70
  Greek, I 104, 109 n.
  Hebrew, I 764, 765, 766
  Latin, I 770
  Mesopotamian, I 459, 780
  metal-work, I 630–1, 637 n., 649
  Moabite Stone, I 764–5
  Nabataean, Arabia (A.D. 26), I 766, 767
  Phoenician canal (691 B.C.), I 531
  pictographic tablets, I 745–6
  Sargon (*c.*2250 B.C.), I 57
  Sennacherib, I 533–4
  Sinaitic script, I 762
  stone door-socket, I 194
  wood, I 684
  writing, invention, I 102
Insects
  instinct and skill, I 1–2
  scavengers, I 333–5
  slave-holders, I 338–9
  symbiosis among, I 328–9

Instinct and use of tools, I 1–6
Institute of Chemistry, VI 153
Institute of Naval Architecture, VI 159
Institute of Physics, VI 161
Institution of Electrical Engineers, VI 159
Instruments, IV 398–415; VII **1204–19**; *see also*
  Machine-tools
  astronomical, III 535–6, 545, 582–94, 614–18,
    625; IV 380, 398–400
  books, III 645
  calculating, *see* Calculating/computing
  cartography, III 553–5; V 443
  chromatography, VII 1210–11
  chronometers, IV 409–15, 598, 615, 619
  computation, III 612–18, 644
  electricity measurement, VII 1208
  gas measurement, VII 1208–10
  geodesy, IV 400–3
  laboratory, IV 403–9
  manufacture, IV 398–415; VII 1206
  materials, III 623, 625
  measuring, I 110–13, 775, 779, plate 31B; III
    336, 375, 386, 594–8, 601–3, 627
  medical, VII 1321–3
  navigation, *see* Navigation
  observatories, III 586–94, 625, 642; IV 380,
    398
  optical, III 229–33, 234, 237, 627, 630–5, 699;
    IV 358–62, 382, 399, 403
  pH measurement, VII 1210
  precision (to 1500), III 325, 337, 345, 400,
    582–619
  production specialization, III 629–31
  scientific (*c.*1500–*c.*1700), III **620–47**
  sighting-tubes, III 375, 593–5
  spectrophotometers, VII 1206, 1207
  surveying, III 510–13, 528, 537–43, 555,
    609–12, 627–8; IV 600–8; V 444–54
  timekeepers, III **648–75**; VII 1212–14
  vacuum-tube, VII 1214–15
  weighing, I 783, 784
Intelligence
  apes, monkeys, I 7–8, 12–13
  brain of man, I 13, 15–17, 19, 35
  invention capacity, I 60–2
  primitive racial groups, I 34–6
Inter-Continental Ballistic Missiles, VI 24
Interchangeable manufacture, IV 438, 440–1
Internal combustion engines, IV 148; V **157–76**;
  VII **997–1024**
  aircraft, VII 819–21, 823–4, 828–30, 1009–13
  diesel, V 120, 163–4, 301; VI 345; VII 345,
    767–70, 988–9, 991, 997–8, 1003–9
  gas-, V 157–60; VII 988
  gas-turbines, VII 760, 990, 991, 998, 1016–23
  gunpowder engine (1680), IV 171; V 119, 157
  historical background, V 119–20, 157

industrial use, 19th century, V 301
motor-vehicles, introduction, V 426
oil-, V 160–3; VII 759–60
petrol (Otto), V 164–76; VI 430–1; VII
    714–17, 997–1003, 1009–13
waterways, IV 548
International Air Transport Association, VII 700,
    701
International Civil Aviation Organization, VII
    699, 851
International collaboration, VI 3, 14, 25
International Council, Exploration of the Sea, VI
    355
International Council, Study of the Sea, VII 1447
International Geographical Congress, Berlin
    (1899), VII 1448
International Hydrographical Organization, VII
    1448
Inventions; see also Discovery, Diffusion,
    Innovation, Patents
  chemical industries, III **676–708**
  definition, I 59, 60; VI 65
  early cultures, I **58–84**
  economic factors, VI 49, 65–7, 115
  ideas, sources, VI 34–6
  independent, I 133, 166, 226, 227–8
  instruments, Royal Society, IV 669–70
  limitations, I 62–6
  mental states, man, I 75
  mutation factors, I 66–74, 83–4
  research, I 75–7; VI 65–7
  scope, technology cycles, VI 29
  slowness, man's progress, I 60–2, 128–9
  time-lag, adoption, VI 52
  U.S.A., 20th century, VI 4
Iodine, IV 236, 255; V 319–20
Ipomoea batatas (sweet potato), III 2–3
Iran (Persia), oil, VI 186
Ireland; see also Europe
  agriculture, III 3
  alcoholic spirits, III 11
  Atlantic cable, IV 661
  canals, waterways, III 455; IV 563, 572
  cartography, IV 623–4; V 438–9
  coins, III 46
  copper, IV 126
  diet changes, III 1–3
  electric railway, V 233
  farm-vehicles, III 140
  fine metal-work, II 464, 471–5
  flax, III 156–8
  linen bleaching, III 175–6
  map, Ptolemy's, III 508
  population, V 1 n. 1
  railway engineering, V 233, 324, 333, 336, 338,
      344, 347
  sericulture, IV 308

ships, shipping, II 581–2; IV 558
spinning-wheels, III 160–2
surveys, IV 601, 608–10, 616
telegraphy, V 218, 225
telescope, Rosse's, IV 399
textiles, V 574, 580
water-supply, V 563
whiskey distilling, V 305
wool, III 152
Iridium, IV 119, 144
Iron, iron industry, I 592–8; III plate 3; IV
    99–117; V **53–71**; VI **462–98**; see also
    Ironworks, Metals, Steel industry
  agricultural implements, II 85–6, 95
  alloys, V 97
  assaying of ores, III 62–3
  bar, III 31–2, 350
  blowing-apparatus, IV 103–5, 111
  books, Renaissance period, III 27–9
  bronze substitute, I 73, 573–4
  building, II 403, 417–18, 437–8, 442; III
      258–9, 267; V 470–7
  cannon, III 34, 361, 363, 365–7; IV 451
  carbon-content, III 33; IV 108–9
  cast-, II 6–7, 56, 72–5, 770; III 31–4, 63, 340,
      343; V 470–7, 499; VI 428, 430, 432–4, 436,
      455–7
  chronological charts, I 574–5
  coal, III 78–81
  coke fuel, IV 99–103, 104–5
  consumption, Middle Ages, III 30–3
  copper extraction process, II 11
  Cowper stove, V 58
  domestic implements, III 31, 38
  Egyptian gold stain, I 243
  electrolytic corrosion, III 45
  expansion, 16th–17th centuries, III 326
  fine metal-work, I 623
  fire-making, I 219–20
  furnaces, III 50; IV 99–106, 111–15, 235, 253;
      VI 489–91
  galvanizing, V 624–5
  glass, colouring, III 218–19
  glazes, ceramics, II 301
  Greek production, II 55–61
  'hard' and 'soft' iron, II 56
  horse-shoes, II 515, 561
  hot-blast process, IV 109–11, 113; V 58
  metal-working, IV 106–7; V 605–7, 610, 611,
      614–15, 620–2, 625
  meteoric, I 593–4, 616
  mining, I 568, 574, 593; II 2, 4, 8–10; IV 65;
      VI 410, 424
  ornamental use, II 429–30, 443
  pig-iron, II 70, 73–4; III 31–2, 78; IV 100,
      106–7, 113, 115; V 53–4, 58–61, 66–7; VI
      462, 465, 469, 479, 482

Iron, iron industry (*cont.*)
  prices, III 30
  production, I 592–8; II 55–61, 69–78; III
    342–3; IV 105, 107; V 53–4
  puddled, III 81; IV 99, 106–7, 113–15; V
    61–2, 66, 499
  Roman production, II 55–61
  rudder fixtures, III 45
  sheet iron, II 75; III 32, 355, 689
  ships, IV 476, 579–80, 583, 587–9, 594
  smelting, I 573, 576–9, 593–8, 617; IV 99–106,
    109–10, 113, plate 7B; VI 469–80
  spiegel, V 55
  sponge iron, III 33–4
  steam-power, IV 104, 116, 161–2
  tinplate, III 42; V 616–17
  tires, II 551
  tools and weapons, I 82, 321, 322, 504, 616–21;
    II 8–10, 228–32, 455–7, 696, 705, 721, 726–7
  trade, medieval, II 76
  tunnels, V 499
  water-wheels, IV 208
  welding, I 654
  wire, II 74, 75; *see also* metal-working
  wrought iron, II 70, 73, 74; III 32–3, 79; V
    472–6, 499, 606, 614, 620–1
Iron and Steel Institute, U.K., VI 489
Ironworks; *see also* Iron
  Bengal, V 70
  Coalbrookdale, Shropshire, IV 99–101, 106
  Ebbw Vale, V 613
  Horsley, Staffordshire, V 351
  Millwall, London, V 352, 362
  Park Gate, Sheffield, V 614
  Shoreditch, London, V 212
  Stanton, Derbyshire, V 211
Iroquois Indians, I 36, 225
Irrigation; *see also* Dams, Water
  agriculture, II 678; IV 19, 34; VI 319, 322
  ancient civilizations, I **520–57**
  Egyptian mechanisms, II 676–7
  Greek period, II 678
  land reclamation, VII 907
  mathematics, I 790
  medieval period, II 680–1
  Nile valley, I 535–9, plate 20B; V 542–4
  perennial, I 542–54, plates 20A, 21B
  projects, 19th century, V 541–5
  Roman machinery, II 637, 670, 678
  town-planning, Italy, III 286
  warfare use, I 554–5
  water conservation, I 522–5, 528–35, 542; VII
    1364, 1366
  wells, water-holes, I 525–8
  windmill pumps, II 617
Isatis tinctoria (woad), I 247; II 364; III 693;
  V 261–3

Islam; *see also* Arabia, Muslims
  calendar, III 580–1
  designs, fabrics, III 200
  paper-making, III 378
  silk, III 191, 197, 199
Israel, *see* Palestine
Italy; *see also* Etruscans, Europe, Romans
  aeronautics, V 400
  agriculture, IV 19, 29, 34–5
  ancient dwellings, I 317, 324
  architecture, IV 446, 470, 478–80
  bauxite, V 93
  books, scientific, III 677
  boric acid, V 320
  bridges, III 418, 424, 428–30
  building, III 245–52; IV 446, 470, 478–80; V
    493, 495
  canals, waterways, III 441–2, 444–51; IV 452,
    555
  ceramics, III 678; IV 329, 334, 336–7; V 662
  chemical industry, V 243
  cinematography, V 736
  coinage, II 490–1; III 46
  dam building, VI 200
  damask, medieval fragments, III 202
  diet changes, III 2, 5–6
  dredging, IV 630, 635–8, 640–1
  dye industry, III 679, 694; V 259
  electricity, V 177, 187, 188, 193
  fulling-stocks, III 170
  geothermal power, VI 219–20
  glass, II 321–2, 326–7; III 206, 209–20, 230–2,
    234, 238, 678, 700; IV 362
  Gregorian calendar, adoption, III 579
  hemp fibre, III 158–9
  hydraulic engineering, V 527, 537, 546
  jewellery, III 677
  land drainage, III 300, 308–15
  mercury, IV 134–5
  navigation, III 526
  oil discoveries, V 102, 103
  paddle-wheels, V 141–2
  paper-making, III 415
  pottery, II 285, 287–8, 293–4
  printing, III 390, 406–7, 409
  railway engineering, V 341; VII 770
  Renaissance, III 245
  roads and streets, IV 520–1, 529, 544; VII 873
  satin, *see* silk industry
  scientific instruments, III 585–6, 621
  silk industry, III 165, 196, 199, 200, 203; IV
    308–11
  soap, III 703, 705
  surveying instruments, V 446
  surveys, IV 610
  telegraphy, IV 652–3; V 227
  vehicles, road, VII 716–18, 735

velvet, III 205
water-supply, V 556, 564, 567
weapons, II 723
weaving, III 159
wheeled vehicles (c.3500 B.C.), I 211, 213–14
Ivory, ivory-work, I **663–83**, plates 23–5
  chryselephantine work, I 661–2, 681
  colouring, I 243, 669–70, 677–8
  Crete and Syria, I 672–3
  Egypt, I 666–70, 685, 691–2
  furniture, I 675–6; II 225–6, 227, 243
  Mesopotamia, I 670–2
  panels, I 671, 673, 676–7, 678
  pyxides, I 674–5, 680
  sculpturing, I 153, 666, 668–9, 673, 681
  supply sources, I 663–4
  toilet objects, I 665, 670, 674
  tools and weapons, I 32–3, 62, 435, 514–17, 665
  Western-Asiatic techniques, I 677–81; II 227

Jacks; *see also* Hydraulic engineering
  clock-, III 652 *and* n.
  door-, III 337–8
  lifting-, III 107–8; VII 938–9
  screw-, for raising cannon, III 368
Jacquard loom, III 167, 187, 201
Japan
  agriculture, mechanization, VI 337
  aircraft, VII 813
  ceramics, I 197, 384; IV 346
  coal, VI 180
  copper production, V 76 n. 1, 80
  fishing, VI 345
  incendiary warfare, II 378
  iron industry, V 70
  language and writing, I 96, 772–3
  mining techniques, II 3–4
  photography, VII 1281–301
  pivoted pestle, II 107
  potter's wheel, I 197
  textile industry, V 828; VI 15, 640–4
  transport, VII 694
  water-mills, II 614
  whaling, IV 61
Javelins, I 156–8, 163–6, 605; II 698, 704
Jersey fabrics, VI 668
Jet engine propulsion, VI 29; VII 817, 819–21, 990, 1017–19, 1021–2
Jewelry, III plate 29
  alloys, precious metals, III 45, 47–9
  assaying, III 60
  bearings, watches, III 670–1
  cloisonné work, II 450–5, 459–61
  enamelling, III 698
  fine gold-work, I 655–9
  gems, artificial, II 11–12
  Greek and Roman, II 449, 450–2

foils under gems, III 49
  glass beads, II 320, 322, 336
  inlays/enamels, I 659–60; II 452–5, 458–68
  repoussé work, II 468–75
  screw-fixing devices, III 629
  weapons, II 455–8
Jews
  calendar, III 572–5
  chronology, III 575
  seven-day week, III 565, 572–3
Joinery, II 237; III 266–7
  carving, II 245–7
  craft emergence, II 242–3
  roofs, construction, V 466–70
  tools and techniques, II 242–3
Journals, *see* Books and Journals
Juglans regia, I 358
Juncus acutus, I 420, 450
J. maritimus, I 245
Jupiter and *Luna 16* probe, VII 868–9
Jupiter Temple, Baalbek, II 30
Jute, I 450–1; V 569–70, 574, 576–80; VI 339, 613

Kalahari Bushmen, I 162, 171–2, 184
Kaolin, IV 339–41, 350, 352–3, 447
Kelp, IV 236 *and* n., 238, 255; V 319
Kerosene (paraffin oil), V 104–6, 110, 120, 160; VI 187
Kersey (cloth), III 179
Keyboard machines, *see* Calculating instruments, Typewriters
Keys, *see* Locking devices
Kilns; *see also* Ceramics
  bottle kilns for cement, IV 449
  brick-kilns, II 388
  corn-drying, II 99–100
  lime-kilns, II 347–8, 355, 407
  muffle-kilns, II 300, 304
  potters', I 392–7; II 266, 276, 294–9, 307–8
Kitāb al-Īdāh ('Book of Explanation'), II 736
Kitāb Sirr al-Asrār ('Book of the Secret of Secrets'), II 737
Kites, V 398, 399
Klismos, II 222, 234
Knitting/knitted fabrics, III **181–6**; VI 661–9
  Arabic colour-knitting, III 182
  coloured fabric, III 185, plate 9A
  embossed, III 186, plate 9C
  Florentine, III 185, plate 9B
  frame-knitting, III 182–5
  knitting-sheaths, III 184
  lace fabrics, III 185–6
  machine, III 185–6
  needle-weaving, III 181
  needles, III 183–5
  Peruvian needle-knitting, III 182
  sprang, III 181–3, plate 8B

Knives; *see also* Daggers, Swords, Weapons
  blades, I 508–9; III 343
  carved handles, I 516–17, 667, 669
  culinary, II 124–5
  draw-knives, II 231, 252–3, 392; III 121
  half-moon, II 166, 167, 168, 172, 178, 188, 189
  long-handled, III 136
  reaping (*c.*5500 B.C.), I 513
  table, II 125, 126, 127; III 20
  tradesmen's, III 110
Kupfernickel (metal ore), V 79, 87
Kuwait, oil, VI 186

Laboratories
  alchemical, II 748–50, plate 43B
  American Dental, VII 1323–6
  Austenal, New York and Chicago, VII 1324
  Ballistics Research, Aberdeen, Maryland, VII 1187, 1189, 1191
  G.E.C. Research, VII 1331
  Geophysical, Washington, D.C., VI 575
  Harvard Computation, VII 1195
  Interchemical Corporation, New York, VII 1351
  Los Alamos, VII 1332
  Max Planck, Mülheim, W. Germany, VI 559
  Naval Research, U.S.A., VI 230
Lace, *see under* Textile industries
Lacquer industry, II 362; III 688, 695–7; IV 133, 356
Lactuca sativa (lettuce), I 288, 355
L. serriola, I 355
Ladders, II 100, 716, 724
Ladles, III 52
Lagenaria siceraria, I 372
Lake-villages, I map 3; *see also* Houses, Town-planning
  Alpine region, I 310
  basketry, matting, I 420–1, plate 9D
  bronze artifacts, I 588
  Iron Age, I 318–23
  Neolithic, Europe, I 311–12, 517, 588
  plant cultivation, I 355, 357–8, 360–2, 369–70, 372–3
  roads, I 713
  textiles, I 414, 435–6, 445, 448, plate 12B
  weapons and tools, I 508, 510, 517
Lamps, *see* Lighting
Land management, III **300–23**
  drainage, *see* Dikes, Drainage
  reclamation, II 680–5; VII 907–12
Language, I **85–109**; *see also* Languages, Speech, Writing
  grammatical systems, I 96–8, 102, 746–7
  Indo-European group, I 103–5, 757
  linguae francae, I 97–8
  primitive structure, I 98–100

social groups, development, I 89–92, 101–3, 104–6
sound values, I 85–90, 93, 95–6, 104, 106, 755
speech, relationship, I 89
spread, Stone Age, I 100–1
syllables, picture-signs, I 747–50
tool-making, development, I 17–19, 27
Languages, specific; *see also* Language
  Arabic, I 86, 97, 98
  Arunta, I 94, 98–100
  Bantu, I 95
  Chinese, I 88, 94, 97, 772
  Egyptian, I 102–3, 287, 755–6
  English, I 85–8, 90, 97–8, 99, 104
  Finno-Ugric, I 98
  French, I 93
  German, I 86, 93
  Greek, I 97, 102, 104–5, 761, 767–8
  Hebrew, I 97, 761, 764–6
  Hittite, I 752, 757–8
  Hottentot, I 90, 95–6
  Japanese, I 96, 772–3
  Korean, I 96
  Latin, I 97, 104–5
  Polish, I 93
  Russian, I 99
  Sanskrit, I 104
  Sumerian, I 101–2, 748–50, 751–2
  Syriac, I 767
Lapland, II 209
Lansium, I 169
Lasso, I 166
Lasur (vase-colour), II 265–6
Latex, VI 602; *see also* Rubber
Lathes, VII 1035–8; *see also* Machine-tools, Turnery
  ancient Greece (*c.*2000 B.C.), I 518
  automatic, V 646–9, plate 40B
  capstan-, V 646
  coins, minting process (1737), III 343
  construction, IV 424–5, 439–41
  earliest forms, I 192–3
  Egypt (*c.*1300 B.C.), I 688–9
  glass lenses, III 235–7
  Greece, 7th century, II 222–3
  India (*c.*2000 B.C.), II 232–3
  light engineering, 17th century, III 345, 365
  milling-cutters, IV 423–5
  Roman times, I 680
  screw-cutting, III 334–7; IV 386–7, 392, 393–4, plate 25A
  tools (*c.*1700–1850), IV 382–8, 417–21, 428–9, 431–3, 436, 439–41, plates 23A, 24C, 25A, 28
  turnery, post-Roman, II 249–51, 643–4, 645
  turret-, V 644–6
  wood-turning, II 232–3, 249–51, 392, 631
Lathyrus, I 362

Laurus cinnamomum, I 288
Laws, *see* Acts of Parliament
Lawsonia inermis, I 246, 294
Lead
  -acetate manufacture (1790), IV 251–2
  ancient times, I 582–5
  assaying, III 29–30, 62
  batteries, electric, V 204, 205–6; VI 410
  bells, III 340
  brass impurity, III 38
  building uses, II 46, 429, 445; III 44–5, 254;
    IV 123
  cast, III 34, 45
  chronological chart, I 574
  coinage, I 585
  density discoveries, III 58
  electric-cable sheathing, V 626–7
  extraction, I 583–5, 590; II 43–5, 50, 67; III
    27, 51; IV 101, 119, 122–3, 137–8; V 72, 74, 96
  -foil, IV 125
  glass-ware, III 221
  glazes, II 300–1, 316–17; IV 332–3
  gold-refining, I 581; IV 136, 142
  Goslar lead ores, III 51
  liquation, precious metals, I 584–5
  Greek and Roman use, II 46
  gutters and pipes, II 46, 404, 444–5, 665; IV
    123
  metal-working, III 342; V 626–7
  mining, II 2, 9, 11; IV 65–6; VI 423
  nickel alloys, IV 146
  niello, III 48
  oxide, IV 359, 376
  petrol additive, VI 410
  poisoning, VI 592
  production (1850–1900), V 73; VI 428
  pumps, III 332
  red-lead, II 361
  seals, III 338
  sheet, III 32, 45; IV 119, 123
  ship-hulls, sheathing, II 569, 572; III 45; IV
    123; V 605
  shot, III 45
  solders, II 46, 47; III 42; VI 410
  speculum-metal, III 40
  sulphide, III 221
  sulphuric acid, IV 231, 243, 245–6, 248
  tin alloys, III 42–4; VI 410
  'white' and 'black', II 47, 747
  window strips, III 41–2, 44, 256, 261
Leather, II **147–90**; *see also* Harness
  book-binding use, II 169, 173–4, 180–1, plate 7
  boxes and caskets, II 170–1, 172, 183, 255–6
  buckets, II 14, 16, 171, 184
  buff leather, II 175, 182
  ceramics, decoration, I 398, 399
  chamois (wash-leather), II 154

chamoising process, II 154
clothes, II 148, 150, 156, 158, 160, 161, 165,
  175–8, plate 5
colouring, II 173–4
cordwain (Cordovan), II 150–1, 155, 177
craft-techniques, II 170–4
dressing, II 173
Egyptian grave finds, II 158
finishing, II 156–7
footwear, II 150, 151, 156, 160, 162–3, 166–8,
  175–7
furniture, II 171, 178–9, 256
gloves, II 163, 177–8
hangings/screens, II 179–80, plate 6A
harness, I 720, 722, 728; II 157, 163–4, 183
lanterns, II 184
miners' kit-bags, II 166
moulding, II 171, 173, 180
musical instruments, II 180, 182
raw-hide, uses, II 160, 162
rope-making (*c.*1500 B.C.), I 453–4
saddles, II 182, 538, 554–7
sedan chairs/carriages, II 182–3
sheaths, daggers, II 159
shields, II 164, 165, 167
stamping and tooling, II 173–4
tanning, II 147–56; IV 231
tawing, II 149–50, 154–5
tools, II 160, 162, 165–6, 169, 350
vessels, II 150, 158–9, 165, 169–70, 171, 174
writing material, I 757; II 164, plate 5C
Lecanora, I 246; III 694
Lens, I 362
Lenses
  'anallatic', IV 602
  achromatic, III 233, 555; IV 358, 380, 389,
    398, 603
  cameras, IV 360; VII 1286–7, 1288–92
  cinematography, V 737, 745; VII 1308
  cutting lathes, IV 382
  glass recipes, III 698–9
  invention (1286), III 230
  microscope, invention, III 232–3
  mounting spring, III 627
  optical glass, IV 358–62; V 673; VI 572, 573
  optical instruments, III 631–5; IV 398–9,
    400–3; V 444–8
  preparation, III 232, 233–7
  spectacles, III 229–31, 627, 630
  subtense principle, V 445–7
  telescope invention (*c.*1580), III 231–3
Lepidium sativum, I 356
Levalloisian culture, *see* Cultures
Levelling instruments
  altitude, III 593; IV 604–5
  Greek and Roman, II 672
  Hero's dioptra, III 609–12

Levelling instruments (*cont.*)
   Middle Ages, III 513, 527–9, 538, 541, 591
   plumb-bob, III 375
   spirit-levels, II 386; III 375
   surveying, IV 600–4; V 444–8
Levers, II 629–30, 632–3, 646; III 18, 173, 177, 236, 330, 359
Leyden jars, III 644; IV 651, 678
Liber Archimedis de ponderibus, III 66
Liber ignium, Marcus Graccus, II 351, 379–80
Liber sacerdotum, II 352
Liber servitoris, Khalaf ibn 'Abbās al-Zahrāwī, II 371–2
Libraries
   Bayerische Staatsbibliothek, Munich, IV 642 *and* n.
   Cairo, III 63
   monastic, III 381
   St Andrews University Library, III 667
   St Bride Technical Library, London, III 410
Libros del Saber Astronomia of Alfonso of Castile (1252–84), III 616, 648
Lifesaving at sea, VII 745–6, 859
Lifts, IV 76–8, 82, 89, 548–50, 569, 571; V 478, 498; VII 930, 938; *see also* Hoists
Light-signals, triangulation (1787), IV 608
Lighthouses
   'Admiralty List of Lights' (from 1895), V 454, 463
   buoys, V 463
   Eddystone, IV 179, 205, 447–8, 468–9; V 463
   electricity generators, V 191–2, 194, plate 10B
   fog-signals, V 464
   gas-lighted, IV 261
   illumination, V 181–3, 191, 210, 463
   lightships, V 463–4
   Pharos of Alexandria, II 521–2; III 275; V 463
   Roman, II 518, 521–4, 573
   wireless telegraphy, V 228
Lighting systems and lamps; *see also* Electricity, Gas industry
   ancient man, I 233–7
   arc lamps, V 182, 208–13; VII 1085
   bulb manufacture, VI 586–7
   carbon-filament lamps, V 98
   cressets, II 299
   Drummond light (lime-light), V 181
   electric filament lamps, IV 274
   engineering work, V 500, 517, 519
   film studio, VII 1304–5
   fluorescent, VII 1085–6
   glasses for, VI 571–3
   hydrogen lamps, IV 260
   illuminants, Middle Ages, II 370
   leather lamps, II 184
   metal filament lamps, V 99, 218; VII 1082–4
   mining lamps, II 2, 3, 9, 24; VI 373

Nernst lamps, VII 1083, 1085
   oil-lamps, I 33, 147, 235–7; II 278; IV 262, 265, 269
   osmium-filament lamps, V 99, 218
   Roman oil-lamps, II 278
   salt, lamps Egypt, I 259
   street, II 531; VI 291, 583
   Swan filament lamps, V 196, 213–17
   thermolamps, IV 258, 263–4
   torches, I 233–4, 566, 567
Lightning conductor, IV 678
Lignite production, 20th century, VI 177
Lime, II 152–3, 189, 355, 363; IV 447, 449, 466, 519, 539
Limestone, II 26, 30, 31–2, 35, 398, 400
Limonite (iron-ore), II 71
Linen, I 448–9; *see also* Spinning, Textile industries, Weaving
   bleaching, II 217–18; III 174–6
   canvas, oil-painting, II 363
   fibre, preparation, II 195–7; III 156–8
   flax, cultivation, I 372–3; II 195; III 156; *see also* Flax
   fulling and beating, II 215–16; III 169
   looms, new design, III 168–9
   mechanization, IV 282, 295, 304; V 569, 574, 579–80
   origin (*c.*3500 B.C.), I 373; II 195
   paper-making, III 411
   pressing, III 177–8
   Scottish shrouds (1666), III 151
   smoothing by rubbing, II 219; III 176–7
   tabby tissue, III 197
   twill, warp threads, III 196
   warping/dressing, III 164
Linguistics, *see* Language
Linseed oil, VI 595, 597–8; *see also* Flax
Linsey-woolsey (fabric cloth), III 179
Linum usitatissimum (flax), I 288, 358, 372–3, 431, 448; II 195
Liquation process, I 584–5; III 27, 33, 42, 51
Liqueurs, II 142
Litharge, III 63–4, 697
Lithography, IV 626; V 706; VII 1268–70
Livestock, VI **298–318**; *see also* Animals
   alternate husbandry, IV 18
   breeding, IV 21, 25–6, 28; VI 303–4, 305–7, 328
   cattle, IV 14, 15–16, 21, 22, 31, 32; V 3–9, 13, 15, 19–20
   dairy industry, IV 16, 25–6; V 32–6
   diseases, V 12–13, 15, 24
   feeding, IV 37, 40–1; VI 307–12
   fodder crops, IV 13, 14, 15, 23, 31; V 4–5, 9, 12, 15–17, 19–20
   herd-books, V 5, 13
   horses, IV 26, 34; V 4–5, 15–16, 18, 20

Middle Ages, III 12, 15
  nutrition, V 17, 24; VI 300–2, 309–11, 316
  pigs, IV 31; V 5, 15, 18, 20
  sheep, I 344, 345, 451; IV 14, 15–16, 21, 26–8,
      34; V 5, 7, 9, 13, 15–16, 20–1; VI 312–13
  underground haulage, VI 421
Lloyd's register of shipping, IV 594
Lock-gates, see under Canals
Locking devices, I 692, 695; II 415–16, 429–30;
    IV 423
Locomotives, see Railway locomotives
Logarithms, III 542, 553
London; see also Index of Place Names
  Clockmakers livery company, III 630
  Great Fire, rebuilding plans, III 297–8, 320
  Metropolitan Board of Works, IV 510, 518; V
      486
  Metropolitan Water Board, V 554, 567; VI 213
  Pewterers' livery company, III 42, 68
  Roman, III 278–9
  sanitation by-laws, L.C.C., IV 507, 516
  seventeenth century, III 295–6
  Spectaclemakers livery company, III 630
  Squares, III 296, 298
Long-bows, II 721–2
Looms, see Machine tools, Textile industries
Low Countries; see also Belgium and Holland
  agriculture, IV 14–15, 17, 22, 28, 31–4, 36–7;
      V 12, 19
  canals and waterways, IV 555–6
  cattle, V 15
  coal, III 72–4
  dredging, IV 636
  farm wagons, III 146
  glazing, IV 345
  iron industry, III 79
  printing, III 401
  roads, IV 529
  scientific instruments, III 586, 621
  windmills, III 89, 93–5, 100–1, 106
Lubricants, I 198; III 41–2; IV 56; V 103, 104,
    118–19; VI 377, 544
Luddite riots, IV 304
Luftwaffe, VI 17, 19; VII 812
Lunar Society, IV 673; see also Moon
Lustre-painting, II 303–4, 342
Luttrell Psalter (c.1338); see also Bible references
  gloves and mittens, II 177
  kitchen, medieval, II 124
  sower's container (seedlip), II 100
  spindle-wheel, II 203
  swing-plough, II 89
  water-budgets, II 184, 187
  watermill, II 596
  wheeled-carts, II 548–9; III 142–3
Lyddite (explosive), VI 548
Lye, for soap, III 704; see also Alkalis

Lygeum spartum, I 423, 450
'Lymbrique stuff', II 76
'Lyngby' culture, I 497, table E

Mace-heads as weapons, I 510, 512, 517
Machine-tools and machinery, II 629–62; III
    324–46; IV 417–41; V 636–57; VII 1035–45
  armour polishing, III 350
  basic components, II 629–30, 660
  bearings, III 41–2, 327, 658, 660
  block-making, Portsmouth, IV 426–7, 437
  boring machines, V 640, 649–50
  cannon/gun-making, III 359, 365, 367
  chain-and-bucket pump, III 325–7
  clockmakers' catalogue, 18th century, IV 421
  cutting, V 640; VI 453; VII 1035, 1037–8,
      1039, 1048
  development, 20th century, VII 986
  definition, IV 150
  drainage, III 82–3
  drilling, V 641–4
  electrification, V 641
  gear-box, V 640–1
  gear-cutting, V 655–6
  Great Exhibition (1851), IV 431–2
  Greek, II 630–6, 675
  grinding, V 654–5; VII 1041–5, 1048–9
  gudgeons, III 41–2
  interchangeable, IV 438, 441; V 638–9
  Johansson blocks, VII 1041
  lathes, see Lathes
  lens-making, III 232, 234–6
  mass-production, IV 427
  materials of construction, IV 420, 424–6, 429
  metals, used, III 326
  Middle Ages, transition, II 638–40
  milling, V 652–4; VII 1038–9
  monastery development, II 649–51
  motion, mechanization, II 640–5, 652–7
  mushet steel, VII 1037
  oil lubrication, see Lubricants
  ore-crushing, III 327–8
  paper-making, III 411–16
  pile-driving, III 422
  planers, V 650–2
  power, mechanization, II 645–9; III 19
  rolling, III 338
  Roman, II 636–8
  scale-cutting, cartography, III 555
  slide-rest, Maudslay's, IV 162, 425
  slitting, III 338
  steam-driven, III 85
  textile, III 152–5, 157–9, 162–78
Mackintosh (waterproof garments), IV 253; V
    761–3
Madder (Rubia tinctorum), I 246, 249, 441;
    V 260–1

Magdalenian culture, *see* Cultures
Magic, *see* Alchemy, Superstition and magic
Maglemosian culture, *see* Cultures
Magnesia (periclase), VI 574
Magnesian limestone, II 31, 33
Magnesite, V 664
Magnesium, V 72, 86, 98; VI 430, 431, 441
Magnetism, II 771; IV 621, 679
MAGNOX atomic reactor, VI 250, 252, 259
Majolica, III 678
Malachite (copper carbonate), I 293; II 362
Malus (wild crab-apple), I 357
Mallets, III 110, 119–20, 126; *see also* Hammers
Man, primitive
    art forms, I 144–53
    carnivorous habits, I 19–21
    brain, development, I 13, 15–17, 19, 35
    discovery, invention, I 58–84
    distribution, industrial activity, I 36–7
    evolution, I 9–10, tables B, C
    hands, I 10–12
    intelligence, I 34–6
    Neanderthal, I 29–31, 35, table C
    Pekin, I 21, 23–5, 31, 218
    skill, I 22–34, 60–1
    speech and language, I 18–19, 85–109
    time-reckoning, I 110–27
    visual power, I 12, 35–6
Man-made fibres, *see* plastic-fibres
Management, VI 77–92
    big-business, VI 86–7
    business education, VI 83–4
    diversification policies, VI 87–90
    industrial psychology, VI 35, 84–6
    origins, VI 77–81
    Taylorism, VI 73–5, 81–3, 93, 115–16, 118–19,
      120; VII 1046–9, 1056
Mandeville: 'Sir John Mandeville's Travels',
    glass-furnace illustration, III 207, plate 16
Mandrels and lathes, III 335–7; V 64
Manganese
    Bessemer converter, V 55, 60
    ceramic pigment, III 702–3; IV 335
    extraction, IV 119, 147; V 97–8
    glass ingredient, III 210, 218, 222, 702–3; IV
      375
    iron ores, Styrian, III 115
    mining, VI 424
    nodules, oceanic, VII 1465
    occurrence, United Kingdom, IV 65
    paint ingredient, VI 595
    steel alloy, V 64–5; VI 410
Manhattan Project, *see under* Nuclear power
Manilla, paper-making, VI 613
Manipuris potters, I 407
Mannheim gold, IV 133
Manometers, IV 406

Maoris, I 318, plate 27
Mappae clavicula de efficiendo auro, II 63, 351, 352
Maps; *see also* Cartography, Charts, Town-
    planning
    ancient world, I maps 1–8
    aqueducts, Roman, II 669
    Belgian canals, IV 557
    Briare canal (1602–42), III 461
    Britain, Roman roads, II 506
    Brussels canal (1550–60), III 451
    Canadian waterways, IV 553; VII 916
    'Catalan Atlas' (1375), III 525
    'Carte Pisane' (*c.*1275), III 523–4
    clay-tablet, Mesopotamia, I 549, 550, plate 21B
    diffusion, metal-technology, I 573
    'Dufour map', IV 612
    economic, IV 622
    England and Wales, waterways (1858), IV 564
    Eratosthenes' (*c.*275–*c.*194 B.C.), III 506–8
    Etzlaub's road-map (1492), III 534–5
    Exeter canal (1564–67), 456
    Fens, Bedford Level (1724), III 316
    flint mines, Europe, I 562
    France (1774–84), III 554
    France, 17th century, IV 605–6
    France, 18th century registry, IV 442
    geological, IV 491, 621–2, 626
    Grand Trunk canal, IV 565
    Hereford *mappa mundi* (*c.*1280), III 519–20
    Holland, 17th century, III 302
    Hong Kong reservoirs, VII 911–12
    hypsometric, IV 614
    isometric, III 555
    Italy, Roman roads, II 501
    Languedoc canal (1661–81), III 465
    magnetic, IV 621
    medical, IV 624
    Milan area, canals (*c.*1500), III 445
    mineralogical, IV 621–2, plate 32B
    Muslim empire (*c.*750 A.D.), II 757
    New River cut, London, IV 492
    ordnance survey maps, IV 491
    pastures, Boston (*c.*1300), III 515
    Persian royal road, II 496
    Peutinger table, III 514
    polders, Zuiderzee area, VII 910
    Pontine Marshes, Italy, II 313, plate 19
    population, IV 623–4
    printing, III 404–10; IV 625–6
    Ptolemaic, III 507–10
    regional, III 537–9; IV 600, 602, 610, 614
    roads, IV 523, plate 34
    roads, Rome and provinces, II 502
    St Lawrence Seaway, VII 916
    Scandinavia, Claudius Clavus's, III 532–3
    Scotland, Telford's roads, IV 533
    star, III 604–5

Stecknitz canal (1391–8), III 443
telephones, Britain (1896), VII 1228
telephones, Europe (1896), VII 1229
Thanet, Isle of (c.1414), III 516
trade routes, Medieval, II 759, 769
traffic, IV 623
United Kingdom, minerals (1851), IV 65
Wey canal (1651–3), III 457
world, III 507, 508, 519–20, 555; IV 596–600,
    625
Marble
    English, II 36–7
    furniture, II 224–5
    Greek, II 24–6, 398–9, 403
    Purbeck 'marble', II 36–7
    quarrying, to 17th century, II 24–7, 28
    Roman, II 27, 406, 411
    white Carrara marble, II 28, 275, 406
Marcasite, III 30
Margarine, IV 61; VI 340
Margarita philosophica (c.1511), Strasbourg, III
    538
Marine works, IV 443, 452, 464–9
Market-gardening, IV 15
Mars space probes, VII 860, 869
Marshall Aid, VI 24, 123
Mass production
    'American system', V 818–19
    bronze castings, III 39
    ceramics, IV 354; V 661
    conveyor-belt system, VII 714–15
    flour, III 19
    industrialization, IV 354; V 822
    invention consequences, V 639–40
    interchangeable concept (1798), IV 438; V
        638–9
    machine-tools, use, IV 427, 440–1; V 636,
        638–9, 818
    obstacles, Greek society, II 604–5
    obstacles, Roman Empire, II 603, 605
    quantity v. quality, III 709–11
    railways factor, V 819
    scientific instruments, III 629
    ship-building, III 711
    social effects, VI 133
    standardization effects (1851), V 818
    tools, III 113–14
Matches, III 355 *and* n. 2; IV 253–5; V 252–3
Materials, strength, IV 149, 443, 471, 478–86; *see
    also* Building materials
Mathematics, I 785–803, plates 33–6; *see also*
    Calculating, Geometry, Tables
    aerodynamics application, V 407
    algebra, I 787–91
    architectural theory, III 257, 432–4
    astronomy, I 793–803; II 675
    Babylonian, I 787–91

decimal arithmetic (1585), III 61, 542
dynamics, III 344; VII 859
early number-symbols, I 785–7; III 501–2
Earth, size estimations, III 505–6
economic problems, I 790–1, 792
education, V 783–4
Egyptian, I 791–3
elasticity, III 58–9; VII 944
equations, I 787–9
fortifications application, III 371–2
fractions, I 792–3
geometrical problems, I 789–90, 791–2
hydrodynamics, applied, V 387–90
instruments, III 521, 627, 636, 644
lever theory, 3rd century B.C., II 630, 632–3
logarithms, III 542
map-making, III 505–10, 536
measures, *see* Measurement, Weights
mechanics, II 632–3, 675; III 250, 252, 433;
    IV 149, 163, 666; V 784
navigation, III 526
numerals, 'Arabic' system, II 766–7, 772; III
    502
planetary theory, Ptolemy's, III 615–16
projectiles, III 374–5
Pythagorean geometry (c.400 B.C.), II 632
ship-building, III 482, 488; V 385–90
survey applications, III 510–13, 537, 539,
    542
tables, I 787
thermodynamics, IV 163–4
trigonometry (1600), III 542; *see also*
    Trigonometry
Mats, I 413–55
    reed and rush-, I 419, 420, 423, 468, 550
Matter
    atomic theory, IV 223–6; VI 223–4
    conservation of, IV 217–18, 220. 223–4, 228
Mattocks, II 5, 99; III 134
MAUD committee, VI 226, 227, 229, 232, 270–3
Maya civilization, I 82–3, 121, 124–6, 189
Measurement, I 774–84; *see also* Weights
    acres, units, I 111; III 528
    altitude, III 555; IV 604, 612, 614
    angle, IV 602–4
    capacity, ancient world, I 781–3
    chain unit, III 542
    Chinese system, I 111–12, 783–4
    compensation bars, IV 600–2
    cubits, I 775, 777–9
    depth, IV 619
    distance, IV 600
    Egyptian, I 775–8
    energy, IV 679–81
    foot measure, I 777–8
    Greek, I 777
    hodometer, III 628; IV 600, 616, plate 23B

Measurement (*cont.*)
  instruments, III 336, 375, 627; IV 395–8,
    432–3; VII 1050–1
  knot unit, III 550
  meaning, I 110–13
  meridians, IV 596, 599
  metric system, IV 600
  mile, German/Italian, III 535–6
  mile, Roman, III 506, 517, 522, 535
  primitive, I 110–12
  sea-miles, III 550
  stade unit, III 505–6, 522, 531
  standardization trend, I 110, 774–5
  survey requirements, III 537–41; *see also*
    Surveying
  temperature, IV 163
  time, I 112–27; III 594–8, 601–3
Mechanica (pseudo-Aristotle), II 630, 632, 658
Mechanical Engineers, Institution of, London, V
    118, 638, 642
Mechanics
  applied science, IV 149, 163, 666
  Archimedes' teaching, II 632–3
  clocks, III 344–5
  motion, Galileo's laws, III 374
  precision, IV **379–416**
  projectiles, III 374–5
  screw invention, II 632
  structural, III 250–1, 252, 256–8, 432–4
  water-lifting methods, II 675–8
Mechanics' Institutes, IV 274, 445; V 714, 777–9,
    783–4, 786
Medallion-lathes, IV 436
Medallions and medals, III 39, 338–9, 341; *see*
    also Coins, Mints
Medical technology, VII **1317–62**
  anaesthesia, VII 1334–40
  artificial limbs and orthopaedics, VII 1359–60
  blood-pressure measurement, VII 1344–6
  dental industry, VII 1323–6
  electrocardiography, VII 1340–2
  electrokymography of heart, VII 1342–3
  hearing, VII 1348–50
  hospitals, *see* Hospitals named
  instruments, VII 1317–23
  microscopy, VII 1350–4
  military, from 16th century, III 373–4
  pharmaceutical industry, VII 1326–7
  polygraphs, VII 1343–4
  radiology, VII 1299–1300, 1327–34
  stethoscopes, VII 1346–8
  surgery, VII 1354–9
Medicines, *see* Pharmaceuticals
Melamine polymer, VI 554–6, 600
Melanesians, I 406, 407
Melinite (picric acid), VI 540, 548
Mentha piperita, I 288

Merchants' metal tokens (*c.*1500), III 46
Mercator charts, III 548, 550–1
Mercurago spoked wheel (*c.*1000 B.C.), I
    213–14
Mercury
  amalgamation process, I 582; II 42; III 33,
    327–8; IV 139–40; V 95
  assaying, III 62
  Castner-Kellner cell, V 250–1
  chemical discoveries (1774), IV 218, 220
  classification, II 737; IV 214
  clocks, III 648–9
  dentistry use, VII 1326
  extraction, III 51; IV 119, 133–5, 136; V 72
  mirrors, 'silvering', III 40; IV 362
  speculum metal, III 41
Meridian, IV 596–9, 605–7, 609, 618
Merino sheep, IV 26–8, 34
Merton College, Oxford
  astronomers, III 585, 593, 617
  medieval equatorium, III 617
Megaliths, I 81, 490–4; *see also* Burial
    mounds/structures
Mesolithic culture, *see* Cultures
Mesopotamia, I maps 4, 7, tables E, F; *see also*
    Sumer
  agriculture, I 360, 362, 364, 365, 367, 545,
    547–54; II 84, 106
  arrow-heads, I 510
  astronomy, I 798–801
  bitumen industry, I 250–6
  boats and ships, I 256, 731, 736–8; II 566
  calendar systems, I 793–4, 795, plates 33–5
  canals, I 469–70, 531–2
  brewing techniques, I 279–80
  bridge, river Euphrates, I 256, 715
  building, I 306, 457, 459–73, plates 18, 20
  ceramics, *see* pottery
  chemical arts, I 239–45, 247
  copper production, I 563–4
  cosmetic arts, I 286–95
  culinary arts, I 272–82
  dams, I 529
  draught animals, I 720–1, 723
  food preservation, I 264–5
  gardens, I 550–4
  glass-ware, II 319–20
  glazing, II 316–17
  iron production, I 594, 597–8
  irrigation, I 532–5, 545–55
  ivory work, I 664, 670–2, 675, 677–8
  land reclamation, I 540
  language and writing, I 80, 101–2, 205, 206,
    252, 280, 747–52
  mathematics, I 785–91
  measures and weights, I 778, 780–2
  metal-casting, II 478, 480

metal-work, I 623–4, 626–7, 632–3, 638–41, 643–6, 648–9, 654–6, 659, 661
ploughs, I 540; II 84
potter's wheel, I 199–200, 202–3
pottery, I 378, 395, 405
qanaats, I 532–4
roads and streets, II 528, 494–5
salt production, I 258
sanitation, II 665
ships, *see* boats and ships
silver and lead trade, I 583
soap and detergents, I 261–2
social organization, Sumerian, I 46, 48–9, 50, 52, 54–7
textiles, I 432
threshing-sled, II 106
transport, I 705–6, 710, 713
water-lifting, I 524–5
watermills, II 614
wells, I 526–7
wheeled vehicles, I 207–8, 209–10, 717–18, 725–7
wood-work, I 684
woollen industry, I 248–9
writing, *see* language
Messenger-services, II 495–6, 508, 514, 524, 527
Metal-mining, IV 64–79; VI 410–26; *see also* Metal-work, Metals and metallurgy
aluminium, VI 410, 413
bauxite, VI 413
boring, IV 67–9, 72
chromium, VI 410
copper, I 563–6, 567; VI 410, 413, 418, 420, 424
drainage, IV 77–9
explosives, IV 70–1, 75
extraction, IV 72–6
gold, I 526, 563, 579–81, 589; II 41–2; VI 410, 413–17, 422, 424
growth of demand, VI 410–11
haulage and hoisting, IV 76–7, plate 5B
iron, I 568, 574, 593; VI 410, 413–16, 424
lead, VI 410, 423
location, IV 64–7
machinery, IV 71–3, 76–9
manganese, VI 410, 424
mica, VI 424
nickel, VI 410
shaft-sinking, IV 69
silver, VI 410, 413, 424
surface methods, VI 411–17
tin, I 563, 589–90; VI 410, 414, 416, 424
tools (hand), IV 67–8, 70, plate 5A
underground, VI 417–25
ventilation, IV 69
zinc, VI 410, 423
zircon, VI 424

Metal-work, I 623–62; II 449–92
Metals and metallurgy, I 572–99; II 41–80; III 27–71; V 605–35; *see also* Metal-mining, Metal-work
aeronautical engineering use, VI 438–43
alchemical equipment, II 731–52
alloys, V 605, 607–8, 626
aluminothermic reduction, V 76, 97–8
annealing, I 624–5, 636–7; III 55–6, 207, 215, 225, 227, 238–9, 393
chronological chart, metallurgy, I 575
assaying, base metals, III 59, 62–5
automobile engineering use, VI 429–38
bearing-metals, III 41–2
bell-casting, II 64
books, II 63–5; III 27–9, 33, 63, 68, 79
cannon, III 361, 363
casting and moulding, I 588, 601, 606–8, 625–35; II 475–80; V 605–9; VI 428
colouring, I 243
copper-casting, I 588, 601, 606, 627–8
copper-steel alloys, IV 108
corrosion, VI 450–3
density, III 57–8
die-cutting and stamping, II 485–92
drawing, cartridge cases, V 629–31
early use, I 46–7, 59–60, 64–5, 73, 503–5, 572–9, 600–1; II 770–1
electrical communications, VI 448–50
electrical engineering, VI 443–8
electroplating, V 633–4
electrolytic refining, V 76, 79, 83–6, 88, 90–4, 96–7
enamels, I 659–62; II 63, 450–5
engineering construction, VI 427
engraving, I 648–9; III 405–8, 623–4
extraction, I 582–5; V 72–101
ferrous, IV 99–117
filigree and granulation, I 654–9
forging, V 610–11
fractures, III 57–8
German pre-eminence, 16th century, III 678
galvanizing, V 624–5
gold refining, I 580–2, 594
heavy plate, V 613–15, plate 38B
inlays and enamels, I 659–62; II 63, 450–5
interaction, metals, III 58–9
machines, III 326, 328, 330, 332, 336–7; VI 453–7
manufacture, metal goods, III 678; VI 457–61
mechanization, II 78; III 30, 340–1
medieval metallurgy, II 61–9
Mond carbonyl plant, V 88–90
nickel-steel alloys, IV 108
niello inlaying, II 453, 454, 480–1
non-ferrous, III 50–1; IV 118–47
ore, concentration methods, V 73–6

Metals and metallurgy (*cont.*)
  ornamentation, weapons, II 455–8
  patination, II 483–4
  plating, III 42; V 632–4
  precious metals, II 41–6; III 63–8
  printing, III 43, 385, 393, 405–8
  refining, II 42–5, 47–50, 56–8; III 42; V 76, 79, 83–6, 88–97
  repoussé work, I 642–8; II 468–75
  Rittinger shaking table, V 74
  riveting, I 649–50; II 469, 483
  rolling, V 611–20
  sheet metal, I 636–9; III 340; V 615–20
  smelting, I 564, 573, 576–9, 585–8, 593–8, 617, 633–4; IV 70–4
  soldering, I 649–54
  stamping, I 648
  structure, III 57
  tensile properties, III 58–9
  thermal expansion, IV 395–6
  transmutation, base metals, II 352, 731–2, 736
  tubes, V 625–9
  utilization, VI 6, **427–61**
  water-power, II 62, 68–9, 73–4
  welding, I 649–54; V 631–4
  Wilfley table, V 74
  wrought metal (statues), I 639–42
Meteorological Office, Air Ministry, IV 491
Meteorology, IV 620–1
Methane, VI 547; *see also* Natural gas
Méthode de Nomenclature Chimique (1787), IV 222
Mexico, oil, VI 185
Mezzotint engraving process, III 406–8
Mica, ore concentration, VI 424
Micrometers, III 640, 644; IV 397, 424, 426
Microphones, VII 1264–5
Microscopes
  balances, use, IV 405
  compound-, invention, III 232–3
  electron microscope, VI 623; VII 1350–3
  food-testing, III 21
  lenses, IV 358; V 673
  metallurgical use, III 41, 57
  petrological, VI 623
  position-finding, IV 394
  precision tools, IV 392–3, 394, 397, 399, 403
  screws, focussing, III 336, 635; IV 386
  seventeenth-18th century, III 630–5
  survey instruments, III 554; IV 401, 402, 604
  twentieth century, VII 1350–4
Milestones, II 507; III 504
Military technology, II **695–730**; III **347–76**; *see also* Warfare, Wars named
  aeronautics, VI 439; VII 793
  air power, VII 811, 816–19
  aircraft; *see also* Aircraft headings
  aircraft design, VII 793–810, 1009–22

aircraft operation, VII 789–96, 800–4, 806–7, 825
atomic bomb, *see* nuclear weapons
ballistic missiles, VI 24; VII 816
Battle of Britain, VI 622; VII 812
Berlin airlift, VII 828
biological warfare, VI 322
Blitzkrieg, VI 18, 20; VII 811, 812
bombing, VI 11, 19–21; VII 795–6, 811–16, 844, 845, 847
dreadnought class battleships, VI 5–6
explosives, VI 547–50
fighter aircraft, VII 793–5; *see also* Aircraft
government's role, VI 42–4, 113–15, 121–5
helicopters, VII 821–3, 834
incendiary bombs, VII 816
Inter-Continental Ballistic Missiles (ICBMs), VI 24
jet aircraft, VII 817–21, 1019–24
jet propulsion, VII 817, 819–21, 990, 1017–19, 1021–2
Luftwaffe, VI 17, 19; VII 812
navies, VI 4–6, 9, 10, 12, 21; VII 751–3, 814, 844–5, 853
navigational aids, VII 816–18
nuclear weapons, VI **268–83**
pyrotechnics, II **374–82**
radar, VI 350, 449–50, 559; VII 812
rocket propulsion, VI 22; VII 812
Royal Air Force, VI 17, 19, 558; VII 796, 811, 812
Royal Navy, *see* Admiralty, Royal Navy
surveillance from space, VII 866
tanks, VI 11, 22
two world wars, VI 8–13, 18–23
Milk industry, VI 302, 312, 584–5; VII 1401, 1414, 1421–2; *see also* Dairy products
Milking machines, VI 312
Mills, miscellaneous; *see also* Millstones
  boring (metals), IV 208, 421–2
  churn-, butter (1660), III 14
  donkey-mills, III 110–11
  drainage (land), III 106, 303–7, 321–2, plate 20
  early history, I 273–4
  floating, II 607–8, 610; IV 209
  flour-, I 72, 273–4, 593, 600–1; II 103–4, 107–11; III 15–19; IV 150, 151; V 29–30; *see also* Corn-mills
  frizing- (1763), III 176
  fulling-mills/stocks, II 216, 355, 609, 610, 611; III 170–3
  gig-, 15th–17th century, III 172–3
  glass-grinding, III 239
  grinding-mills, II 349; *see also* flour *above*
  hammer-mills, II 609–10, 611–12, 770
  hand-mills, I 274; II 109, 111, 653–4
  horse-, III 688, 702

machine-tools, IV 428–9, 434–6, 439, 441; VII 1038–9

mud-mills (dredging), IV 638–40, plates 45B, 46A

paper-making, III 412–13, 415

post-, II 536, 623–4, 649

roller-mills (crushing), II 112, 382; III 16–17, 19; V 30

rolling (metals), III 32, 41, 45, 47, 341–3; IV 161; V 611–20

rotary-querns, II 107–11, 127

saw-mills, II 62; III 106–7, 308; IV 202

slitting-, III 32

spinning-mills, IV 154

stamp-mills, II 66, 69, 107, 642; III 412–13, 415; IV 161

treadmills, II 16, 591, 602, 636–7, 642, 659–60; III 178, 366, 422

tide-mills, II 610, 614, 648; IV 209

twisting-mills (silk yarn), II 206–7

watermills, II 69, 349, 381, 395, 593–614, 647–8; *see also* Water-power, Watermills

windmills, II 349, 395, 396, 536, 610, 614–20, 623–8, 649; *see also* Windpower

winnowing-, III 14

Millstones (querns), I 72, 273–4; II 108–12; III 15–16; IV 200, 212

Millwrights, III 100–1, 107–8, 111, 114, 345

Mineral-water industry, IV 228

Mineralogy, treatise, III 28

Mines, named or places; *see also* Mining
  Bloomfield colliery, Staffordshire, IV 183
  Comstock, Nevada, IV 66
  Consolidated, Cornwall, IV 79
  Cook's Kitchen, Cornwall, IV 78
  Cornwall, England, V 127, 129
  Dolcoath, Cornwall, IV 190, plate 6B
  Ferndale colliery, Rhondda Valley, IV 91
  Fowey Consols, Cornwall, IV 194
  Griff colliery, near Coventry, IV 176
  Hebburn colliery, Durham, IV 93, 97
  Killingworth colliery, Gateshead, IV 189
  Long Benton, Northumberland, IV 179–80
  Murray, Sudbury, Cananda, V 87
  Parys mountain, Anglesea, IV 64
  Peacock, Hyde, Cheshire, IV 84
  Rio Tinto, Spain, V 77
  Risca, Newport, IV 91
  Walker colliery, Newcastle-upon-Tyne, IV 178
  Wallsend colliery, IV plate 6A

Mining, I 558–71; II 1–40; IV 64–98; *see also* Coal, Metal-mining
  coal, III 72–88
  chronology of techniques (chart), I 574
  flint, I 514–15, 558–63
  German pre-eminence, 16th century, III 678
  Greek, II 1–6
  legal aspects, II 53, 66; III 30
  medieval, II 10–24
  metals, VI **410–26**
  pumping devices, II 7–8, 16–22, 654–5
  regulations, miners, III 30
  Renaissance writers, III 27–8
  Roman, II 6–10, 41
  salt, I 566–7
  shafts, III 30
  tools, II 3, 5, 8–10, 12, 14–19, 24, 110
  ventilation, II 19, 20, 23

Minium (red lead), II 361

Minoan civilization, *see* Crete

Mints; *see also* Coins, Medallions, Presses
  Cassel, Sweden, III 343
  England, III 46; IV 143, 270, V 269
  France, III 47, 60, 384; IV 405, 436
  Rome, III 341
  Scotland, III 341

Mirrors, IV 362–4
  arsenicum metallicum, III 29
  bronze, II 449, 460, 463
  Celtic, II 460, 463
  glass materials, III 228
  Greek and Roman, II 449
  light velocity test, IV 409
  mercurial process, IV 362–3
  optic curvatures, III 232, 237; IV 361
  plate-glass, III 238; IV 362, 363, plate 21A
  reflective properties (1269), III 230
  silver salt process, IV 363
  solar furnaces, III 55
  speculum metal, III 38, 40–1, 237
  surveying, III 537, 555
  telescopes, III 237; IV 398, 400

Mispickel (arsenical pyrites), IV 121

Missile-engines, II 698–701, 707–14

Mittelalterliche Hausbuch, Das (*c.*1480), II 66, 67, 68; IV 425

Mixers, gold-amalgamation, III 327

Moabite Stone, I 764–5

Mockadoes (goat-hair fabric), III 179

Mohole Project, VII 1456–7

Mohs's scale, IV 339, 346

Molecules, Dalton's discovery, IV 224–6

Molybdenum, IV 119; VI 448

Money, II 773 n.; III 379–80; *see also* Coins, Economics

Mongol Empire
  European travel, III 523
  printing, III 378, 380

Monodermon, II 167

Moon
  Babylonian astronomy, I 798–9
  calendars, I 119–20, 293–6, 797; III 558–75, 577–81
  geology, comparisons, VII 1449

Moon (*cont.*)
  Lunar Society, IV 673
  moon-count notion, I 117–18
  navigation tables, III 552–3
  primitive time-reckoning, I 112–13, 115, 117–20
  space technology, VI 26, 118, 132; VII 857–9, 866–9
  tidal power, VI 216–18
  tide-tables, III 553
Mordants, II 350, 359, 364, 366–9; IV 231, 233–4, 244, 249–51
Moringa oleifera, I 288
Mortar, *see* Cement
Mortising, IV 462–8, 437, plate 26A
Morus alba, II 197
Mosaic decoration, I 462, 464, 465
  mosaic 'standard', Ur, I 705, 718
  'town mosaic', Knossos, I 485
Moseley Education Commission, U.S.A. (1903), VI 140, 148
Mossynoeci (people), II 54
Mother-of-pearl, I 465, 665, 670
Motion, *see* Mechanics, Rotary motion
Motor cars (road-vehicles), V 414–37; VII 711–35; *see also* Bicycles, Goods vehicles, Motorcycles
  Benz, V 167–8, 170, 426, 428, plates 16B, 17A
  Cannstatt-Daimler (Mercédès), V 428, plate 17B
  Daimler, V 428
  de Dion-Bouton, V 169, 170, 434
  design, VI 172–5; VII 711–22
  electric, V 419–20; VII 731–3
  engineering, VI 429–38
  engines, I.C., V 164–76
  Ford Models, V 433; VI 4; VII 714, 716, 718
  gear boxes, V 640–1; VI 431–2
  Isotta-Fraschini Tipo 8, VII 716
  Lancia Lambda, VII 716–17
  Lanchester, V 430–2, plate 19A
  Mercédès (*see* Canstatt-Daimler)
  Oldsmobile Curved Dash Runabout, VII 713
  petroleum motor-carriages, V 426–34
  Peugeot, V 171–2, 429
  production, VII 704–6, 709–10, 714–15, 726
  radiators, VI 429–30
  Renault, V 433–4
  road carriages, V 418–33
  road surfaces, requirement, IV 536–7
  Rolls-Royce Silver Ghost, VII 716
  Rolls-Royce Silver Shadow, VII 718
  Rover Ten, VI 431
  shows, VII 872
  Siegfried-Markus, V 425, 426, plate 18A
  steam, V 420–6; VII 720–2
  transmission, VI 431–2, 437–8; VII 715, 719–20

  tyres, VI 339, 563; *see also* Tyres
  Wolseley, V 432–3
Motor-launches, V 165
Motor-tricycles, V 169, 424, 429, 434
Motorcycles, V 434–6; VII 733–5
Motorways, VII 875–6
Moulds and processes
  assaying, III 60–1
  casting metal, III 39, 43–5, 47, 53; *see also under* Metals
  paper-making, III 412–13, 415
  glass bottle-making, V 674–7; VI 580–5
  glass television-tubes, VI 588
  injection-moulding, plastics, VI 552
  plastics, thermosetting, VI 555
  printing, III 385, 387, 391–3, 402
  rubber extrusion, V 764–5, 772
Mousterian culture, *see* Cultures
Movement, *see* Rotary motion
Mummification, I 260, 266–70, 277, 756
Munich Academy of Sciences, IV 654
Muntz-metal, IV 130, 133; V 607–8
Murex purple, I 247–8
Museums
  Athens, National Museum, III 618
  Blaise Castle House, Folk Museum, Bristol, IV 209–10
  Bristol City, V 423
  Chadwick Museum, Bolton, IV 279
  Conservatoire des arts et métiers, Paris, V 742
  Curtis Museum, Alton, Hampshire, III 116
  Danish Technical, V 429
  Greenwich National Maritime Museum, III 497
  Hôtel des Monnaies, Paris, III 340
  Leiden Museum for the History of Science, III 663
  Modern Art, New York, VII 953
  Munich National Museum, III 598
  Oceanographical, Monaco, VII 1447
  Plantin-Moretus, Antwerp, III 401 n.
  Reading Museum of English Rural Life, III 149 n.
  Science Museum, London, IV 318, 423, 456, 639, 652; V 139 n. 1, 143, 181, 432, 742; VII 1069, 1174–6
  Sheffield Art Gallery and Museum, V 281
  Smithsonian, Washington, VII 862
  Stockholm City Museum, III 224
  Swedish Commission on Mines, III 342
  Technical, Vienna, V 426
  Teylers Museum, Haarlem, IV 408
  Victoria and Albert Museum, London, III 186
  Wolseley Company's private museum, V 433
Music
  instruments, II 149, 180, 182, 256, 614, 634
  origin claim, III 58 n. 1

printing, III 405–6, 406–10
wire instruments, V 623
Muslim people
 ceramics, IV 334
 dials, III 596, 600
 globes, III 612–13

Nagas, pottery, I 406, 407
Nails
 brass, II 255–6
 building construction, II 403, 443
 cranes, avoidance, II 661
 furniture, Roman periods, I 234, 242 *and* n.,
  255
 -making, II 76; III 32, 116, 342; V 623
 Middle Ages, II 76, 394, plate 30A
 nail-lifters, II 231
 wrought iron, II 443
Naphtha, I 251; II 374–5, 376; V 105; VI 467
'Napier's bones' or 'rods', III 627, 636
National Academy of Sciences, U.S.A., VI 271
National Aeronautical and Space Administration,
 VII 857, 858, 869
National Association of Master Bakers, V 28
National Bureau of Standards, VI 230
National Coal Board, U.K., VII 1083
Natron (sodium carbonate), I 257, 259–60,
 268–70; II 354, 355, 737
Natufians of Palestine, *see* Cultures
Natural gas, IV 258; VI **376–409**
 ancient times, importance, I 251
 Baku, Marco Polo's visit (1272), V 102, 104
 consumption, VI 176
 handling, VI 405–8; VII 1481
 supply, VI 191–2, 407–8
Natural power, VI **195–222**
 dams/reservoirs, VI 200, 212; VII 1367–70
 geothermal, VI 219–21
 hydro-electric, VI 172, 195–7, 200–1, 212
 pipelines, valves and turbines, VI 201–3, 203–8
 solar, VI 221–2
 tidal, VI 216–18
 water, VI 172, 183, 195, 213
 wind, VI 172, 214–16
Nautical Almanac, the, IV 597–8
Naval Architects, Institution of, Britain, V 372–3,
 386
Navigation and equipment, III **501–29, 530–57**; V
 438–65; VII **837–56**
 air safety, VII 854, 856
 airborne aids, VII 849, 851–3
 artificial horizons, V 457–8
 astro-navigation, VII 849, 852
 automatic instruments, VII 841–4, 851
 buoys, V 463
 charts, III 510, 517, 526, 536, 544, 548, 550–1,
  555–7

chronology, 20th century, VII 855
chronometers, III 662, 672; V 458
collision avoidance, VII 849–51
compasses, V 454–6, plate 31C
direction finding, VII 838–41
electronic, VII 845–8, 850–1
fog-signals, V 464
gyroscopes, VII 838, 842–3, 853–4
inertial navigation, VII 853
instruments, III 523–9, 546, 549, 555–7, 627;
 IV 400; V 454–64; VII 841–5, 849, 852–3
lead-lines, V 460–2
lighthouses, *see* Lighthouses
logs, marine, III 549–50; V 458–60
Marteloio, Italian method, III 526
methods (1400–1750), III 544–53
missiles, VII 848–9
pilots, shipping, III 547–9
quadrants, V 457
radar, VII 844, 847, 850
radio and radio beacons, VI 349–50, 838–9
sextants, V 456–7; VII 853
sonar, VI 350; VII 844
sounding machines, V 460–2
station-pointers, V 462–3
tables, III 526–7; IV 597
Navigations, *see* Canals, Waterways
Nazarites, I 285
Needles; *see also* Compasses, magnetic
 etching, III 407
 knitting, III 183–5
 magnetic, III 518, 540–1, 523, 548–9
 manufacture, II 75; V 624
 sewing-machine, V 589; VI 679–82, 683
Needle-shuttles, I 432–4
'Neglect of Science Committee' (1916), VI 151,
 154, 157
Neolithic, *see* Agriculture, Cultures
Netherlands, *see* Holland
Nets, netting, I 166, 167, 170, 176–7, 178; III
 181–2; VI 345–9
Nettles, fibre use, I 355–6, 449
New Atlantis (1627), Francis Bacon, III 718
New Zealand
 agriculture, V 2–3, 7, 9; VI 307, 311, 312
 Cook's chart and voyages, IV 575, 598, 617
 food production, V 2–3, 7 9
 geothermal power, VI 220
 railway engineering, V 347
 refrigeration, V 49
 telegraphy, V 227
Newcastle-upon-Tyne Chemical Society, England,
 V 214
Newcomen Society, IV 206–7
Newspapers; *see also* Books, Paper, Printing
 hand-presses, V 693
 illustrations, V 701–3, 704, 705–6; VII 1269

Newspapers (*cont.*)
  Linotype machine, V 686–7, 688
  paper-making machine, VI 615
  phototypesetting, VI 29; VII 1279
  rotary-presses, V 697–701; VII 1278–9
  steam-presses, V 696
  type-casting machines (*c.*1851), V 683
  typesetting, VII 1272–5
Nickel, III 30
  alloys, IV 146; V 608; VI 430, 443, 630
  discovery, Canada, IV 66
  extraction, IV 119, 145–6; V 76, 79, 86–90, 96, 97
  glass colouring pigment, IV 374
  mining, VI 410
  Morfa Works, Swansea, V 86
  nodules, oceanic, VII 1465
  plating, electrolytic, IV 146; V 86
  production of, V 72–3
  steel alloys, V 64, 66, 88, 608
Nicobar Islanders, I 170
Nicotiana, III 8–9
Niello, I 623–4; II 453, 454, 480–1; III 48, 406
Niobium (columbium), IV 119
Nitre (potassium nitrate), I 261–2; *see also* Saltpetre
Nitric acid
  aqua fortis, II 695, 706
  aqua regia, II 356–7, 706; IV 144
  brass-casting process, II 39
  brass surface effect, IV 133
  caput mortuum residue, II 63 n.
  cochineal dye, II 695
  early descriptions, II 356
  explosives, V 286–7, 290; VI 522, 528, 547
  manufacture, II 12; IV 233, 242; V 255; VI 522–3, 527–8
  nitrobenzene synthesis, V 270, 274
  'parting' acid, gold, II 65; IV 141, 142–3
Nitro-celluloses, V 290–6, 729; VI 547–8, 600
Nitrogen fixation, V 255; VI 501, 504–5, 521–8; *see also* Ammonia, Fertilizers, Nitric acid
Nobel Prizes, VI 3, 557, 559; VII 1340
Nocturlabes (nocturnals), III 546, 614–15, 625
Nogg *or* Moot (tool), III 119
Noose weapon, I 166
Nootka peoples, I 168, 176, 177
Norfolk four-course rotation, *see* Agriculture
North America
  copper production, V 77, 79
  farming and food, V 2–3, 24, 30, 51
  oil, V 112
  population growth, V 1
North Atlantic Treaty Organization (NATO), VI 24
North of Scotland Hydro-Electric Board, VII 1072

Norway; *see also* Europe
  agriculture, V 20
  coal imports, VI 181
  fish industry, IV 48–9, 54
  hydro-electric power, VI 196
  nickel, IV 145
  nitrogen fixation, VI 522–3
  ships, ship-building, IV 593; V 380–1
  surveys, IV 610
  whaling, IV 60–1; VI 346–7
Nuclear power, VI **223–67**; *see also* Nuclear weapons, Reactors named
  Argonne, VI 255
  atomic fission, VI 224–5, 269–70
  atomic structure, VI 223–4
  Calder Hall, VI 66, 249, 251
  chemical processes, VI 256–67
  costs, development, VI 66
  Dounreay power-station, VI 248
  energy resources, VII 1481–2
  Hanford power station, Columbia, VI 239, 240, 243, 245, 252, 257–8, 261–2
  Harwell, VI 67, 243–6, 248–50, 265, 281
  Manhattan Project, VI 132, 227, 230, 231, 233, 238, 241, 245, 252, 258, 259, 271; VII 857–8, 1468
  moderators, VI 226, 238, 241, 243, 268
  Oak Ridge power station, VI 230–4, 239, 252, 260–1; VII 1333
  plutonium, VI 227, 238, 260, 271–3
  power stations, construction, VI 248–56, 265
  reactors, construction, VI 236–56
  Risley power station, VI 245–7, 249–50, 263, 265
  Springfields power station, VI 257–8
  uranium, VI 228–36
  Windscale power station, VI 247
Nuclear Research Office, Germany, VI 268
Nuclear weapons, VI **268–83**; *see also* Nuclear power
  atomic bombs, VI 23–4, 226–8, 260–71, 279–80, 548; VII 816
  atmospheric pollution, VI 283
  British programme, VI 281–3
  deuterium research, VI 278
  fall-out, VI 283
  gun technique, VI 273–4, 276
  implosion technique, VI 274–5, 276, 281
  international control, VI 277
  nuclear explosions, VI 276, 280–1
  research before World War II, VI 268–70
  Soviet programme, VI 279–80
  thermo-nuclear tests, VI 282
  U.S.A. programme, VI 277; VII 1333
Numerals, Arabic, II 766–7, 772; III 502
Nut-milling machines, IV 429
Nuts (metal), III 339

Nuts and oil-seeds, *see* Agriculture, Plants
Nylon, VI 349, 501–2, 560–1, 646, 650–1, 672

Observatories; *see also* Astronomy
  clocks, III 665, 667, 671
  Greenwich, England, IV 607; V 448, 458
  instruments, III 584–5, 586–94, 625, 639, 642;
    IV 380, 398–400
  Montpellier, France, V 398
  Paris, IV 394, 597, 607; V 739
Obsolescence, VI 74, 124
Occam's razor, IV 224
Ocean depths, VII **1446–66**
  depth sounding, V 460–3; VII 1446–8, 1452,
    1455, 1458
  gravity measurements, VII 1462–3
  heat flow, VII 1464
  magnetic measurements, VII 1462
  minerals, VII 1464–6
  oceanographic surveys, VII 1447–8
  sea-floor exploration, VII 1448–55, 1458
  sediments, VII 1455–8
  seismic research methods, VII 1458–62
Oceanographic Institute, Paris, VII 1447
Octants, III 555–6, 636; IV 400
Odyssey, I 444
Oil wells and oilfields, named; *see also* Petroleum
  oil
  Caspian oilfield, VI 408
  Drake Well, U.S.A., V 105, 106, 110; VI 377
  Duck Creek, Ohio, U.S.A., V 108–9
  Galicia, Spain, V 110–11
  Huntingdon Beach oilfield, VI 392
  Sarata-Monteoru, Rumania, V 111
  Spindle Top Well, Texas, VI 381, 397
  Summerland oilfield, VI 386, 389
  Wietze, Hanover, V 110, 111
Oils; *see also* Petroleum oil
  'coal-', V 104, 110
  cosmetic, I 286, 287–9, 290–2; V 301, 316–17
  diesel, V 120
  -engines, V 160–3
  essential oils, III 12; V 301, 316–17
  explosives, V 294 n. 1
  fat, conversion, IV 61
  inks, oil-varnish, III 405, 408
  lamp-, *see under* Lighting
  linseed-, III 381, 697; *see also* Flax
  lubricating, *see* Lubricants
  mineral-, V 102–6, 111–12; VI 185–91
  'Oil of Petre' ('British Oil'), III 687–8
  olive-, *see* Olives
  ore flotation process, V 74–5
  -painting, III 242
  paraffin- (kerosene), V 104, 105, 116, 160
  soap manufacture, II 356 *and* n.; III 703–5

  vegetable-, I 354, 360; III 5; V 20, 23–4
  whale-, IV 55–7, 58, 61–2, 264, 272, plate 4
Olea europaea, I 288, 359
Olives
  consumption, medieval, II 127
  cultivation, I 359–60; II 121
  oil-mills, II 112
  oil-presses, II 112–18
  oil production, II 106, 121–3
  soap manufacture, II 356
Olympiads, III 570, 572, 578
Olympic games, III 569, 572
Onagers (ancient catapults), II 713–14, 723–4
Opium poppy, I 358–9
Optical instruments, *see under* Lenses
'Optics', of Alhazen and Ptolemy, III 230
Opuntia coccinellifera, I 245; II 366
Ordnance, Board of, England, IV 604, 618
Ordnance Survey (Great Britain), IV 601–2,
  607–8, 622, 626; V 439, 444, 452
Ores, assaying, III 62–4; *see also* Mining
Organization for Economic Co-operation and
  Development (OECD), VI 49, 114
Organs, *see* Music, instruments
'Orlon' (acrylic fibre), VI 651
Ornamental turning, IV 419–20, 425; *see also*
  Machine-tools
Orphreys, III 196, 200, plate 12C
Orpiment (arsenic mineral), II 362, 737, 747; III
  29; IV 122; *see also* Arsenic
Orreries (planetaria), III 616, 618, 638, plate 27B
Osmium, IV 119, 144
Osmund (Swedish iron), II 70 *and* n., 76
Otto engines, *see* Internal combustion engines
Ovens, I 272–3, 321; II 118, 119, 127; VII
  1136–9
Oxalic acid, synthesis, V 311
Oxen, III 148; *see also* Livestock
Oxides, metal, III 29, 218–19, 234, 237, 241
Oxygen, Lavoisier's discovery, IV 220

Pack-horses, III 139, 141–2, 417, 438
Packaging industry, VI 502
Painting, paints, VI **590–606**; *see also* Art,
  Chemical industries, Colours
  air-drying, VI 594–8
  application methods, VI 604–6
  Cennino's handbook (*c.*1370–*c.*1440), II 352–3
  classification, 20th century, VI 595–604
  decorative, VI 594
  distemper, VI 602
  early man, I 33–4, 146–9, 235, 451, 744, 745,
    plate 2
  eye/face paints, ancient, I 292–4
  emulsion, VI 602
  enamels, II 462; *see also* Enamels
  finishes, VI 594, 599, 600, 603

Painting, paints (*cont.*)
  fossil resins/lacquers, VI 600–1
  frescoes, I 462–3, 470–1, 486
  gesso, II 249, 363
  glass, II 323, 327–8, 341–2, 444
  gold effect, I 49
  grounds, I 240–3; II 359–64
  gums and resins, I 244–5; VI 595–600, 603
  inks, I 245, 247
  lacquers, VI 100–1; *see also* Lacquers
  layering of paints, II 362
  lustre-painting, II 303–4, 342
  manufacture, IV 256; VI 590–4
  media, I 243–4
  minai technique, II 306–7
  mural-, I 241–3; II 363; IV 447
  pigments, I 238–40, 244, 502, 514; II 359–64; VI 591–4
  pottery, I 241, 383–4, 408, II 264–5, 268, 301–4
  reproductions, III 408
  resins, VI 595–600, 603; *see also* gums
  stained-glass, III 241–2
  tempera, I 243; II 363–4
  varnishes, I 243–5; II 359–64; III 381, 405, 687, 695–6
  water-borne, VI 594, 601–2
  wood, II 242, 247–9, 363, 443
Paiute Indians, I 172–3, 183, 184, 300
Palace of Minos, Knosses, I 486–7, 672, 714, 715
Palanquins, I 705
Palestine, I maps 4, 8, table F; *see also* Bible references, Jews
  agriculture, primitive, I 520
  alum, I 263
  basketry and matting, I 419–20
  beer trade, II 140
  ceramics, I 200–2, 204, 387, 390
  copper production, I 564
  cosmetic arts, I 289, 292
  culinary arts, I 271–2, 274, 277, 281–2, 285
  dyeing, I 250
  food preservation, I 265
  fuller's soap, I 261
  glass-ware, II 319
  iron production, I 597
  ivory-work, I 672, 675
  lamps, earthenware, I 236–7
  metal-work, I 633–4
  Nastufian, *see under* Cultures
  Phoenician alphabet, I 764–6
  Phoenicians, *see under* Cultures
  potter's wheel, I 200–2, 204
  salt production, I 258–9
  textiles, I 432, 434
  threshing methods, II 106
  water conservation, I 530–1

Palladium, IV 119, 144
Palm-oil, V 23
Paludina (fresh-water snail), II 36
Panicum (millet), I 369–70
Papaver somniferum (poppy), I 358
Paper, VI **607–21**; *see also* Parchment, Printing
  chlorine bleaching process, IV 255
  'deckle', III 413, 415
  esparto-grass, I 450; V 713
  European introduction, III 378
  Foudrinier machine, III 411, 416
  'Hollander', III 412, 415
  invention, China, II 771
  machines, paper-making, III 382; VI 614–20
  manufacturing, III 411–16; IV 255; VI 610–12
  manuscript printing, III 383
  money, III 379–80
  paper-mills, II 608, 611, 643
  papyrus, *see* Papyrus-reed
  pulps, III 411, 415; V 713; VI 612–13
  raw materials, III 412; V 712–13; VI 607–14
  sizing, III 415
  stamping-mill, III 412–13, 415
  wallpaper, III 415
  watermarks, III 413–14
Papyrus Carlsberg 9 calendar (*c.*A.D. 114), III 563
Papyrus-reed, I 756, 757; II 188, 197
Para rubber, *see* Rubber
Parachutes, V 398–9
Paraffin, *see* Petroleum oil, kerosene
Parchment; *see also* Paper
  bookbinding, II 180
  Egypt, I 757; II 164, plate 5C
  manufacture, II 187–8, 189–90
  rolls and codices, II 188–9, 190
Paris Convention (1919), VII 699
Parliament
  Acts, *see* Acts of Parliament
  Commission on Secondary Education (1899), VI 144
Parsons and Marine Engineering Turbine Research and Development Association, VII 747
Parthians, II 706
Pasteurization, milk, VII 1417, 1421–2
Paston letters and ships, III 477
Patagonian Indians, I 175
Patents, IV 149–50; VI 44–5, 72, 115; *see also* Innovation, Inventions
  alkalis, IV 237–9
  bleaching, IV 247–8
  building construction, IV 448, 457, 459, 461
  canals, IV 570
  ceramics, IV 340–2, 347, 352, 356
  dredging, IV 630, 635–6, 640–2
  engines, IV 171–2, 174–5, 177–8, 183, 185–6, 188, 191, 193, 499

gas industry, IV 261, 271
glass-making, IV 358, 360, 362, 368, 375
locks, IV 423
machine tools, IV 426, 432, 434, 437
paper-making, IV 255
road-making, IV 538
sulphuric acid, IV 244
superphosphates, IV 254
tar, IV 252
textile-machinery, IV 278, 280, 286, 288, 290,
    292–4, 299–300, 302–3, 311, 313–15, 317,
    321–5
water-works, IV 499, 501, 503
Patterns, III 182, 186–7; *see also* Fabrics, Textiles
Pavements, paving, II 28, 528–30, 531–3; IV 505–
    6; *see also* Roads
Paviotso seed-gatherers, I 154–5
Paviours, II 525–6, 532–3
Peat, salt resource, II 683
Pekin man, *see* Man
Pendulums, III 267, 344–5, 553; IV 395, 408–9,
    411; *see also* Clocks
Pennisetum (pearl millet), I 369
Peperino stone, II 406, 410, 529
Perfumes, I 289–91; V 300, 317
Persea, I 542
Persia, I map 5
  architecture, III 245
  brass production, II 54
  ceramics, II 285–6, 302, 306–7
  figured-silk weaving, III 193, 195
  furniture, II 224
  measurement, I 779, 782–3
  metal-work, I 660, 661; II 174, 453, 472, 481–2
  potter's wheel, I 203
  roads, I 713; II 495–7
  spinning and weaving, II 191
  steel production, II 57
  syllabary, 5th century B.C., I 752
  trade routes, III 380
  water-wheels, II 676
  weaving, III 191, 193, 195
  wheeled vehicles, I 204–9
  windmills, II 615–17
Peru
  basketry, I 414, 415, 421–2
  early cultures, I 82–4
  reed rafts, I 731–2
  textiles, I 424, 436, 445, 449
Pestles and mortars, I 273, 502, 514; II 106–7,
    380, 382, 747, 749
Pests and pesticides, VI 315–16, 322–5, 333–4,
    674
Petroleum oil, V **102–23**; VI **376–409**; *see also*
    Asphalt, Bitumens, Natural gas
  atomizing processes, V 117
  books, V 115

  chemical investigations, V 106, 109–11, 115
  consumption and demand, VI 172–5, 176–7,
      759
  cracking processes, V 112, 120–1; VI 29, 55–7
  drilling, V 106–12; VI 376–94; VII 1481
  gasoline, *see* petrol
  geology, applications, V 111–12
  handling, VI 405–8; VII 1481
  internal combustion engines, V 160–75, 426–36;
      VII **997–1034**
  kerosene (paraffin), V 104–6, 110, 116–17, 120,
      160; VI 187
  marine and off-shore, VI 386–90; VII 1464–5
  marketing, VI 408, 543–4
  oil-fields, V 102, 110–12; VI 390–42, 402–4,
      405; *see also* Oil wells
  petrol, VI 533–5; VII 1002, 1014
  petrol, first use of term, V 429
  products, I 250–6; II 374–6
  refining, V 112–16; VI 408, 543–7
  storage, V 122; VI 406–7
  transport, V 103, 120–2; VI 408
  uses, V 102–3, 116–21
  water removal, VI 407
Pewter; *see also* Type-metal
  antimony ingredient, III 29, 42, 43
  assay methods, III 43, 68
  casting, III 386, 393
  cochineal scarlet process, III 695
  composition, II 47; III 42–3
  engraved plates, printing, III 408–10
  table-ware, 17th century, III 20
  zinc, name 17th century, III 30, 690–1
Phaistos disk, I 759
Pharmaceuticals, I 258, 260, 287; V 317–18; VII
    1326–7
Phaseolus (grams), I 361
Philosopher's stone, II 732
Phlogiston theory, III 57, 720; IV 218, 220
Phoenicians
  alphabetic script, I 104, 763–8
  dyeing, I 247
  glass-making, II 321
  harbour-works, II 516–17
  irrigation, I 531
  ivory work, I 672, 674–6, 678–81
  measurement, I 782
  shipbuilding, II 568
  traders, I 56, 664; II 568
  weapons, I 512
Phoenix dactylifera (date-palm), I 371, 450
Phosphates, IV 246, 253–4; *see also* Fertilizers
Phosphorus, V 54, 60, 253–5
Photography, V **716–33**; VII **1281–1301**; *see also*
    Cinematography
  ambrotype, V 725, plate 42B
  apparatus, VII 1287–8

Photography (*cont.*)
  camera obscura, V 716–17
  cameras, V 717, 722, 723, 726–7, 730–2; VII 1287–94
  colour, VII 1295–8
  daguerrotype system, V 720–2
  exposure meters, VII 1284–5
  film materials, V 729–32, 747–51; VI 553, 554; VII 1283
  glass base, V 724–9
  lenses, VII 1281, 1286–7, 1289
  metal base, V 719–22
  miniature cameras, VII 1288–92
  panchromatic materials, VII 1282–3, 1303–4
  paper base, V 722–4
  photo-chemistry, V 717–19; VII 1296–7
  photo-etching, V 719, plate 42A
  Polaroid system, VII 1294–5
  printing industry, V 703–8; VII 1275–9
  printing paper, VI 554–5
  radiology, VII 1299–1300
  sensitometry, VII 1283–4
  shutters, VII 1285–6
  studio equipment, VII 1292–4
  surveying use, V 448–9
Phragmites communis (reed grass), I 450
Physical Therapy, Council on, VII 1349
Physika kai Mystika (Demokritos), II 732
Picks, I 514–15, 558, 561, 563, 620; II 3, 5, 8–10, 12, 16, 99; *see also* Mattocks
Picric acid, V 296; VI 548
Pigments; *see also* Colours, Dyes
  ancient civilizations, I 238–40, 502
  beeswax, varnish fixative, I 244
  eye/face paints, I 292–4
  industrial, II 359–64; III 692–5; IV 256; V 257–83; VI 563–8
  paint, II 249, 359–64; VII 591–4
  preparation, I 502, 514
Pile-driving and piles, II 410, 436; III 420–7
Pimpinella anisum (aniseed), I 288
Pincers, II 630; III 110–11, 113, 120, 212; *see also* Pliers, Tongs
Pins; *see also* Nails, Needles
  banking-, clocks, III 657
  firearms, III 354, 359
  manufacture, V 624
Pipes; *see also* Tubes
  drainage, IV 508, 512–13, 630–1; VII 1386–7
  gas, IV 123, 512
  hollow tree-trunks, III 329, 330–2
  stoneware, V 669; VII 1386
  suction-pump systems, III 329, 330–4
  water, IV 123, 512; V 559–60, plate 33A; VII 1371
PIPPA atomic reactors, VI 249, 250

Pirotechnia, Vanoccio Biringuccio (1540), II 475; III 27, 210, 214–15, 364
Pistons
  fire-piston (fire-making), I 226–8
  Greek machinery, II 634
  I.C. engines, V 157, 163–6, 169–76; VII 1004
  pumps, III 330–3
  steam-engines, IV 169–71, 176–87, 196; V 124, 126–8, 130–7
Pisum (cultivated pea), I 361–2
Pitch, III 679, 685; V 102; *see also* Asphalt, Bitumens
Pithecanthropus, I 22–3, table C; *see also* Man, Pekin
Pitjendadjara tribe, I 23
Plane-table (survey), III 540–3, 553, 556; IV 600, 603, 616
Planes, carpenter's
  machine-tools, IV 426, 429
  Middle Ages, II 243, 250, 253, 392–3, 641
  Roman periods, II 230–2, 243
  tradesmen's tools (*c.*1500–1850), III 111, 113, 115, 118, 121, 130, 131
Planetary models/calculations, III 614–18
  nocturnals, III 546, 614–15, 625
Planing machines, metal, IV 424–5, 429, 432–5, 441, plate 25B; V 650–2
Planispherium, Ptolemy, III 603, 605–6
Planning, *see* Management, Town planning
Plant breeding, VII 1407–8
Plant nutrition, IV 37–40, 42; *see also* Agriculture, Fertilizers
Plantago lanceolata, I 370
Plants, cultivation, I 353–75; *see also* Agriculture, Gardens
  fibre plants, I 372–4, 424, 447–51, 522
  gourds, I 372
  mustard and sugar, I 370–1
  origins, I 353–4
  reeds, I 371–2, 449–50
  trees, I 371, 450, 551–4
Plaster of Paris, I 482 n.; IV 212, 353, 449–50
Plastic-fibres, VI 299, 501–2, 560–3, 648–51, 659
Plastics, V 729, 747; VI 501–3, 551–63, 602, 643–4; VII 991–2
Plates, printing processes, III 406–10
Plates and dishes, metal, III 343
Platinum
  ceramic lustre-ware, IV 356
  extraction, IV 119, 144–5; V 90, 96–7
  glass textile process, VI 589
  measuring instruments, IV 397, 601
  printing patent (1810), IV 356
  vessels, refining, IV 141, 142–3, 145
Playing-cards, III 380–1, 385
  'Master of the Playing Cards', III 406
Pliers (glass-making), II 331

Plinth instrument, observatory, III 587–9, 599
Ploughs, ploughing, II 81–93, 94; III 134–5, 148;
    IV 2–4, 21
  ancient Egypt, I 539–40
  Babylonian (c.2000 B.C.), I 550
  Chinese (c.85 B.C.), II 770
  contour-ploughing, VI 319, 322
  drainage, III 135, 136
  evolution, I 73, 74, 504
  France (1755), IV 30
  mechanization (tractors), VI 334, 336–7
  North America (c.1880), V 6
  steam-ploughing, IV 24–5; V 11–12
Plumb-lines
  astronomy, instruments, I 122–3; III 582–3,
    587–8, 589, 590, 593, 594, 608
  carpenter's tool, II 252, 386
  gradient level, aqueducts, II 672
  mathematical instruments, III 512, 614–15
  medieval, II 386
  navigation instruments, III 526; V 452, 460–1
  plinth (observatory) instrument, III 587–8, 589
  sun-dials, III 598–9
  survey equipment, III 512, 538, 541, 611
  timing device (c.525 B.C.), I 122, 123
  tradesman's tool (c.1500–1850), III 110
Plumbago (graphite), IV 133
Plutonium, VI 227, 238, 260, 271–3
Pneumatic apparatus, see 'Gasometers', Pumps,
    IV
'Pneumatics', Hero's, IV 672
Pneumoconiosis, II 24 n.
Poison used in hunting, I 158, 163, 166, 169, 172,
    173
Poland
  coal, VI 180
  oil, V 103, 110–11, 113–14, 115, 116
  shipping, IV 558
  textile-machinery, IV 292
Polarimeters, IV 403
Polders, see Dikes and polders, Drainage
Police, R. Thames (1798), IV 467
Polimetrum, surveying instrument, III 538, 541
Polishes, polishing
  engraving plates, III 408
  furniture, Roman period, II 234
  lenses, III 233–7
  metal, II 449, 459; III 343, 350
  speculum-metal, III 40–1, 237
Pollution
  air-, IV 241; V 78, 236; VI 133; VII 821
  coal-gas impurities, IV 270–1, 272–3
  control legislation, VI 133–4
  domestic appliances, VII 1138–9
  house-drainage, IV 507–10
  nuclear weapons tests, VI 283
  pesticides, VI 133, 322, 324–5, 327

  sewage disposal, IV 518; V 552; VII 1395–7
  smoke, London, 17th century, III 77
  street refuse, II 532–3; IV 504–6
  water, IV 492–3, 495, 501–3, 509–10, 518; V
    552, 561–2, 566–7
Polverine (fused wood-ash), II 354, 356, 358
Polyester fibres, VI 501, 646, 650–1
Polygonum lapathifolium, I 370
Polymers, VI 501–2, 506, 551–63, 602, 643, 645–6
Polystyrene, VI 501–2, 555–6
Polytechnic Institute, Milan, VI 560
Polyvinylchloride (PVC), VI 556
Pontil (punty), glass-making, II 331, 332
Ponts et Chaussées, Corps des, IV 521, 525
Ponts et Chaussées, Ecole des, IV 442–4, 453–4,
    479–80, 526
Poor Law Commissioners (1843), IV 509
Population changes
  Australia/New Zealand, growth, V 7
  Britain, growth, IV 489, 506; V 1, 10, 823
  cartography, developments, IV 622–4
  Europe, growth (1800–1950), V 1, 26, 823
  explosion, V 822–5; VII 962, 968, 1409–10
  India, growth, V 1
  movements, V 825–6; VII 967, 976
  town-planning effects, VII 962–5, 968, 976
  U.S.A., growth, V 26, 823
  wartime losses, 20th century, VI 11, 23–4
Porcelain, see under Ceramics
Porphyrogenitus title, II 30
Porphyry ('Roman stone'), II 24, 29, 406
Portland cement, see under Cement
Portland stone, II 32, 34
Portland vase, II 323, 338, plate 26A
Ports and harbours named; see also Civil
    engineering, Docks
  Dover, VII 920, 922–3
  Fishguard, VII 920, 923–5
  Port Talbot, VII 920, 927–9
  Tema, Ghana, VII 920, 925–7
Portugal; see also Europe
  agriculture, IV 16
  cartography and navigation, III 530, 544, 547–8
  ceramic trade, IV 328, 337
  Gregorian calendar, adoption, III 579
  mining, II 8, 12
  ships, III 476–8, 480–1
  snuff, III 10
  surveys, IV 610
  tobacco, III 8
  windmills, II 618
Potash (potassium carbonate)
  ancient sources, II 354–5
  assaying flux, ingredient, III 63
  glass-making, III 206–7, 221–2, 229, 685, 700;
    IV 235, 359, 376
  plant ash, II 354

Potash (potassium carbonate) (*cont.*)
    potash-lime crystal, III 221–2
    production, IV 231–2
    soap-making, II 354, 356; III 703–5
    uses, II 354; IV 232, 235
Potassium
    carbonate, *see* Potash
    cyanide, V 95
    extraction, IV 118; V 72
    metals, refining process, V 97
    nitrate, II 377–8, 380; V 255
    nitrite, III 63
Potatoes, *see under* Food
Potter's wheels
    Americas, non-discovery, I 83
    early civilizations, I 80, 198–202
    fifteenth century onwards, IV 330–1, 333
    function and operation, I 195–7, 378, 388–90;
      IV 330–1
    Greek pottery, II 260, 261, 262–3
    invention, I 65, 71–2, 80, 202–4
    medieval period, II 284–5, 288–93
    Megarian bowls, II 269
    origin, I 65, 202–4, 404–5
    rotary motion, discussion, I 187, 195–204
    varieties, I 197–8
Pottery, *see* Ceramics
Power, II **589–628**; IV **148–67**; *see also*
    Natural/Nuclear-power, Power stations
    animals and slaves, II 590–3, 600–1
    machinery, power-operation, III 18–19, 158,
      162, 167–9, 239, 324, 330, 340–1
    prime-movers, II 589–90
    steam, II 1, 590, 608
    water, *see* Water-power, Watermills
    wind, II 614–20, 623–8
Power stations, named; *see also* Electricity,
    Nuclear power stations
    Battersea, VII 1071
    Blyth, VII 1071
    Carville Station, Newcastle, VII 1071
    Mockfjärd, Sweden, VI 200
    Neptune Bank, Newcastle, VII 1068, 1070,
      1071
    Niagara, VI 200
    North Tees, VII 1071
Pozzolana (*pulvis puteolanis*), II 407, 410, 412
Precipitation records, *see* Rainfall records
Precision instruments, V **582–619**
Precision mechanics, IV **379–416**
Press, printing-house
    Blaeuw, P. and J., III 402
    Clarendon Press, Oxford, III 402
    Imprimerie Royale, III 402
    Plantin and Moretus, III 401–2
    Ripoli, III 393
    Stamperia Vaticana, III 402

Presses
    bag-press, II 112
    balance-, III 340–1
    beam-press, II 112–16
    coinage, III 338–40
    linen-, III 177, 382, 389, 397
    printing, *see under* Printing
    screw-press, II 113, 116–17, 218–19, 638, 647;
      III 41, 177, 334, 337–9, 382, 414
    tread-mills, operation, II 637
    wedge-press, II 117–18
Pressure
    atmospheric, IV 168–71, 173, 182, 605, 621
    hydraulic engineering, V **522–51**; VII 1060–7
    machines, 16th–17th centuries, III 324–44
Prices rise, 16th century, III 30
Prime movers, IV 159; *see also* Energy, Power,
    Steam-power, Water-power, Wind,
    Windmills
Primitive man, *see* Man
'Prince Rupert's metal', III 37
Prince's metal, IV 133
Principia, Newton's, IV 223
Printing, III **377–416**; V **683–715**, plate 39; VII
    1268–80
    block-, III 379–381, 404–5, plate 2B
    books, III 377, 676; V 711–14
    'chase', III 396–7
    colour-, III 408; V 704–8
    compositor's work, III 388–91, 394–6, 403
    developments, 15th–18th centuries, III 386–404
    die-casting, III 385
    illustration, III 379, 380, 404–10; V 701–8
    ink, III 381–2, 390, 398, 403–8, 415
    intaglio printing, III 379, 406–10; VII 1268
    invention, II 771; III 377–8, 379–80, 676
    jobbing, V 708–11
    letterpress method, VII 1268, 1279
    Linotype machine, V 686–8
    music, III 405–6, 409–10
    offset method, VI 29; VII 1275–9
    photocomposition, VI 29; VII 1275–9
    photogravure, VII 1271–2
    presses, III 337, 382, 389–90, 396–401, 408–9;
      V 690–701
    printing-offices, III 386–90, 395, 401–4
    production rates, III 387–8, 393, 404
    readers, III 389, 403
    relief method, III 379–80; VII 1268, 1279
    techniques, outside Europe, III 378–80
    tools and equipment, III 386–8, 391–2, 394,
      398–9, 404–7
    training, 19th century, V 714
    type-casting, V 683–4
    type-composing, III 395; V 684–8
    type-metals, III 43–4, 385–6, 392–3, 406
    type-moulds, III 385, 387, 391–3, 402

typefounding, III 383–6, 390–4
typesetting, VII 1272–5
typewriters, VII 1270–1
wages, 19th century, V 714
Prints, earliest block method
  Brussels 'Virgin' (1418), III 380
  Buxheim 'St Christopher' (1423), III 380, plate
    2B
Probierbüchlein ('Essay on Assaying'), III 27, 59,
    63
Proconsul (Miocene ape), I 9, table C
Production engineering, VII **1046–59**
  equipment, VII 1049–51
  machine factors, VII 1049–52
  processes and techniques, VII 1049
  productivity factors, VII 1053–5, 1056–6
  standardization, VII 1051–3
  Taylorism, VI 73–5, 81–5, 93, 115–16, 118–19,
    120; VII 1046–9, 1056
  tool materials, VII 1047–9
Projectiles
  bombs, III 38, 347, 368–9, 374
  cannon-balls, III 362, 368, 374
  grenades, III 347, 369
  musket-balls, III 34
  scientific principles, III 374–5
  shot, III 357, 359–61, 368, 375
  stones, III 353
Prosimians, I 6–10
Protractors, III 593, 623
Prunus, I 288, 357–8
Prussia; *see also* Germany
  diet, III 2
  geodesy, V 442
  iron industry, IV 105
  railways, V 340
  silver, IV 138
  surveys, IV 601, 610
  windmills, III 97
Przewalski horse, Mongolia, I 342, 346–7, 721
Public service vehicles, VII 706, 728–30
Pudding-stone, II 35
Pulley-blocks, IV 426–7, 437
Pulleys, I 523; II 630, 646–7, 658–62, 723; III
    334, 337, 432
Pumps and pumping; *see also* Hydraulic
    engineering
  air-pumps, IV 93
  ancient devices, I 522–5
  bridge-building, III 420, 422–3, 425–6
  centrifugal, VII 1061–2
  chain, II 17, 19, 22, 617, 626, 770; III 484
  chain-and-bucket, III 325–7
  Cornish beam engine, V 126–30
  dockyards, IV 581
  drainage, III 307–8
  dredging, IV 630–1

force-pumps, II 376; III 333
  hand-, III 484
  Hubbard rotary-gear pump, IV 438
  innovation, 15th–16th centuries, II 654, 691
  mines, IV 69, 76–9, 93, 671–2, plate 7A
  pneumatic, III 636–8; IV 403
  polders, drainage, II 688–9
  pumping-stations, IV 443, 511, 517; V 128–9
  Roman devices, II 7–8
  sand-pump (Burt and Freeman), IV 631
  sewage-disposal, IV 517
  siphon type, II 376
  steam injector, III 52
  steam-power, IV 171–87, 192–4, 197, 499
  suction-, II 16–17, 633–4, 654; III 324,
    329–33, 344, 432; IV 93, 169
  types, VII 1060–3
  vacuum, III 637, plate 27A; IV 170
  water-lifting methods, I 522–5; II 675–8
  water-powered, IV 491
  water-supply service, IV 497
  Woodford, pump, IV 631
Punan people, Borneo, I 164
Punched card machines, VII 1153–4, 1157
Punches; *see also* Hammers
  Assyrian, 8th century B.C., II 229
  block-printing, relief, III 405
  coin-making, II 486, 490–1; III 341
  fine metal-work, II 468, 469, 486
  machine-tools, IV 429, 441
  music printing, III 410
  typefounding, III 385–7, 391–2, 402
Punica granatum (pomegranate), I 372; II 151
P.W.R. (Pressurized Water Reactor), VI 244, 253,
    254, 255
Pygmies, I 170–1
Pyramids, I 473, 475–6, 481–2, 714, 792, 793
Pyrites, IV 246
Pyrus (apples, pears), I 357; II 139

Qanaats, I 532–4; II 666–7
Quadrants
  altitude, sun-dial use, III 598–600
  arc of the meridian (1669), III 544
  astrolabes, III 614–15; *see also* Astrolabes
  cannon, III 362, 374–5
  geodesy use, IV 400–1
  horary quadrant (1399), III 597
  hydrographic survey, IV 615–18
  instrument improvements, III 555–6, 586, 625,
    636, 640–1; IV 399–41
  latitude measurement, III 522
  navigation use, III 526, 528, 536–7, 545–6, 625;
    IV 598; V 456–7
  shadow-square scale, III 608
  survey use, III 536–7; IV 603
  telescope fitment, IV 380

Quarrying, I 558–71; II 1–40
  alluvial mining, VI 414–17
  ancient Egypt, I 477–9, 526, 569–70
  Greek, II 24–7, 398–9
  opencast mining, VI 411–14, 418
  Roman, II 27, 32, 406, plate 2A
  Saxon and Norman, England, II 32–8
  Stone Age, I 558—60
  tools used, II 36–7, 384
Quartz, II 348
Quebec Agreement (1943), VI 227
Quercus aesculus, II 413
Q. cerris, II 413
Q. coccifera, I 245
Q. infectoria, II 151
Q. robur, II 413
Querns, *see* Mills, Millstones
Quicklime, III 29
'Quincunx', IV 618

Radar, VI 17, 349–50, 449–50, 559; VII 812
Radio; *see also* Radar, Television
  aerials/antennas, VII 1243–4
  beginnings, 19th century, V 227–8; VII 1230–6
  broadcasting, VII 1253–6
  land systems, VII 1253
  liberal era (1920s), VI 13
  marine use, VI 349–50; VII 1252–3
  modulation, VII 1109–11, 1238–40
  radio-telephony, VII 1250–1, 1252
  receivers, developments, VII 1256–7
  solid-state devices, VII 1116–23
  telegraph, cable competition, VII 1248–50
  theoretical developments, VII 1247–8
  thermionic valves, VII 1092–1106, 1237–8
Radiology, VII 1299–1300, 1325, 1327–34; *see also*
  X-rays
Radiology, Fifth International Congress of (1937),
  VII 1329
Radium, isolation of, VI 224–5
Railway and Canal Commission, U.K., VII 702
Railway locomotives and classes named
  '567' diesel, VII 768–9
  'Ae 4/4', VII 773
  'Atlantic', VII 763
  'Big Boys', VII 766
  'Castle', VII 764
  'Deltic', VII 769
  'George the Fifth', VII 763
  'King', VII 764
  'Mallet', VII 766
  'Pacific', VII 766
  Rocket (1829), VII 762
  'Star', VII 764
Railways, general, V 322–49; VII 762–88; *see also*
  Railways, lines and systems
  amenities, V 341–4

brakes, V 344
bridges, IV 452–3, 460–1; V 503, 505, 513, 643
canal links, IV 549, 569
diesel traction, VII 767–9
early history, V 322–9, plate 20A; *see also*
  tramways
electricity generation, V 194, 197
electric traction, V 230, 232 n., 233, 346–8; VII
  770–3
freight rolling stock, VII 775–6
fuel, V 327
guages, rail, V 324
locomotives, IV 189; V 323–9, 333–41, 345,
  348, 349, plates 20, 21; VII 762–7; *see also*
  Railway locomotives
networks, V 804–5; VII 702–4, 787–8
permanent way, V 331
points and crossings, V 65
Pullman cars, V 342–3, plate 20B, C
rails, V 53, 61–2, 323, 331, 804–5
road competition, IV 537; VII 706–9
rolling-stock, V 62, 341–6; VII 773–6
signalling, V 329–31; VII 778–87
social consequences, V 819–20
stations, V 194, 197, 211, 218, 332–3, 468–70,
  473–50, plate 22
track construction, VII 776–8
traffic marshalling, VII 778–87
tramways, mining, IV 77, 79, 85, 89, 90
underground (subways), V 346–8, 520; VII
  770–1
Railways, lines and systems, VII 787–8; *see also*
  Railways, general
Australia, V 8
Baghdad, VI 7
Benguela, Angola, VII 788
Bern-Lötschberg-Simplon, VII 773
Blackwall, U.K., IV 658
British Railways, VII 771, 772, 775, 779
Buenos Aires Great Southern, VII 767
Canadian Northern, VII 788
Canadian Pacific, V 87; VII 769, 775, 784
Chester and Holyhead, U.K., IV 460
French Railways, VII 772
German State Railways, V 820; VII 768
Grand Trunk Pacific, Canada, VII 788
Great Central, U.K., VII 785
Great Indian Peninsular, VII 771
Great Western, U.K., IV 461, 657; VII 763–4,
  774, 775, 779, 780
Indian, V 61, 819–20
Inner Circle, London, VII 770
Ireland (1838), IV 623
Lancashire and Yorkshire, U.K., VI 296
Liverpool and Manchester, U.K., IV 657
London, Brighton and South Coast, U.K., VII
  771

London, Midland and Scottish, U.K., VII 892
London and Birmingham, U.K., IV 657
London and North-Eastern, U.K., VI 465; VII 768, 771
London and North-Western, U.K., VII 762, 763, 778
London and South-Western, U.K., VII 771
London Transport, VII 771
Metropolitan District Underground, London, VII 779
National Trans-Continental, Canada, VII 788
Netherlands State Railways, VII 771
New South Wales, VII 775
Norfolk and Western, U.S.A., VII 772
North America, V 3–4
Paris–Orleans, VII 764, 766, 771
Pennsylvania Railroad, U.S.A., VII 780, 785
Rockefeller Railroad and Steamship Lines, U.S.A., VI 464
Russian, V 61; VI 8
Siberian, VI 6
South African, VII 782
Stockton and Darlington, U.K., IV 460
Sudan Government, VII 770
Swiss Federal, VII 772
Teheran-Djolfa, Persia, VII 788
Trans-Iranian, VII 788
Union Pacific, V 3
Western Railway, France, VII 776
Rainfall records, IV 490–1; V 558; VII 1364–5, 1366
Rainwater disposal, IV 505, 514, 517
Rakes, II 94, 331; III 134, 137, 138
Ramie (China grass), I 449
Raphanus sativus (radish), I 288, 356
Raphia vinifera, I 277
Rasps, II 228–9, 230, 231
Ratine (woollen cloth), III 179
Rayon (cellulose acetate), V 579; VI 554, 643–4, 648–9
Reactors named; *see also* Nuclear power
  A.G.R. (Advanced Gas-Cooled), VI 251, 259
  BEPO, VI 245, 246
  B.W.R. (Boiling Water Reactor), VI 244, 254, 255
  CANDU, VI 241, 242
  Gleep, VI 243
  MAGNOX, VI 250, 252, 259
  PIPPA, VI 249, 250
  P.W.R. (Pressurized Water Reactor), VI 244, 253, 254, 255
Realgar (arsenic disulphide), III 29; IV 122
Reamers, III 111
Reapers, mechanical, III 136
Reaping machines, II 96–7
Rectangulus, III 593–4
Reduction-house, zinc, IV 131–2

Reeds, rushes, sedges, grasses, I 371–2, 415–18, 423, 449–50, 468, 473, 550
Reeling yarn, reels, II 206, 208–10, 213; III 162–4, 167
Reflection-circle, IV 618–19
Refractory materials, III 39; V 664–5; VI 630–6
Refrigeration, food
  bacon-curing (from 1850s), V 32
  chilled meat, VII 1400, 1425
  dairy products, V 8, 33
  'deep freeze' storage, VII 1423–5
  domestic appliances, VII 989, 1144–6
  early history, V 45–51; VII 1422
  fish, IV 51, 53–4; VI 350–4; VII 1400–1, 1424
  freeze-drying process, VII, 1415, 1416
  ice storage, Roman Empire, II 121
  liquid air, VI 991
  methods, improvement, VII 989–90, 1422–5
  'modern' *v.* 'traditional', V 29
Refuse disposal, II 532–3; IV 505–6; *see also* Sewage
Reliefs, terracotta, II 278–9
Religion, religious practices
  animal sacrifice, I 327
  astrology, I 121, 125, 802
  cosmetics, I 286–7, 289
  early Egyptians, I 123–4, 266, 267
  geometry, Greek, I 121, 126
  lunar calendar, I 794
  Mayan number system, I 125–6
  taboos, I 241, 331
  temples, early societies, I 47–9
  time, early beliefs, I 114, 120–1, 123–6
Religious houses, drainage, III 309, 316, 318
  prints, 15th century, III 380
Renaissance, effects, III 151, 378, 609
Repeating-circle instrument, IV 401, 604
Research and development, VI 27, 33–6, 38–40, 44–5, 62–4, 114, 123–4, 127–32
Research institutions
  Agricultural Research Council, VI 128
  Commonwealth Scientific and Industrial Organization of Australia, VI 128
  Medical Research Council, VI 128
  National Institute of Oceanography (U.K.), VII 1451
  National Research Council of Canada, VI 128, 241
  National Science Foundation (U.S.A.), VII 1456, 1457
  Science Research Council, U.K., VI 127
  Scientific Research and Development Office, U.S.A., VI 271
Reseda luteola (weld), yellow dyes, II 365; III 692
Resin, III 685
Rhamnus catharticus (green dye), III 692
R. saxatilis (yellow dye), II 362

Rhodium, IV 119, 144
Rhus (sumac sources), I 247; II 151
R. cotinus (yellow dye), II 366
R. vernicifera (varnish shrub), III 696
Rhymes for apprentices, III 112–13
Ricinus communis (castor), I 288
Rig-Veda, I 449, 728
River navigations, III **438–70**
Rivets, riveting, II 469, 483; III 354; V 368–9, 502
Road-vehicles, *see* Motor-cars, Wheeled vehicles
Roads and streets, general, II **493–536**; IV **520–47**; *see also* Roads . . . named
  ancient, I 712–14; II 493–8; III 269–71
  asphalt, IV 539–41
  city streets, II 527–33, 690, plate 38A; IV 542–6
  concrete, IV 538–9
  construction methods, IV 520–42, plate 32A
  drainage, IV 536
  equipment and tools, IV 537–8
  foundations, IV 527–8, 531, 536
  Greek, II 498–9
  Hellenic, III 272–5
  labour statutes, road-building, IV 524, 526–7, 536
  London, III 283, 296–8
  macadam system, IV 532–6
  maps, II 496, 501–2, 506
  medieval, II 525–7, 537; III 282–4
  metalled, IV 535–6
  milestones, II 507
  motorways, U.K., VII 875–6
  pavements (side-walks), *see* Pavements
  Renaissance, III 285–93
  road-books, IV 524–5
  road trusts, IV 532
  Roman, II 500–15, 524; III 276–82, 504
  rut-roads, II 499, 511–12, 539, 548
  street-cleaning, IV 505–6
  street-lighting, II 531; III 282
  surfaces, IV 531–2, 536
  tolls and imposts, II 524–5 533
  trenched roads, IV 525
  turnpikes, IV 530–1, 536–7
  vehicles, 20th century, VII 711–35
  wood-paving, IV 545, plate 33C
Roads and streets named; *see also* Roads . . . , general
  Autobahn, Germany, VII 872
  Autostrada, Italy, VII 873
  Banihal Cart, Kashmir, VII 874
  California Freeway, VII 972
  Charles Street, London, IV 545
  Commercial Road, London, IV 544
  Edgware Road, London, IV plate 33B
  Germany to Rome, 15th century, III 534

  Grand Trunk, India, VII 871, 882, 883
  Great North Road, IV 530
  London–Birmingham Motorway, England, VII 876
  London to Holyhead, IV 531
  Montgenèvre–Mont Cenis, IV 529
  Ontario Provincial Highway, Canada, VII 919
  Oxford Street, London, IV 545
  Persian Royal Road, II 496
  Rowanduz, Kurdistan, VII 894
  Simplon road, IV 530
  Stane Street (London to Chichester), II 505–7
  'Street called Straight', Damascus, III 275
  Tottenham Court Road, London, IV 545
  Via Aemilia, II 501, 505, 679
  Via Altinata, II 679
  Via Appia, II 28, 500–1, 504, 505, 507, 510
  Via Cassia, II 501, 510
  Via dell' Abundanza, II 531
  Via Domitiana, II 500, 507
  Via Flaminia, II 500–1, 504, 508, 511, 513
  Via Praenestina, II 28, 504, 513
  Via Sacra, II 28
  Via Traiana, II 508, 510
  Wigan road, England, IV 530
  Winchester by-pass, England, VII 875
Roccella (archil dye), I 246; II 366
Rock asphalt, *see* Asphalt, Bitumens
Rockefeller Institute for Medical Research, VII 1341
Rockets, II 771; VII 859–66; *see also* Space technology
Rollers, rolling-mills, III 340–3; IV 104, 161; V 30, 611–12
  grooved (patented 1783), IV 106–7
  steam-rollers, roads, IV 535, 537–8
Roma Instaurata (1444–6), Flavio Biondo, III 288
Romania, oil, VI 185
Rome, Roman Empire/Republic
  aqueducts, II 7, 10, 47, 404, 405, 418–19, 669 (map), 670–1
  beer, Romans' aversion, II 140
  brass production, II 55
  bridges, II 508, 510–11; III 417–18, 420, 432
  Britain, and Celtic art, I 82
  building construction, II 404–20, 446
  calendar and chronology, III 575–8
  canals/river navigation, III 439; IV 550
  carpenter's tools, II 230–3
  cement, II 407; IV 448
  ceramics, I 378, 393, 394; II 272–4
  chemical techniques, II 350–2, 355
  chronology, Italy, II table 1, lx
  concrete, II 405, 407, 410–13
  copper production, II 52–3
  cornmills, IV 199
  crafts, IV 664

fire-making, discussion, I 219
foods and diet, II 103–5, 108–12, 114–16, 118–23, 600–1
food plants, I 355, 356, 362
furniture, II 224–5, 233–8
glass-ware, II 322–5, 340
harbour installations, II 516–17, 518–21
harness, II 515, 592
harvesting implements, II 95–7
heating braziers/stoves, I 231
horse-riding, II 556, 560–1
iron and steel production, II 55–7, 59–61
irrigation and drainage, II 678–80
laws, II 53
leather, use, II 150, 166, 168
lighthouses, II 518, 521–3
lighting, I 234, 235, 237
machinery, II 636–8
maps, III 502–4, 507–10, 518
mathematics, III 501–2
measurement, linear, I 778
metal-work, II 449–50
metals, II 41, 42–5, 47–50
mining techniques, II 6–10
mural paintings, II 363
pavements, III 194
pigments, II 359, 361–2
ploughs, II 86–8
pottery, *see* ceramics
quarries, II 27–32, 406, plate 2A
roads, II 500–8, 511–12, 529–31; III 276–82, 504; IV 521, 529, 539
sanitation, II 530–1
scripts, Latin, I 770–1
sewers, IV 504
ships, II 569–76
signalling methods, IV 645
statuary, terracotta, II 276–9
sun-dials, III 596–8, 600
surveys, III 502, 504, 511
textiles, II 191–7, 201–2, 211–12
viticulture, II 133–7
water-engineering, II 663–4, 671–3
water-supply, II 671–4
watermills, II 595–600
weapons and armour, II 703–20
wheeled vehicles, II 515, 538, 541, 545, 550, 554, 592
Roofing, *see under* Building construction
Roofing-sheets, III 343
Rope-making, I 451–4
Rope-tackle, III 107–8, 139
Rosa sancta, I 288
Rosetta Stone, I 754
Rosmarinus officinalis (rosemary), I 288
Rotary fan, III 20–1

Rotary motion, I 71, 187–215, 274
partial, I 187–94
potters' wheels, I 194–204
reciprocating motion, conversion, II 652–5
wheeled vehicles, I 204–14
Rotary querns, II 107–10
Routers, coach-makers', III 113, 115
Royal Academy, IV 443; V 217
Royal Agricultural Society, Britain, V 161
Royal Air Force, VI 17, 19, 558; VII 796, 811, 812
Royal Astronomical Society, VII 1173
Royal College of Chemistry, London, V 269, 273, 782 and n. 1, 785, 821
Royal Commissions
Compensation and Betterment (1942), VII 968
Conditions of the Towns (1844), VII 962
Distribution of Industry (1938), VII 967
Scientific Instruction and Advancement of Science, V 792–3
Sewage Disposal (1898), VII 1388, 1396
Small-Arms, V 639, 653
Technical Instruction (1884), V 786–7, 795; VI 142
Water-supply, V 554
Royal Institution, London
battery, Humphry Davy (1806), IV 227
'coal-oil' and James Young, V 104
fire-damp, explanation (1815), IV 95
food adulteration (1820), V 834
founding (1799), IV 674–6; V 831
optical glass research (1827), IV 361
photography on paper (1839), V 722
Rumford's project, IV 674–6; V 831
steel, Faraday's experiments (1819), V 65
Royal Laboratory, Woolwich, England, V 630
Royal Mint, London, V 269
Royal Naval College, Greenwich, England, V 372
Royal Navy, British; *see also* Admiralty
compasses (from 1837), V 455–6
composite ships, 19th century, IV 592
copper sheathing, hulls, IV 580
dockyard mechanization, IV 581–2
draughts, 18th and 19th centuries, IV 577–8
Dreadnought class battleships, VI 5–6
fuel-oil, introduction (c.1903), V 118
Henry VII and VIII (1546), III 477–82
hydrographic surveying, V 449–54
Indiamen convoy, IV 576
iron cables and rigging (1808), IV 585–6
ironclads (1859), IV 588–90
list (1872), IV 576
marine boilers (from 1835), V 153–5
marine steam-engines, 19th century, V 148
Massey's Patent Log (c.1840), V 459–60
Naval Architecture and Marine Engineering, Royal School of, V 372
Queen Victoria's hurry, IV 590

Royal Navy, British (*cont.*)
Quiberon Bay, battle (1759), III 499
'rates' classification, III 490; IV 576
sailing warships, IV 582–5, 586–8, 590
seventeenth century, III 484
shipbuilding, VII 751–3
steam-turbines (1897), V 151–2, 156
steel plates, adoption (1876), V 615
survey-equipment, 18th century, IV 616
surveying ships (1814), IV 618; V 449
Thomson's sounding machine (*c.*1878), V 462
Victory (1765), IV 576–7
weapons (1628), III 679
wire rigging (1838), IV 593
yachts, Charles II (*c.*1660), III 497
Royal Observatory, III 552, 557
Royal Society, London
achromatic lens, patent, IV 358
agricultural research (from 1660), III 15
bridge drawing, Pont-y-Ty-Pridd, III 436
chemical interests, 17th century, IV 234
coal-gas (*c.*1684), IV 259
craft-technologies, survey (1667), IV 669, 677
crystal-glass, colouring materials, III 703
dyeing experiments (1660s), IV 248
dynamo-electric generators (1866–7), V 184–6
early history, IV 667, 668–71
electromagnetic induction (1831), V 179
explosives, III 374
founding (1660), III 631, 681; IV 234, 667–71
gas-lighting (1808), IV 262
'Histories of Nature, Arts or Works' (1667), IV 668, 669, 674–5
inventions, IV 669–70
lathes, Ramsden's, IV 388
metals, strength tests, III 58–9
mirrors, telescope- (1672), III 41
natural gas (1618), IV 258
octant (1731), III 556
optical-glass (1824), IV 360–1
'Philosophical Transactions', III 22, 682
photo-etching process (1827), V 720
photography on paper (1839), V 722
pistol claim (*c.*1631), III 358
rigging, sailing ships, 17th century, III 486
science, Boyle's views (*c.*1659), III 718
science education, 20th century, VI 153–4, 162
steam-engines, IV 172, 189 n.
Swan lamps (1882), V 217
tar/pitch, wood source, III 685
telescope description, Hooke (1684), IV 645
Volta's pile (1800), V 177
water wheels, Smeaton's, IV 204, 206
wind-pressure experiments (1759), V 502
Royal Society of Arts
building (1772–4), IV 473, 474
cellulose nitrate (1865), VI 552

coal-gas, purification method, IV 268
Coalbrookdale, Darby's bridge, IV 456
copper smelting (1879), V 80
education, 19th century, V 28, 778–9, 782, 783–4, 789, 794
electromagnetic invention (1825), V 178
grassland improvement, IV 20
gutta-percha, uses (1843), V 221
meat preservation (1884), V 42
reaper, prize entry (1812), IV 7
roller-milling, wheat (1891), V 30
Royal Technological Institute, Stockholm, VII 1174
Rubber ('caoutchouc'), V **752–75**
calendar, V 763–6, plate 5
cut sheet and thread, V 759–60
extrusion process, V 764–5, 772
'india-', V 756
latex processing, V 752–3
manufacture, V 755, 757–70; VI 339
masticator, V 758–9
plantation, V 773–5
proofing, V 761–3
raw, V 752–4, 770, 773–5
synthetic, VI 506, 510
tapping methods, V 752–3; VI 339–40
tyres, *see* Tyres
uses, V 44, 220–1, 756–7, 764–5, 770–2
vulcanization (curing), V 221, 765–70
waterproof fabric (1822), IV 253; V 761–3
Rubia tinctorum (madder), I 246; II 366; III 693
Rubus, I 353
Rumex spp. (docks), I 370
Russia, *see* Union of Soviet Socialist Republics
'Russian' leather, II 156, 177
Ruthenium, IV 144

Saccharum officinarum (sugar-cane), I 370–1
Saddle-querns, I 273–4, 502, 514
Saddles, II 182, 538, 554–7
Safety measures, V 284, 289–90, 314, 330–1, 378, 480–3, 523, 636–7, 832
Safflower/saffron (yellow dye), I 246–7, 249, 441
Saggars, II 297
Sails, *see under* Ships
Sakai people, Malaya, I 164–5, 169–70
Sal alkali (glass-salt), III 210
Sal ammoniac (ammonium chloride), I 262; III 48, 225, 689, 706; IV 232
Salicornia (marsh samphire), II 682
Salicylic acid, V 300, 316
Salsola kali (soda plant), I 261
S. soda, II 354
Salt (sodium chloride)
chlorine production, IV 248
early civilizations, I 256–9
food additive, II 125, 127

food preservation, I 264–6; II 148–9, 155; IV 46–7, 49–50
glaze, II 287, 307
Leblanc's process, IV 239–42
mining, I 566–7; II 166
peat source, Zeeland, II 683
purification, IV 230, 232–3
salt-pans, III 76
salt-works, III 82
saltpetre, separation, II 381
soda manufacture, IV 236–9
tax duties, IV 237, 239
transport, II 527
Saltpetre (nitre, potassium nitrate)
  Akkadian description, I 251–2
  alchemical price (1501–8), II 747
  aqua fortis preparation, III 695, 706
  assay, metal ores, III 62–3
  Chile-(NaNO₃), IV 246
  Chinese origin, II 370
  dyeing process, III 695
  explosive, II 370; IV 235; V 285
  gun-cotton manufacture, V 290–1
  gunpowder ingredient, II 371, 379–82; V 285
  manufacture, IV 255; V 285
  matches, hand-guns, III 355 n. 2
  metallurgical fluxes, III 39, 41
  military pyrotechnics, II 374, 376–82
  preparation, II 369, 370–1, 380–1; IV 232
Sand
  building, II 407
  glass-making, II 314, 320, 323, 328
  metal-casting, II 475
Sand-glass, III 526, 550
Sandals, *see under* Footwear
Sandiver, III 62 *and* n., 63
Sanitary engineering, II 663–94; *see also* Water supply, Sewage
  drainage systems, IV 504–13; VII 1382–98
  sanitation, II 527–33, 689–90, 691; III 273, 282–3; IV 504–19; VII 1382–97
  water-closets, IV 423, 507–9
  water-supply, II 663–74, 689–92; IV 489–519; V 561–7; VII 1365–6, 1373–81
Sanson-Flamsteed projection, IV 625
Santalum album (sandalwood), I 288
Saponaria officinalis (soapwort), I 249; II 215
Sarsen rocks, I 491–3; II 35
Satellites, Earth, VII 848, 857
Satin, III 192, 199–200, 203, 205; *see also* Textiles
Saturn rockets, VII 848, 861, 869
Saudi Arabia, oil, VI 186
Saws
  ancient origin, I 142, 143, 207–8, 603
  band-saws, IV 437
  bow-saws, III 112
  bronze, I 603, 613

carpenter's, II 228–30, 390–2, 393
circular-, II 385; IV 427, 437
copper-, I 207–8, 570, 603, 688
fire-saws (fire-making), I 220–2
flint-, I 603
frame-saws, II 391–2
iron-, I 619–20
joiner's (c.1703), II 253
machine-saws, II 643–4; III 365
mechanization (1808), IV 427–8, 437, 581–2
medieval, II 389, 390–3
Neolithic stone-workers, I 500
quarrying/stone-masonry, II 29, 30–1, 385
ship-building, IV 580
tradesmen's tools (c.1500–1850), III 110, 112
two-handled, II 230, 389, 391
water-powered (c.1250), II 644
Saxons, II 32–3, 93
Say (cloth), III 179
Scaffolding, II 386; VII 938
Scandinavia, N.W. Europe
  metals, IV 120
  planks, ready sawn, IV 580
  roads, IV 529
  shipping, IV 558, 574
  whaling, IV 57–8, 62
  wooden pipes, IV 500
Sceptical Chymist, The, Robert Boyle, IV 215
Schedula diversarum artium, Theophilus Presbyter, III 206
School of Mines, Britain, V 782, 785
Science and technology, IV 663–81; VII 1467–82
  agriculture, IV 14, 621, 674; V 24
  ancient world, II 603–4, 606, 638
  arch theory, III 433–4
  assaying, III 59–68
  atomism, III 56–7
  bridge-building, III 432
  British government departments, V 550, 782–4, 790, 794
  building and engineering, III 258; IV 442–6, 471, 478–86
  cables, transatlantic, IV 674
  ceramics, IV 328, 352
  chemical industries, III 676–708; IV 214–57
  dimensional theory, III 432
  economics, factors, VI 48–76
  education, V 776–98; VI 138–71
  engineering, III 344–5
  Europe after 17th century, III 717–21
  food management, V 26–52
  gas, IV 678
  glass-making, IV 358, 360–1, 374–5, 674
  gunpowder, IV 674
  hydraulics, III 313–15
  industrial revolution, IV 148–51, 230–57
  instruments, III 620–1; IV 398–415

Science and technology (*cont.*)
  metallurgy, III 56–9, 68; IV 674
  navigation instruments, IV 674
  optical glass, III 230; IV 358–62
  philanthropic schemes, IV 675–7
  phlogiston theory, III 57, 720; IV 218, 220
  quantitative analysis, III 59
  road construction, IV 532–6, 539
  ship-building, III 482–3; IV 577, 579, 584
  social consequences, V 814–41
  steam-engines, development, IV 162–6, 179,
    181–8
  strength of materials, III 58–9, 250–1, 432
  structural theory, III 432–4
  town-planning, III 285–8
  warfare, III 373–5; IV 665–7
  water-supply, IV 490–1, 674
  water-wheel experiments, IV 603–9
Scissors, II 100
Scirpus (mats and basketry), I 420, 421, 450
Scotland, *see also* Britain, Europe
  agriculture, III 3, 142; IV 2, 3, 7, 10, 17
  alcoholic spirits, III 11
  bleaching industry, IV 243, 244, 246–8
  Board of Trustees (1727), IV 244, 246
  canals and waterways, IV 565, 569, 571–2
  chemical industry, IV 238–9, 241, 244–5,
    251
  coal-mining, IV 88
  coinage, III 341
  dam-building, VI 199–200
  diet changes, III 3
  dredging, IV 641
  dyeing, IV 248–9
  farm-vehicles, III 142–4
  fish industry, IV 47, 53–4
  flax, III 157–8
  furnaces, IV 122
  Highlands mapped, III 554
  hydro-electric engineering, VI 212
  iron industry, IV 105, 109–11
  kelp, IV 236
  linen, III 151, 176
  New Year's Day, date, III 580
  patent law, IV 244
  roads, IV 530, 532–4, 539
  sensitometry, VII 1283–4
  shipping, III 478; IV 558, 592–3
  soap, III 703
  spinning-wheels, III 151
  steam-engines, IV 162, 172, 178
  sulphuric acid manufacture, IV 245
  surveys, IV 607, 610, 613, 616
  telegraphy, IV 647, 649–50
  textile-industry, III 163; IV 283, 292, 299–300,
    315
  water-supply, IV 497

  windows, III 237
  wool, III 152
Screws
  accuracy, IV 424–6
  advances, 16th–17th centuries, III 325
  Archimedean screws, II 633, 676–7; III 106,
    304–5, 423; IV 157
  armour, III 629
  automatic production (1879), IV 647–8
  'Barrow' screw regulator, III 667
  clocks, III 657, 659
  Greek, II 631–2
  instruments, use, III 610, 629
  lathes, screw-cutting, III 334–7; IV 425, 429,
    432–3, 439, 441
  lead-screws, III 335–6
  linen-press (printing), III 382
  machine-tool parts, III 343, 345; IV 384–5,
    418–19, 432
  machinery (*c.*1556), III 326–7
  medieval, II 293–4, 394–5
  micrometers, III 554; IV 395–7, 403, 423, 426;
    VII 1041
  origin, II 242 n.
  printing-press, 16th century, III 398–400
  screw-cutting, II 631–2, 655; III 334; IV
    385–8, 392, 419–20, 424–5, 434, plate 25A
  screw dies/taps, 19th century, III 113, 114
  screw-jacks, II 647; IV 463, 480
  screw-presses, III 41, 177, 334, 337–9, 340,
    382, 414
  screw-stoppers, II 293–4
  standardization (1841), IV 433, 637
  tangent screws, IV 391–3, 395, 402–3
  worm screws, III 610, 629; IV 384, 393–4
Scripts, *see* Writing
Sculpture and statuary
  casting statues, III 38–9
  Colossus of Rhodes, II 470
  Greek temple figures, II 470
  ivory, I 153, 661–2, 666, 668–9, 673, 681
  stone, I 469, 470
  terracotta, II 274–9
  wood, I 689–70, 699, 702
Scythes, I 514; II 75, 95–6, 396; III 136–8; IV 7
Scythians, I 373
Seals, hydraulic, IV 266–7
Seasons, cycle, III 559–60, 569, 573; *see also*
  Calendars
Seaweed, *see* Kelp
Secale (rye), I 368
Sector instrument, III 627–8, 636
Seedlip (seedcot), II 100
Selenium, V 97
Semang people, Malaya, I 162, 168–70, 183, 184
Sericulture, IV 308–9, 319–20; *see also* Silk
  industry

Sesamum (sesame), I 288
Serbia, VI 7, 8
Serge (cloth), III 178–9
Serpentine quarries, II 30
Set-squares, II 253, 386–7
Setaria (a millet), I 369, 370
Sewage and sewerage, IV **504–19**; VII **1382–98**;
    see also Sanitary engineering
  Akkadian closet, I 467
  ancient cities, II 527–9
  'Bill of Sewers', England (1531), IV 506
  City of London (from 1349), IV 505–6
  construction, 20th century, VII 1386–7
  design progress, VII 1385–6
  disposal, IV 517–19
  health hazards, sewer-men, IV 516–17
  house-drainage, IV 507–9
  London, drainage-works (from 1850), IV
    509–12; VII 1383–4
  medieval, II 689–90, 691
  Mesopotamia, third millennium B.C., I 466–7,
    528
  Middle Ages, II 532–3; IV 504–5
  Minos, Crete (*c.*1500 B.C.), IV 504
  Paris, drainage works (from 1663), IV 512, 513
  Roman period, III 282; IV 504
  sewers, designs and repairs, IV 512–17
  treatment, VII 1387–95
  urban systems, VII 1383–5
  water pollution, VII 1395–7
Sewing-machines, V 588–91, 639–40, 655, 819; VI
  679–84, 685–9
Sextants, III 556–7, 639; IV 390, 399, 401, 598,
  615–18, plate 22B
Shag (fabric, goat's hair), III 179
Shagreen (leather), II 157
Shalloon (woven worsted), III 179
Shaper-machines/engines, IV 428, 430, plate 26B
Sharpening-stones, II 97
Shearing-machines, III 173–5
Shears, II 100, 192, 193, 215, 217–18, 331; III
  340, 352
Sheep, *see under* Livestock
Sheffield plate, III 48; IV 130; V 632–4
Shellac, II 362; III 696; IV 133, 163; *see also*
  Lacquer
Shells
  burnishing tools, I 390
  currency, I 174
  ivory-work, I 670–1
  lamps, I 236
  metal-work, I 661
Shields
  bronze, II 460, 462, 468, 696
  leather, II 164, 165, 167
  'tortoise', II 716
  wood, II 169

Shimosite (picric acid explosive), VI 548
Shipping lines; *see also* Ships . . . headings
  Australian Mail Company, V 362
  Blue Funnel, V 378–9
  Cunard, V 145, 352, 374, 378
  East India Company, III 481, 496; IV 576,
    578–9, 590
  Eastern Steam Navigation Company, V 362
  General Steam Navigation Company, IV 591
  Inman, V 217
  National, V 149
  Orient Line, IV 593
  Pacific Steam Navigation Company, V 372
  White Star, V 378
Ships and ship-building, II **563–88**; III **471–500**;
  IV **574–95**; V **350–90**; VII **736–61**; *see also*
  Shipping lines, Ships named
  ancient man, I **730–43**
  Armada campaign, III 480–2
  barges, II 576–7, 686–7; III 478–9
  bells, III 484
  block-making project, IV 427–8, 437, 581–2
  brigantines, III 491, 499
  bulk carriers, VII 737, 740
  cable-ships, V 225
  cables, IV 580
  caravels, III 476–8
  cargo-ships, V 378–83; VII 737, 739–41
  carracks, II 586–7; III 475–9
  'cat-built barks', IV 574–5
  cataphracts, Greek, II 566–7
  chain anchor-cable, IV 585
  clippers, IV 590–3
  coal consumption, V 363, 365
  cogs (types of sailing ships), III 476
  colliers, IV 574–5; V 371, 380
  decoration, III 493–5, 498
  designs, IV 577–8, 584; VII 736, 746–7, 756
  developments, from 17th century, III 482–6,
    495–9
  docks, III 492
  dredgers, II 656–7; IV 630
  fishing-vessels, IV 46–8, 52–4
  frigates, III 496
  full-rigged, III 474–8, 499
  galleasses, III 473, 478–80, plate 21B
  galleons, III 479–81, 483, 487
  galleys, III 353, 471–4, 478–80
  Greek, II 517–18, 563–7
  guns, III 473, 478–81, 499
  herring-busses, III 499
  H.M. Victualling Office, Deptford, IV 208
  hoys (sailing vessels), III 498
  hydrofoil craft, VII 744–5
  industry, 20th century, VII 693–4
  international trade, VII 692, 694, 696
  Irish, II 581–2

Ships and ship-building (*cont.*)

iron ships, IV 476, 579–80, 583, 587–92, 594; V
  62, 350–3, 355–72
ironclads, IV 588–90
Liberty ships, VII 753–4, 877
Lloyd's Register, V 373
medieval, II 390, 582–8
merchantmen, II 567–9, 573–6, 585–6
models, III 482, 490–2, 497, plate 24
mould-loft, III 489–90
naval architecture, IV 584
oars, III 471–3
oil-burning, V 117–18; VII 759–60
oil-tankers, V 121, 380–3; VII 696–7, 737,
  740–3
ownership, world fleet, VII 694–5, 757–8
paddle-propulsion, II 606, 607, 651, 719; V
  141–6, 372
passenger ships, VII 737, 743–4
performance, IV 575–9, 592
pinks, III 485
pinnaces, III 478–9
propulsion, 20th century, VII 758–60
rigging, III 476–7, 483, 485–7, 495–7; IV
  582–6, 593–4
Roman, II 569–76
round-ships, III 473–4, 479–80
rudders, III 475–6
rust prevention, III 679
safety, VII 745–6, 748
sails, II 567, 571, 574, 583–4, 771; III 472,
  474–7, 481, 485–7; IV 582–6, 593–4; V 353,
  362–3, 378, 454
Scandinavian, II 578–81
screw-propulsion, IV 431, 587; V 146–9
sheathing, IV 123, 130, 579–80, 591–2
ship-burials, II 578–80, 581
shipyards, II 517–18; III 486
sloops, III 491
snows (snaauw), III 499
steamships, IV 548, 571–2, 587–8, 590, 593–4;
  V 352–3, 362–5, 381, 383
steel ships, V 62, 372–8
steering-gear, III 497–8; V 383–6
stores, III 487
submarines, V 117, 234; VI 254
textbooks, III 482, 488
theory, III 432; V 353, 386–90
timber, III 487–9, 481, 492–3
tonnage, III 480–1, 493; VII 757–8
trading-ships, III 472–8, 481–2, 496, 498–9
turbines, V 149–52; VII 758–9
Victory ships, VII 753, 877
warships, III 471–4, 478–82, 497, 499; IV
  576–9, 582–8; V 141, 147, 152, 155; VII 752
whaling-ships, III 497
Whitby colliers, IV 574

wooden ships, IV 574–88, 590–2; V 350, 353,
  355 n. 1, 365, 375
yachts, III 486, 497–8
Ships named; *see also* Shipping lines, Ships and
  shipbuilding
Aaron Manby, V 146, 351
Agamemnon, IV 575
Ajax, V 148
Alarm, IV 580
Albatross, VII 1468
America, VII 744
Amphion, V 148
Annette, V 361
Antarctic, IV 60
Antelope, III 479
Anthracite, V 154
Aquitania, VII 743, 759
Arcadia, VII 744
Archimedes, V 147
Arctic Prince, VI 352
Arctic Queen, VI 352
Ariel, IV 592, plate 43A
Ark Royal, III 481
Arrow, IV 579
Atalanta, IV plate 41A
Atlas, IV 42B
Augustus B. Wolvin, VI 464
Bacchante, V 148
Banshee, V 62
Bellona, IV plate 41B
Birkenhead, V 441
Blenheim, IV 590
Blue Peter, VI 352
Boadicea, V 148
Brandon, V 149
Britannic, VII 743
Bull, III 479
Caliph, IV 593
Campania, V 374–7, plate 25
Candia, V 148
Captain, IV 589, plate 43B
Challenger, V 453, 465; VII 1451–3, 1447–8,
  1456, 1465
Charles, III 491
Charlotte Dundas, IV 572; V 143
Circassia, V 48
City of New York, V 378
City of Paris, V 378
City of Richmond, V 217
City of Rome, V 369–71
Clermont, V 143
Clytemnestra, V 62
Comet, V 143, 152
Conte di Savoia, VII 744
Cordelière, III 478
Coronia, VII 759
Cutty Sark, IV 592

Dart, IV 579
Defiance, IV 548
Devastation, IV 590
Dreadnought, IV 577; VI 5
Duke of Wellington, IV 587–8
Dunedin, V 49
Élégante Messagère, IV 559
Elizabeth Watts, V 121
Endeavour, IV 575, plate 40A; V 146; VII 1446
Erebus, V 41
Etruria, V 378
Europe, VII 744
Fairfree, VI 353
Flamboro, V 149, plate 15B
Foreland, VI 353
Franconia, VII 759
Frigorifique, V 48
Fullagar, VII 736
Garry Owen, V 351, 362
George (1546), III 478
Gettysberg, VII 742
Glatton, IV 590
Glomar Challenger, VII 1457
Glückauf, V 121; VII 740
Golden Hind, III 477, 494
Goliah, IV 661
Gorgon, V 145
Grâce Dieu (1418), II 587, 588
Grand Souterrain, IV 556
Great Britain, IV 431, 476; V 147, 362, 375, plate 14A
Great Eastern, IV 661, plate 46B; V frontispiece, 146, 148, 361, 363–5, 372, 374–5, 377, 383–5, plates 13A, 15A, 24; VII 743
Great Michael (1506), III 478
Great Western, IV plate 44A; V 362, 372
Greyhound, V 389–90
Hamburg, VI 354
Harry/Harry Grâce à Dieu, see *Henry* . . .
Hart, III 479
Henry Grâce à Dieu, III 478, 487, plate 21A
Holland, V 149
Inflexible, V 217
Investigator, V 452
Iris, V 373
Isle of Arran, V 146
James Watt, IV 594
Janot, VI 352
Janus, V 153
John Sergeant, VII 760
Lady of the Lake, IV 548
La Gloire, IV 588–9
Lancing, IV 593
Lenin, VI 244
Leviathan, V 540
Lindesnaes, V 380–1
London Engineer, V 144

Lucania, V 374–7
Lusitania, VII 743, 759, 1071
Madre de Dios, III 481
Majestic, V 378
Manhattan, VII 744
Marlborough, IV 587, plate 40B
Mauretania, VII 693, 743, 759, 924, 1033, 1071
Mercury, V 373
Merrimac, V 822
Minotaur, V 148
Minx, V 147
Monitor, IV 589; V 819
Nautilus, VI 254
Neverita, VII 737
New Zealand, V 155
Newcombia, VII 737
Nina, III 477
Normandie, VII 744
Northumberland, V 148, plate 14B
Oceanic, VII 743
Olympic, VII 743
Oswestry Grange, V 48
Paraguay, V 48
Pereire, IV 593
Powerful, V 154
Preussen, IV 594
Princess Henriette, V 146, plate 13B
Princeton, V 147
Propontis, V 149
Pyroscaphe, V 142
Queen, IV 587–8
Queen Elizabeth, VII 743
Queen Mary, V 539; VII 743
Rattler, V 147
Regent, III 478
Reliance, V 449
Resolution, IV 575
Rising Star, V 144
Royal George, III 499; IV 580, 590
Royal Prince, III 485–6, 494, plate 22
San Juan, III 480
San Martin, III 480–1
São João, III 480
Savannah, V 144
Scotia, V 145
Selandia, VII 759
Shark sloop, III 491
Sharpshooter, V 154
Sheldrake, V 155
Sir Lancelot, IV 592
Sirius, V 153
Sovereign of the Seas, III 482, 485–6, 494, plate 23
Speedy, V 155
Strathleven, V 49
Subtile, III 479
Taeping, IV 592, plate 43A

Ships named (*cont.*)
Terrible, V 154
Terror, V 41
Teutonic, V 378
Thermopylae, IV 592
Thetis, V 154
Thunderer, V 153
Tiger, III 479
Titanic, VII 743, 745
Transit, IV 579
Trial, V 350
Trident, IV 590
Trieste (bathyscope), VII 1453
Turbinia, V 150–1, 156; VII 758, 1025
Umbria, V 378
Vaderland, V 381–2
Vespasian, VII 759
Victoria (1518), III 477
Victoria (1859), IV 587
Victory, III 499; IV 576–7, 582, 585, plate 42A
Virginia, VII 759
Vityaz, VII 1453
Volkswohl, VI 352
Vulcan, V 350, 355–6, 358, 366, 368–71
Warrior, IV 588–9; V 148
Zoroaster, V 121
Shipworm (*Teredo navalis*), II 569
Shoe-last celts, Danubian period, I 507
Shoemakers, II 166–8, 170
Shoes, *see* Footwear, *under* Horses
Shuttles, III 161, 165–70, 187
Siamese spoke-reel, II 206
Sickles, I 502, 503, 513–14, 520, 541–2, 698; II
     94–5; III 134, 136–7
Siege-engines; *see also* Military technology
     battering-rams, II 702, 715–16
     cannon, 14th century, II 726–7
     fighting car ('tank'), II 726
     scaling-ladders, II 716, 724
     seventeenth century, III 370
     towers, II 716–17, 725–6
     wall defences, II 717–19
Sieves, I 275, 422, 566; II 5, 104; III 16–18
Sighting-tubes, III 375, 593–5
Silex, II 28, 529
Silica, III 206, 218; IV 332–3, 347, 359; VI 430
Silk industry, IV **308–27**; *see also* Sericulture,
     Textile industries
     ancient textiles (*c.*1000 B.C.), I 445
     bleaching and dyeing, II 365
     British, Irish and Colonial Silk Company, IV
         308
     Byzantine influence, II 756, 761
     Chinese, III 195, 201–2
     combing processes, IV 321–2
     damasks, III 201–3
     dyeing and weighting, IV 320–1

early sericulture, II 197–9
     fabrics and cloths, types, III 178–9
     figured silks, III 190–1, 193
     figured twill, III 195–6, 197–8
     French loom (*c.*1605–20), II 165
     Islamic, II 762–3; III 191, 197, 199
     Italian, III 165, 196, 199, 203
     machinery, IV 309–19
     medieval, III 190–1, 196–8, 200, 202
     'old silk-route', II 759, 763
     raw-silk trade, III 195
     reeling process, IV 309–10
     Renaissance period, III 151
     satin weave, III 200
     sericulture, III 195
     spinning techniques, II 205–7, 644–5, plate 8;
         IV 313–16, 321
     tabby fabrics, III 195, 201–2
     throwing, IV 311–13, plate 15
     varnished for gas-pipes, IV 262
     warping, II 210
     watered effects, III 178
     weaving, IV 316–19
Siloam tunnel, I 764–5
Silver
     aqua regia, reaction, II 356–7
     assaying, II 45–6; III 59, 63–8
     blanching, III 46, 56
     chloride, III 241
     chronological chart, I 575
     coinage, I 585; II 13 n., 46, 485; III 46–7; V
         94; VI 411
     cupellation, I 581, 584; III 64–5
     cups, II 449–50, 463
     early civilizations, I 582–5
     electrum, I 623, 644, 645, 661; IV 146
     extraction, I 572, 574, 584; II 356–7; III 33,
         42; IV 135–40; V 72, 76, 79, 95–7; VI 424
     fine metal-work, I 623–4, 626
     furnaces, melting, III 53
     galena source, II 43–4
     German silver, IV 146
     gold, separation, I 356–7; II 65
     ground, III 49
     liquation process, I 584–5
     metallurgy, III 27, 28
     mines, II 1–4, 5, 65, 110; III 80–1, 82; IV 66;
         VI 411, 413
     mirrors (from *c.*1840), IV 363
     niello ingredient, III 48
     paint/ink, III 49
     plating, II 227; IV 119; V 95
     price (A.D. 1501), II 747
     refining techniques, II 42 n., 43–5, 65–7; IV
         142–4
     repoussé work, II 468, 470–2
     rolling-mills, III 342

Roman period, II 63
Sheffield plate, III 48; IV 130
soldering, I 650, 651, 652; III 48
speculum-metal, III 38, 40–1
sulphide, IV 66
tape, III 48
touchstone assay method, II 45–6; III 65–6
Vase of Entemena, I 652, 653
wire, II 452–3
yellow glaze, I 359
Sinapis (mustards), I 370
Sinopia (red pigment), II 361
Siphons, II 668–9, 671
Skill, I 1–37
    animals, use of tools, I 1–6
    cultural factors, I 34–6
    hereditary influences, I 34–6
    human-, evolution, I 22–34
    primates, I 6–19
    tool-making, origins, I 19–22
Skis, Stone Age, I 708–9
Slates, roofing, II 34, 415; III 260–1
Slave-labour, II 590–2, 600–1, 605–6, 638–9,
    671–2
Sledges
    cave pictograph (*c*.3500 B.C.), I 205
    farm vehicles, III 140–1
    Egyptian building methods, I 479, 481
    origin, I 206, 210, 704
    runners, types, I 516, 708
    wheel-less vehicles, I 707–11; II 19; III 537
Sleeve and socket principle, I 71
Slide-car, III 141
Slide-rules, 17th century, III 542, 628
Slings, I 156, 157, 502, 512.
Sluices, III 441–2, 444, 454, 462
Smalt, II 361; III 678, 698; IV 145
Smelting, III 27–33, 51, 74, 79; *see also under*
    Metals
Smiths; *see also* Machine-tools
    anchor-, III 116
    ancient crafts, I 579, 596
    blacksmiths, II 58–61, 69–70, 236, 242, 256,
        396; III 111–14, 116, 126–7, 487
    blade-, III 343
    bronzesmiths, I 591; II 472; III 116
    chain-, III 116
    coppersmiths, II 53
    goldsmiths, III 623
    locksmiths, II 60
    London Blacksmiths' Company, II 243
    nail-, III 116
    needlesmiths, II 75
    shear-, III 116
    silversmiths, II 482
    swordsmiths, II 479–80; III 115, 350
    tinsmiths, III 116

Snuff, III 10
Soap
    alkali supply, IV 232, 235, 238, 241, 246
    animal-oils, IV 61
    cloth, finishing, III 169–71, 175
    detergents, 20th century, VI 547
    Egyptian detergents, I 249, 260, 261
    essential oils, use, V 317
    manufacture, from 18th century, IV 253-4; VI
        547
    origin, II 354, 355–6
    palm oil, I 828; V 23
    soap-making, II 356 n.; III 703–5; IV 232
    Sumerian recipes, I 261
    -tax, abolished (1853), V 834
Soapstone (steatite), IV 343
Soay (Hebrides) sheep, I 344, 345
Société d'Encouragement, Paris, V 528
Société des Ingénieurs Civils de France, V 489–90
Societies, forms; *see also* Cultures
    domestication, animals, I 327–8
    early periods, I 38–57
    economic classes, emergence, I 53
    economic imperialism, I 56–7
    food-gatherers, I 40–2, 184–6
    industrialization, consequences, V 814–41
    ivory workers, I 664–6
    language, I 89–92, 102–3, 104–6, 745–6, 755–7
    Neolithic revolution, I 42–5, 46–7, 79, 340–1,
        349–50
    science, effects, IV 663–81
    tribute state, I 50–5
    urban revolution, I 45, 49–50, 52–3, 80, 101–2,
        377
Society of Arts, *see* Royal Society of Arts
Society of Chemical Industry, London, V 802; VI
    153
Society of Dyers and Colourists (Bradford), VI
    672
Society of Glass Technology in Britain (1916), VI
    589
Soda (sodium carbonate); *see also* Natron
    alchemical description, II 737
    caustic- (hydroxide), II 737; IV 227; V 244; VI
        519, 521
    cleansing agent, II 355
    glass industry, I 261; III 206, 234, 700; IV
        256
    glazes, II 354
    manufacture, IV 231–2, 236–42, 245, 248; V
        236–44
    mummification process, I 268, 270
    nitrun, Nile Delta, II 354
    plant source, I 261
    soap manufacture, II 354; III 703; IV 253; V
        244, 671–2
    wood source, III 685; IV 231–2

Sodium
  aeroengine coolant (1940s), VI 442
  carbonate, *see* Soda
  Castner process, V 91, 248–9; VI 518
  cyanide, V 95, 249; VI 518–19
  extraction, IV 118, 227; V 72, 91, 248; VI 519
  nuclear reactor coolant, VI 247–8
  oxide, III 234
Soil erosion, VI 319
Soil fertility studies, VI 311–12
Soil mechanics, VII 946
Solanum (potato), III 1–2
Solar power, VI 221–2
Soldering, solders; *see also* Welding
  alloys, equilibrium diagrams, I 651
  ancient metal-work, I 649–53
  blowpipe, use, III 55
  boilers, steam-engines, IV 172–3
  gold and silver, III 48
  inlaid metal-work, II 453
  lead, Greek and Roman, II 46, 47
  pipework, Middle Ages, II 665–6
  recipes, 13th–15th centuries, II 352
  tin alloys, III 42; IV 124
  window-leads, III 42
Solutreans (stone-workers), I 500
Sorghum, I 369
Sound recording, VII 1262–5, 1302, 1304–5,
  1309–15
South Africa
  cartography, V 441
  food, V 2, 51
  geodesy, V 442
  gold mining, V 95, 249
South America; *see also* America
  barbed wire fencing, V 625
  copper, V 76 n. 1, 79
  dyestuffs, V 264
  food, V 23–4, 51
  precious metals, V 85, 95
  rubber, V 221, 752, 754, 773–4
Space technology, VII 857–70; *see also* Space
  vehicles
  Apollo space flights, VII 848, 857, 858–9, 865,
    868–9
  astronavigation, VII 849, 952
  manned space flight, VI 26; VII 865–6
  Moon, VII 857–9, 866–7
  rocket flights, VI 22; VII 859–65, 869
  satellites, VII 848, 857, 866–9
  Saturn rockets, VII 848, 861, 869
  unmanned space vehicles, VII 866–9
  V-2 missiles (1940s), VII 848–9
Space vehicles, manned
  Apollo series, VII 848, 857, 866, 868–9
  Apollo–Soyuz joint flight, VII 866
  Gemini series, VII 866

  Mercury series, VII 866
  Saturn V, launch vehicle, VII 861
  Shuttle, VII 869
  Skylab, VII 866
  Soyuz, VII 857, 866
  Voskhod, VII 866
  Vostok, VII 865, 866
Space vehicles, unmanned
  Early Bird, VII 867
  Luna series, VII 867
  Pioneer, VII 869
  Sputnik, VII 857, 867
  Titan-Centaur, launch vehicle, VII 860
  Viking, VII 860, 869
Spades and shovels, II 9, 12, 16, 98–9; III 115,
  134–6, 138
Spain; *see also* Europe
  aeronautics, V 405
  agriculture, IV 13, 16, 26–9, 33–4
  aqueducts, II 47, 670, 671
  canals and waterways, IV 551, 554
  cannon, III 367
  ceramics, IV 329, 334–6
  chemical industry, III 677; V 243
  coinage, III 341
  copper, V 77, 95
  cotton, II 199
  damasks, III 201–2
  diet changes, III 1–3
  dyeing industry, III 695; V 258, 264, 266
  esparto-grass, V 713
  figured silks, III 196, 198–9
  flax, II 195
  food, V 3, 21
  furnaces, IV 122, 134
  geodesy, V 442
  glassware, II 323
  gold production, II 42
  Gregorian calendar, adoption, III 579
  infantry, III 348
  iron production, II 61; V 60
  knitting, III 185
  lead, 2nd century A.D., II 45
  leather, II 150–1, 180
  looms, medieval, III 192
  mercury, IV 134
  mining, II 5, 11–12, 13, 46–8; IV 65
  navigation, III 526, 548
  non-ferrous metals, IV 119, 123
  paper-making, III 412
  platinum metals, IV 119, 144
  population increase (1800–1950), V 823
  precious metals, V 85, 95
  roads, IV 529
  satin, III 200
  ships, III 477–8, 480–1, 499; IV 576, 586
  silver, II 44; IV 138; V 85, 95

soap, III 703, 705
'Spanish leather', II 150–1, 179
steam-engines, IV 178
sulphur, IV 232
telegraphy, IV 651, 653
textile industry, III 151
tiles, ornamental, II 307
tin mines, I 589, 591–2
tobacco, III 8
viticulture, II 134
water-supply, V 557
windmills, III 89, 95–6, 106
wool, II 192; III 152, 155
Spanners, III 112, 114, 338
Spears; *see also* Harpoons, Javelins, Weapons
   all-wooden, I 30, 141
   bronze heads, I 615–16
   copper heads, I 605
   food-gatherers, I 157–61
   hafted, I 32–3, 142
   heads, bone/flint, I 33, 498
   iron heads, I 618; II 696
   multiple-point heads, I 160–1, 173, 498
   origin, I 73, 157–8
   Roman *pila*, II 703–5
   -throwers, I 157–60
   war-weapon, II 696, 698
Spectacles, III 229–31, 630–1, 633, 699, plate 1
Speculum-metal, III 38, 40–1, 237
Speech, I **85–109**
   animal sounds, difference, I 88–9
   children and aphasiacs, I 92–4
   click-sounds, African, I 90, 95–6
   consonant sounds, I 86–8, 93
   early communication, I 18–19
   gesture language, I 94–5
   imitative words/sounds, I 89–90, 91–2
   language, relationship, I 89
   linguistics, terms, I 85–8
   vowel sounds, I 85–6, 93
Spelter, *see* Pewter
Spergula arvensis, I 370
Sphagnum, I 370
Spherometers, IV 361
Sphyrelaton work, II 469–70
Spinacia oleracea (spinach), I 355
Spindle-whorls, I 433, 434, 435, 438; II 201–2
Spindles, I 70, 187, 194, 424–5, 433, 437–8; III
   160, 162–3, 223–4
Spinning, I plates 14–17, II **191–220**; III **151–86**;
   see also Textiles, Weaving
   competition, 20th century, VI 639–40
   Egypt (*c*.3000 B.C.), I 434
   factory system, England (from 1500), III 151–3,
      160–2
   machinery (from 1760s), III 153; IV 277–94,
      299; V 577–9

primitive processes, I 424–5
   rope-making, I 453
   silk, II 205–7, 644–5; IV 311–16, 321–2
   spinners, Greek vase (*c*.560 B.C.), I 444
   tomb remains, Egypt (*c*.1900 B.C.), I 437–8
   Troy settlements (*c*.2500 B.C.), I 435
   -wheels, I 72, 83; II 201–5, 644–5, 656; III
      151, 160–2
   wool and vegetable fibres, II 200–5
Spirit-levels, IV 603–4
Spits, III 20–1
Spoons, II 125, 127; III 20
Sprang fabrics, III 181–3, plate 8B
Springs
   balance-spring, III 344
   mechanisms, II 629–30, 643–4, 712–13
   -winding machine, IV 423
Squares, geometric, III 608
'Stadiometer', IV 602
Stament (worsted fabric), III 179
Stamping-mills, *see under* Mills
Standardization
   interchangeability principle, IV 437–9; V
      638–9, 818
   International Annealed Copper Standard, V 86
   machine-tools, V 818
   measurement, I 110, 774–5
   muskets, IV 437–8
   plane surfaces, IV 424–5; V 818
   Portland cement (1904), V 486–7
   screws, IV 424, 426, 433
   weights, *see* Weights
Stars and navigation, III 521, 523, 545–7, 552–4,
   556
State intervention, III 30, 347, 352, 361; *see also*
   Government, 20th century
Statics, III 433
Statistical records
   agricultural machinery (1862–92), V 16
   arable farming, Germany, V 14
   cereal yields (1855–99), V 8–9
   cocoa, V 22
   copper production, V 76, 84, 605
   dyestuffs, V 258, 279, 281
   frozen meat, V 49
   ground-nuts, V 24
   heavy chemicals, V 239, 244, 247, 255
   metals, miscellaneous, V 73, 96
   oil drilling, V 108
   palm oil, V 23
   pins, V 624
   population, V 1, 7, 823
   postage, V 816
   rail exports (1870s), V 612
   rubber, V 770
   sheep, V 9
   ship-building, V 352, 373, 539–40

Statistical records (*cont.*)
    small-arms, V 639
    steel industry, V 53, 58, 61, 67
    sugar-cane, V 9
    tea crop, V 21
    techniques, China to West, II 770–1
    textile industry, V 816–17
    tinplate, British, 19th century, V 617
    water-supply, V 552
    wine production, V 9
Statuary, *see* Sculpture and Statuary
Steam, general; *see also* Steam-engines
    bridge-building machinery, IV 452–3
    distillation technique, III 12
    heating scheme (1720), III 687
    latent heat, IV 182
    rock-boring machines, IV 71
    steam-hammer (from 1784), IV 116
Steam-boats, IV 548, 571–2; *see also under* Ships
Steam-engines, IV 168–98; *see also* Aeolipiles,
        Engines, Railway locomotives
    atmospheric, III 344; IV 175–7, 180
    beam-engine, IV 183, 186–9, 191–2, 197, 499
    boilers, V 137–8; *see also* Boilers, steam
    ceramic industry, IV 348, 352
    coal consumption, IV 166
    chronological chart (1600–1850), IV 165
    compound, IV 187, 191, 194
    Cornish beam, V 126–30
    cylinders, boring-machines, III 368; IV 103–4,
        409
    direct-action, IV 195–7
    double-acting, IV 185–7
    factories, IV 130, 148, 150–1, 161–6, 196–7
    glass-working machinery (1789), IV 372, 377
    grasshopper, IV 195–7
    high pressure, IV 188–94, 197, 499
    horizontal, V 130–3
    injector-pumps (aeolipiles), III 52–3
    inventions, to 17th century, IV 168–71
    locomotives, IV 188–9, 197
    machine-tools, IV 380, 441
    marine, III 83–4; IV 587–8, 590, 593–4; V
        141–56, 301, 374, plates 12–15
    marine boilers, V 152–5
    Maudslay's table-engine, IV 196–7
    mining, IV 77–80, 81–2, 84, 161–2, 671–3
    paddle-steamers, V 143–6
    paddle-wheel propulsion, IV 196–7; V 141–2
    'parallel motion', IV 186–7, 191, 195–6
    plunge-pole, IV 192–4
    power measurement, IV 177, 680–1
    prime movers, 17th–19th centuries, IV 161–6;
        VII 997
    rotative, IV 185–7, 195, 499
    screw-propulsion, V 146–9
    steam-pressure, IV 162–3, 165, 188, 190–3, 195

    shipyards (1799), IV 581–2
    stationary, V 124–40, plates 7–9, 10A
    suction-pump principle, III 330, 344
    textile industry, IV 283, 288, 319
    thermal efficiency, IV 163–5, 680–1
    triple expansion, IV 164
    turbines, V 138–40, 149–52; VII 758–9, 989,
        990, 997, 1025–34
    vertical high-pressure, V 133–7
Steatite in porcelain (1700), IV 343
Steel industry, IV 99–117; V 53–71; VI 462–98;
        see also Iron
    alloys, IV 108; V 64–6
    ancient forms, I 573, 576
    basic process, phosphoric ores, V 60
    Bessemer converter, IV 678; V 53–7, 60, 373;
        VI 462–3, 478–82, 484–5, 490
    Chalybean, II 5, 59
    crucible-cast steel, II 57; III 35; IV 107–8
    damask-, II 57
    electricity, application, V 66, 73
    furnaces, V 53–60, 66–70; VI 489–91
    German production, 15th century, II 75
    Greek production, II 60
    hardening process, III 34–7
    heat treatment, III 35
    instruments, IV 381, 404–5, 601
    metal-working, III 343; V 609, 611–13, 617,
        621–2, 623, 625
    metallurgy, III 57
    production, 16th–18th centuries, III 28–9, 31,
        32–4, 79; IV 107–8
    production (1900–50), VI 462
    quenching, III 35–6
    Roman production, II 60
    screws (*c*.1570), III 338
    Siemens-Martin process, IV 678; V 57–60, 373
    structural use, V 476–7
    temper-colours, III 36
    tool-engraving, III 407
    tool-making, II 453; III 115
    trade, Middle Ages, II 76
    tube-making, V 63–4
    uses, V 61–3, 640
    weapons, II 63; III 350
    -works, V 56, 59, 63, 66, 68–9
Stellaria media, I 370
Stibnite (antimony trisulphide), I 293, 582, 588;
        II 12, 47, 74, 737; III 29; IV 120
Stills, *see* Distillation
Stipa tenacissima (esparto grass), I 421, 450; II
        197
Stock market crash (1929), VI 15
Stones: *see also* Building materials, Quarrying,
        Sarsen rocks
    ancient buildings, I 456–94
    beads, I 139–40

breaking, IV 534
bridges, III 419–20, 423–4, 426, 428, 436
canal-building, III 446, 450, 453, 455, 462, 468
cannon-balls, III 34
crushing, IV 538
door-sockets, I 193–4, 306, 472
dressing, IV 212
flaking techniques, I 129–39
hammerstone technique, I 129–30
lamps, I 33, 147, 235–6
microliths, I 496, 502
missiles, I 157, 512
paving-stones, IV 543–4
polishing & grinding, I 139–40, 506
precious/semi-precious, I 243, 572, 574, 659, 661
quarrying, Egypt, I 477–9, 526, 560, 569–70, 714
roads, I 714
sculpture, I 469, 470
techniques for working, I 128–40, 496–7, 504, 506, plates 1, 21A
tools and weapons, I 19–32, 67–9, 73, 128–9, 501–15, 601
vessels, I 398–9
Stonehenge, I 81, 195, 315, 490–4, 712, plate 21A
Stoves, Cowper, V 58, 67
Straw, ancient basketry, I 450
Straw, paper-making, III 411, 415
Streets, see Roads and streets
Strikes, industrial, VI 14, 98–9, 102, 104, 106
Stuff (worsted fabrics), III 179
Sublimation equipment, II 740–1, 742, 750
Succinic acid, preparation, II 741
Suffrage, VI 4, 14
Sugar, see under Food production, Sweetening-agents
Sulphides; see also Sulphur
  arsenical pigments, IV 122
  definition, IV 222
  glass, discoloration, III 221
  hydrogen-, isolation, IV 216
  metal ores, II 42, 43–4, 47–50; IV 120, 126, 129, 133–4; V 74, 80, 88
  metals, parting process, III 65
  niello ingredient, III 48, 406
Sulphur
  alchemical theory, II 352, 736–7
  copper alloys, 16th–17th centuries, III 42
  cupellation process, II 42 n. 2, 44
  dioxide pollution, IV 78
  distillation, 16th century, II 371
  gunpowder ingredient, see Gunpowder
  metals, parting process, III 65
  military pyrotechnics, II 376, 380–2
  Mond recovery process, V 238
  names, compounds, IV 222
niello ingredient, III 48, 406
refining, II 371, 740–1; IV 232
sources, IV 232; VI 515–16
sublimation, II 740–1; IV 232
value (1501–8), II 747
Sulphuric acid
  balloons, hydrogen synthesis, IV 255
  brass 'pickling' process, IV 133
  definition, IV 222
  lead, resistance, IV 231
  manufacture, III 12; IV 231, 237, 239, 242–6, 248; V 244–8; VI 514–18
  matches, manufacture, IV 254–5
  nitric acid synthesis, V 255; VI 517
  nitrogen fertilizers, VI 522, 529
  ore refining process, IV 124
  origin and isolation (c.1535), II 357
  phosphate fertilizers, IV 246, 253, 254; V 254–5; VI 529–30
  platinum laboratory ware, IV 145
  production, 20th century, VI 515
  silver, parting process, IV 142–3
  soda manufacture (Leblanc), IV 239–42; V 244
  textiles, bleaching, III 176; IV 246
  value (c.1800–30), IV 245–6
Sumac, tanning-agent, II 151, 155–6
Sumer, I 48–9, 101–2, 203, 464–9, 545, 670, 745–52, 777; see also Mesopotamia
Sumerians, II 164
Summa perfectionis (Geber), II 356, 738–9
Sun; see also Sun-dials, Time
  astrolabes and latitude, III 521
  calendars, I 121–3; III 558, 560, 567, 577
  daily journey (c.130 B.C.), III 517
  instrument theory, III 582–3
  navigation, III 523, 545–7, 549, 551–3, 556
  shadow survey methods, III 513, 528
  size of earth measurement, III 505–6
  solar furnaces, III 55
  space technology, VII 858, 866
Sun-dials
  altitude dials, III 594–600
  direction dials, III 600
  equatorial dials, III 598
  Greek time-recorders, I 126
  'ham' dials, III 596–7
  instruments, 16th century, III 586
  portable, III 533, 535, 594–601
  Saxon scratch-dial, III 594
  shadow-clocks, I 113, 123
  'shepherd's dial'/'chilindre', III 596
  ship-dials, III 597–8
  skaphe (hemicycle), III 594, 596
  universal rectilinear dials, III 597–8
Supersonic aircraft named; see also Aircraft headings
  Bell XS-1, VII 830

Supersonic aircraft named (*cont.*)
  Concorde, VII 834, 835, 1469
  Douglas Skyrocket, VII 832
  F-86, VII 830
  F-100, VII 830
  Tu-144, VII 834
Superstition and magic; *see also* Alchemy
  alum, magical powers, I 262
  cosmetics, I 286–7, 289
  magical 'knife', I 666
  Mayan number mysticism, I 124–6
  metal artifact taboos, I 504, 592
  mummification, I 266, 267, 270, 756
  natural-gas myths, I 251
  pottery shapes, I 400–1
  primitive art, I 144, 146
  salt, magical powers, I 257, 259, 260
  temple city society, I 47–9
  time, magico-religious, I 114
Surfactants, VI 547; VII 1415
Surgery, *see under* Medical technology
Surveying and surveys, III 510–29, 537–57; *see*
      also Cartography, Maps, Navigation
  aerial photography techniques, V 449
  ancient maps, III 503–4
  base-measuring equipment, V 447–8
  cadastral, IV 608, 614
  cartographical (1400–1750), III 537–44
  civil engineers, IV 443
  classical times, III 510–12
  coastal, IV 616, 618–19
  Fens, Great Level scheme (1588–1630), III 317
  fortification, IV 603
  French geodetic, 18th century, III 554
  geodetic, IV 604–5; V 441–4
  geological, IV 622
  hydrography, V 449–54; VII 1448–55
  instruments, III 510–13, 528, 537–43, 555,
      609–12, 627–8; V 444–54
  levelling, III 542–3
  marine, IV 618
  Middle Ages, III 513–17
  military requirements, III 373
  national, IV 605–12, 625
  Ordnance Surveys (from 1791), III 554
  Roman Empire, III 504
  space technology, VII 866
  'stasimetric', IV 616–17
  theodolites, *see* Theodolites
  topographical, IV 600–11, 613
  training, surveyors, IV 442
  triangulations, III 503, 511, 517, 539–40, 544,
      554, 556, 626; IV 397, 402, 599–610; V
      450–2
  trigonometrical, IV 599, 603, 607, 610–11
Sweden; *see also* Europe, Scandinavia
  agriculture, III 156; IV 26, 35, 39; V 20

asphalt, IV 539
blast-furnaces, IV 129
canals and waterways, III 455; IV 554
ceramics, V 662
chemistry, III 682
cobalt, IV 145, 335
computers, VII 1174
copper, IV 129
domestic appliances, VII 1146
dyestuffs, V 268
electricity, V 191
flax, processing, III 156–7
glass, III 217
Gregorian calendar, adoption, III 579
hydro-electric power, VI 200–1
iron industry, III 31, 79, 342–4; IV 106, 108,
    115–16; V 60; VI 490
machine-tools, V 648
marine steam-propulsion, V 150
matches, V 253
metal-mining, IV 65
oil, V 108
paper, VI 612–13
railway engineering, V 333, 338
roads, IV 528
steam turbines, V 140
steel industry, V 55, 63, 70; VI 490
surveying, IV 610; V 447
tinplate, III 689
town-planning, VII 976–7
weaving, traditional, III 187
windmills, III 97, 99
Sweetening-agents, II 127, 139, 372; *see also* sugar
    under Food production
Swiss lake-villages, *see* Lake-villages
Switzerland; *see also* Europe
  aluminium, V 93
  asphalt, IV 539–40
  bridges, III 427–8; IV 459
  bronze, I 614–15
  building, structural theory, V 493
  cartography, V 440
  ceramics, IV 329; V 663
  chronometers, IV 413–15
  clock-making, III 674
  dyestuffs, V 265
  electricity, V 190; VII 1071
  glass industry, IV 359–60; VI 577
  hydraulic engineering, V 533
  hydro-electric engineering, VI 213
  linen, finishing, II 219
  metal working, V 608
  milk, condensed, V 35
  nickel, V 86
  oils, lubrication, V 118–19
  pikemen, III 348–9
  plastics, VI 551–2

population growth (1800–1950), V 823
railway engineering, V 345–6
relief-tiles, II 306
roads, IV 528–30, 536, 540
stained-glass windows, III 241
steel industry, IV 108; V 66
surveying equipment, V 449
surveys, IV 610, 612
water-supply, IV 498
weaving, III 159
Swords; *see also* Daggers, Knives, Weapons
bronze, I 614–15
heat treatment, steel, III 35
iron, I 617–18
Lydian steel, medieval, II 60
Middle Ages, II 69, 75
moulds, II 479–80
ornamented, II 451, 453, 455–8, 474
pattern-welded, II 457, 460
Renaissance variants, III 350–1
Roman, II 57, 455, 704
sharpening wheel, 9th century, II 651
steel, outer layer, II 57
Syria, I map 8
alphabetic script, I 766–7
glass industry, II 322, 324–5, 327–8, 337, 340
ivory work, I 672–3, 675–6
merchant ship model, II 575–6
Roman roads, II 504
water-wheels, II 614

Tabby weave, III 187–8, 194–5, 197–205
Table engines, IV 426
Table-ware, V 634, 661–3; *see also* Cutlery
Tables, furniture, *see under* Furniture
Tables, numerical; *see also* Mathematics
'Alfonsine', III 535, 585, 617; *see also*
Ephemerides
amplitude, III 549
astronomical, III 503, 547, 557, 584–5; IV
597–8
Babylonian mathematics, I 787
cartographical, IV 596
conversion, distance, III 536
engine parts, Smeaton's proportions (1772), IV
179
engine-power, IV 177
gravities, III 22
logarithms, III 553
lunar, III 552, 557
mathematical, III 57–8, 521, 537, 541–2
nautical, III 552–3, 557
navigation, III 526–7
Peutinger (Roman Empire), III 503–4, 514–15
range, III 361, 369, 374
'Seaman's Kalendar' (1600), III 552–3
shadow-lengths, III 595

stellar (Flamsteed's), III 667
tide-, III 553
trigonometrical, III 542, 551, 553
Tacheometry, IV 602
Taffeta (silk fabric), III 179
Talmud, II 117
Tamis (wool fabric), III 179
Tammies (worsted), III 178–9
Tannin compounds, II 151, 155; IV 321
Tanning, *see under* Leather
Tantalum, IV 119; V 99; VI 448
Tape recording, *see* Sound recording
Tapestry, III 193–4
Tapers, I 234
Tappets, III 328, 412
Tar, III 685; IV 252–3, 541–2; VI 467; *see also*
Asphalt, Bitumen, Pitch, Wood-tar
Tartar (argol), II 354–5, 358; III 29, 41, 217,
218 n.; V 308
Tasmanians, I 183, 184, 185
Tawing of leather, II 149–50, 154–5
Taylorism, VI 73–5, 81–5, 93, 115–16, 118–19,
120; VII 1046–9, 1056
Taynton stones, II 35
Tea, III 1, 5, 23; V 2, 21–2
Teazels, fuller's, II 193, 217; III 172
Technology; *see also* Discovery, Invention
accidental discovery, I 60–2
diffusion factor, I 77–84; *see also* Diffusion
economics, VI **48–76**
education, industrialized societies, VI **138–71**
environmental factors, I 62–6
government function, VI **113–37**
human needs, I 38–40
novelty, VI 52; *see also* Innovation
quality of life, VI 132–3, 626; VII **1467–82**
Teeth, cutting-machines, IV 381, 383, 385, 391–3,
395, 420–1
Teeth evolution, man, I 20–1; *see also* Dentistry
Telegraphy, IV **644–62**; V 218–28; VII 1220,
1236
cables, IV 660–1; V 218–20, 222; VII 1242
electric, IV 647, 649–52, 657–61, plate 47; V
330, 448
Electric Telegraph Company, IV 659; V 218–19
electrochemical, IV 652–4
electromagnetic, IV 654–6
insulation, cables, V 220–3
radio-, V 227; VII 1236, 1248–50, 1266
semaphore, Chappe method, IV 645–7, 651
submarine, V 224–6
surveying technique, V 448
Western Union, IV 660
Telephony, V 226–7
beginnings (c.1876–1915), VII 1220–9
cables, VII 1242–3
exchanges, VII 1223–6, 1244–7

Telephony (*cont.*)
  inter-urban networks (1886), VII 1226–7
  long-distance, VII 1227–30, 1238–41
  radio-, VII 1236, 1239, 1250–2, 1265–6
Telescopes; *see also* Lenses
  achromatic lens invention (1758), IV 358
  astronomical, IV 398–400
  Cassegrain, IV 398
  dilatometer (1781), IV 396
  geodetic instruments, IV 400–3
  Gregorian, IV 398
  instruments, 17th century, III 234
  invention (*c.*1580), III 231–3, 631
  lens improvements, 18th–19th centuries, IV
    359–61
  navigation, III 552, 641; IV 400–3
  precision mechanics, 18th century, IV 380, 382
  reflecting, III 41, 233, 237, 632, 633–4; IV
    398–9
  scientific instruments, III 630–6
  sighting-tubes, III 594, 595
  speculum metal, III 41, 237, 634
  surveying, III 543, 553–5, 641; IV 603–4
Television, VI 25; VII 1257–62, 1264
Tellurium, IV 119; V 97
Temperature; *see also* Thermometers
  alchemical classification, II 740
  assay methods, III 64
  automatic furnace, Drebbel's, III 680
  bimetallic devices, IV 395, 412–14
  bread-making, V 31
  canning, V 40, 42–3
  ceramics, IV 330–1, 334, 336–7, 346, 350–1
  clocks and watches, III 672–3
  furnaces, metallurgical, III 33; IV 109–10; V
    58, 65
  glass-making, III 210, 238
  Gulf Stream, IV 619
  I.C. engines, V 162–3
  milk pasteurizing, V 33
  'pyrometric beads', IV 350–1
  refrigeration, food, V 45–51
  solders, equilibrium diagrams, I 651
  steam-heating methods, III 687
  survey wires, V 447–8
  tempering steel, III 36
  thermal expansion, measurement, IV 395–6
  thermodynamics, IV 163
Temples
  architectural features, I 457–8, 460–1, 462–4,
    482–4
  Painted Temple, Uqair, I 462–3
  Red Temple, Erech, I 463
  temple city societies, I 47–9
  White Temple, Erech, I 462
Templets (pottery tools), II 289
Tène culture, *see* Cultures, La Tène

Tenoning-machine, IV 437
Teredo navalis (shipworm), II 569
Terracotta, II 262, 269, 274–9
Terrestrial magnetism, IV 621
'Terylene' (plastic fibre), VI 646, 651
Textile industries, processes and textiles, I
    413–55; II **191–220**; III **151–205**; IV
    277–327; V **569–604**; VI **638–74**; *see
    also* Textile machinery
  ancient fabrics, I plates 11–17
  bleaching, III 174–6, plate 31; V 587–8
  block-printing, V 588, 594
  calendering, III 177–8
  carding, III 153–4; IV 277, 280, 284–5, 298
  carpets, V 569, 580–5
  chemistry, IV 228, 233, 244, 248–9, 255
  chronological development, I 431–45
  cloth types, III 178–9
  clothing manufacture, VI **675–90**; VII 1475–7
  combing, III 154–6; IV 295–8, 321–2; V 574–6
  cotton, II 199–200; III 159–60; IV 277–91,
    315; V 569–70, 572–4, 576–80; VI 640–2
  creasing, permanent, VI 673
  cylindrical printing, IV 246
  dyeing, V 588, 594; VI 644–5, 671–4; *see also*
    Dyeing
  fibre webs, VI 670–1
  fibres, heat-setting, VI 672
  figured fabrics, III **187–205**, plates 10–15
  finishing, II 214–19; III 169–78; IV 304–5; V
    587–8; VI 671–4
  flax, II 195–7; III 156–8; IV 291–4, 315; V
    580; *see also* linen
  fulling, III 169–72
  fustian, III 159
  heckling process, IV 293–4
  hemp, II 195–7; III 158–9
  hosiery, V 595–600; VI 666–8
  jute, V 569–70, 574, 580
  knitting, VI 661–9; *see also* Knitting
  lace, III 185–6; V 595, 600–4, plate 37
  linen, V 569, 574, 579–80
  man-made fibres, VI 643–6, 649–51
  mercerizing process, V 587
  natural fibres, VI 640–2, 651–8
  nylon, VI 345, 348–9, 501, 560–1, 646, 650–1
  origins, I 413–15
  Paisley shawls, V 583
  patterns, III 164–6
  pile fabrics, VI 671
  pressing, III 176–8; VI 684–5
  raising, cropping and frizing, III 172–5
  rayon, V 579
  reeling, II 208–10; III 162–4
  scientific study, 17th century, III 681
  sewing, V 588–91
  shearing process, III 173–5; IV 304

shoddy, V 569–70
silk, II 197–9, 205–7; IV **308–27**; V 570, 578
spinning, *see* Spinning
textured yarns, VI 659–60
tow conversion, VI 658–9
washing and scouring, III 153, 162–3, 169, 171
waterproof fabrics, IV 253
weaving, *see* Weaving
wool, II 192–5; III 152–6; IV 282, 295–305, 315; V 569–72, 574–80; VI 640; *see also* Wool
worsteds, III 178; IV 282, 295, 298, 304; V 569, 572, 574, 576–7; VI 641
Textile machinery; *see also* Textile industries
   Arkwright's water-frame, IV 278
   backstrap looms, I 445
   Barmen lace machine, V 604
   Blackburn loom, IV 302
   Bramwell's automatic machine, V 571
   Brussell's power-loom, V 582–3
   carding machines, V 571–3
   carpet looms, V 580–7
   Chinese, II 770
   circular knitting frame, V 598–9
   condensers, V 571–2
   Cotton's knitting frame, V 598–9
   Crompton's mule, IV 279, 314
   cropping, III 174
   'dobby', IV 303; V 579
   draw-looms, III 164–7, 187–9, 191–5, 202
   factory system, from 16th century, III 151–2; VI 677–85
   flax preparation, III 157–8
   Garnett machine, V 570
   gig-mills, III 172–3
   hand-looms, IV 299, 304
   Hargreave's spinning-jenny, IV 278–9
   Heathcoat lace machine, V 600–3
   Heilmann-Schlumberger comber, V 574
   Holder-Lister machine, V 575–6
   horizontal looms, I 426, 427, 437–8; II 212–14, plate 9
   Jacquard loom, IV 299, 303, 316–19; V 579–80, 603
   jenny, IV 278–82, 299
   Kay's fly-shuttle, IV 277
   Leavers lace machine, V 595, 602
   Lee stocking frame, V 596
   lighting, textile mills, IV 246, 252
   Lister-Donisthorpe comber, V 574
   Livesey lace-curtain machine, V 603
   looms, I 425–8, 432; II 770; VI 29
   mules, IV 279–80, 287–91, 295, 299, 429, 560, plate 14B; V 577
   Nasmith cotton comber, V 574
   Noble comber, V 574–5
   Northrop automatic loom, V 584–7

   power-looms, III 167–9; IV 299–304; V 580
   reed looms, V 579
   ring spinning-frame, V 577–8
   rotary milling-machines, V 587
   rotary shearing-machines, IV 305
   roving-frames, IV 286–7, 295
   Royal Axminster power-loom, V 584–5
   screw-gill, IV 293, 315
   sewing-machines, V 588–91, 655, 819
   throstles, IV 291, 295, 314
   Vaucanson's loom (1750), III 166–7, 169, plate 8A
   vertical looms, I 426, 427, 438–40; II 211–12
   warp-weighted looms, I 426, 427–8, 442, 443–5; II 209, 211
   water-frames, IV 277–9, 285, 291
Théâtre des Instrumens Mathématiques (Besson), II 626
Thenard's blue, IV 145
Theodolites; *see also* Surveying
   alternative instruments, IV 598, 617; V 451
   Cook's surveys (1763–73), IV 616
   dioptra instrument (*c.*A.D. 62), III 609–12
   eighteenth century, III 553; IV 602–4
   Everest's (*c.*1840), V 444
   hydrographic surveys, V 451
   manufacture, III 636; IV 403
   origin, III 541, 542, 543, 627; IV 402
   polimetrum prototype, III 538–9
   Ramsden's (1784 and 1990), IV 402–3, 604, 608; V 444
   subtense principle, V 445–7
Theodosian Code (A.D. 438), II 63, 539–40
Théorie des proportions chimiques (1819), Berzelius, IV 226
Thermal expansion, metals, IV 395–6
Thermodynamics, IV 163–4, 679–81
Thermometers; *see also* Temperatures
   bimetallic concept, Lenoir's, IV 397
   designs, Florentine (*c.*1660), III 636–7
   eighteenth century, IV 403
   Fahrenheit's (1775–7), IV 619
   Lavoisier's 'gasometer' scale (1781–2), IV 406
   scientific revolution (from 1650), III 630
   thermostatic device (1663), III 680
Thermostats, cookers (1930), IV 274
Thlaspi arvense, I 370
Thracians, I 373
Threshing-sled, II 106
Throwing-sticks (boomerangs), I 155–7, 499, 517, 518
Tickhill Psalter (*c.*1314), II 177
Tidal power, VI 216–18
Tiles, *see* Bricks and tiles
Timber; *see also* Carpentry, Wood
   bridge-making, III 423, 425, 427
   building, III 248–50, 253–7, 262–6

Timber (*cont.*)
  canal construction, III 447, 450–1, 453, 455, 463
  iron industry, III 31
  ship-building, III 487–9, 481, 492–3
Time, measurement; *see also* Astronomy, Clocks, Sun-dials, Watches
  calendar systems, I 123–4, 793–5; *see also* Calendars
  chronological structure, II 753–4
  early civilizations, I 120–7
  primitive peoples, I 110–27
  temporal and equinoctial hours, I 113, 796–7
  time-dimension, I 114–15, 126–7
  timekeepers, mechanical, III **648–75**
  zodiacal, I 112–13, 797
Time and motion work-study, VII 1056
Timekeepers, *see under* Time, measurement
Tin; *see also* Tinplate
  alloys, III 42–3, 58
  ancient mines, I 563, 589–90
  assaying, ores, III 62
  bell metal, III 38
  chronology tables, I 574–5
  -copper alloys, I 588–92; II 47; III 38, 42; *see also* Bronzes
  density studies, Middle Ages, III 58
  enamel, IV 333–4, 336, 338
  extraction, I 590; II 47; III 27, 29; IV 119, 123–5; V 72
  foil, IV 124–5
  iron casting process, III 34
  -lead alloys, II 47, 63; *see also* Pewter
  Malaccah-, IV 119
  mineral map, U.K. (1851), IV 65
  mines, II 10, 12–13, 46–7; VI 410, 414, 416–17
  ore flotation process, VI 424
  oxide, IV 334
  paint/ink, III 49
  solders, I 651–2; *see also* Soldering
  speculum-metal, III 38, 40–1
  tinning process, III 342; V 616; *see also* Tinplate
  solar furnaces, III 55
  type-metal, III 43–4
  value (1501–8), II 747
  -wares, III 343
  world production (1850–1900), V 73; VI 428
Tincal, III 228 *and* n.
'Tinglass', III 43
Tinplate, III 343, 689–91; IV 104, 124–5; V 39, 43–4, 615–17; *see also* Tin
Tire-bender, wheelwright's, III 113, 127
Tires, iron, III 113, 114, 127–8
Tissued fabrics, III 196–200
Titanium, IV 119, 221; V 97
TNT, VI 548, 550

Tobacco, III 1, 8–10
Tollund (Iron Age) Man, I 355, 359, 370, plate 6B
Tombs; *see also* Burial mounds
  Agamemnon, I 486, 488
  Baqt, I 437–8
  Hetepet, I 540
  Khety, I 437–8
  Khnemhetep, I 437, 438
  Mereruka, I 341
  Tehutihetep, I 438
  Tutankhamen, *see* Tutankhamen's tomb
Tongs; *see also* Pincers
  glass-making, III 210, 219
  hearth accessory, I 233
  hinged-, smith's, I 620; II 58, 60, 61
  holders, coin dies (1552), III 341
  metal-founders (*c.*1500 B.C.), I 578, 579, 635–6
  moulds, cannon-balls, III 34
  origin, II 630
  tradesmen's, 19th century, III 113
  weeding-tongs, III 135–6
Tool-making/tools; *see also* Machine-tools, Weapons
  agricultural, I 513–14, 620; II 81–102; III 14–15, 34, 112, 115, **134–50**
  ancient man, I 23, **128–43**
  animals, use, I 1–6, 13–15
  bamboo, I 164, 169, 233
  bone, I 31, 32–3, 62, 141, 172, 188–9, 432–3
  boring, I 188–9
  bridge-building, III 420
  bronze, I 81–2, 504–5, 608–16
  cannon manufacture, III 365
  carnivorous habits, influence, I 19–21
  carpentry, I 307, 311–12, 517, 518, 611–12, 613, 685, 687–90, 696; II 228–33, 252, 389–95, 641–4, 652, plate 12
  catalogues and lists, III 116–18, 122–3
  centres, tool-making, III 115
  coach-making, III 113, 115
  coin-making, II 488; III 341
  cooking, II 127–8; III 20
  copper, I 46, 59–60, 600–7
  core- and flake-tools, I 23, 28–30, 132–9
  cutting-tools, III 35–6, 336
  discovery factors, I 67–74
  distilling use, III 11–12
  drilling, I 189–93, 612–13, 679
  early man, I 10–13, 15, 19–34
  engineering, IV 386
  engraving, I 642, 648–9; III 407–8, 624
  farm use, *see* agricultural
  flake-, *see* core- and flake-tools
  food disintegrating, II 106–18; III 16–19
  food-gatherers, I 154–68
  glass-making, II 331, 332–3; III 208, 210–12, 217, 219–20, 222–9, 231–2, 234–6

hafting, I 31, 33, 69, 70
hand-tools, III 110, 334; IV 381–4
home-made, III 112–13
iron, I 82, 321, 322, 618–21
iron smelting, I 596
ivory, I 140–1, 188–9
ivory-work, I 679–80
joinery, II 242–3, 250, 253
language factors, I 17–19, 27
leather-work, II 160, 162, 165–6, 169, 350
machine-tools, III 334–7, 345, 375; IV 385,
    417–41
masonry use, I 481
mass-produced, III 113–14
mechanical principles, I 33
mechanically guided, IV 384
metal, I **600–22**; III 19, 37
metallurgy and fine metal-work, II 58–61, 66,
    67, 396, 453, 469
mining, I 514–15, 560–3, 567, 574; II 3, 5,
    8–10, 12, 14–19, 24
Neolithic trade, I 500–1
non-metallic, I **495–519**
pottery, I 390; II 274, 289–90, 294
precision instruments, III **582–619**
printing, III 386–8, 391–2, 394, 398–9, 402–7
quarrying and stone-masonry, I 478, 479; II
    36–7, 384–6
rope-making, I 454
scientific (c.1500–c.1700), III **620–47**
specialization, I 46–7, 49, 52, 53; III 116–17
spirits, testing, III 22
standardized implements, origin, I 25, 27, 29,
    68, 79, 188
stone, I 23–5, 128–40, 496
sugar refining, III 7–8
tanning, II 147–9, 152–4
tool-using, difference, I 14, 128
tradesmen (c.1500–1850), III 110–**33**
wooden, III 114–15
Torches, I 233–4, 566, 567
Torquetum (or 'turketum'), III 585, 592–3, 627
Torricellian tube, III 543
Tortoise-core technique, I 29–30, 31, 35, 132–4
'Tortoise-shell' varnish finish, III 696–7
Tortoise-shell veneer, II 227
Touchstones, II 45–6; III 65–6
Town-planning, III **269–99**; VII **958–82**; see also
    Towns/cities named
  advertising control, VII 970, 976
  analysis of problems, VII 968, 979–81
  ancient, III 269–72
  compensation and betterment, VII 967, 970
  Development Corporations, VII 970
  education and planning, VII 978
  garden city concept, VII 964–5
  Green Belt concept, VII 962, 968, 969

Hellenic and Hellenistic, III 272–5
housing reform, VII 962
Industrial Revolution legacy, VII 962
land, acquisition, VII 970
listed buildings, VII 976
medieval, III 282–5
New Towns, VII 968, 970, 975
One-industry towns, VII 967, 976
population growth, VII 962, 965, 968
regional planning, VII 965, 968, 976, 977
Renaissance, III 285–93
ribbon development, VII 965
Roman, III 275–82
rural land, utilization, VII 965
satellite towns, III 285–6
treatises, III 285–6, 288, 290
urban problems, VII 962, 965, 968
war damage, VII 974, 976
Towns/cities named (planning); see also Town-
    planning; Index of Place Names
  Alexandria, III 275
  Amsterdam, III 293–5
  Antioch, III 275, 282
  Aosta, III 280
  Athens, III 272–3
  Babylon, III 270–1
  Bath, England, VII 960
  Bournville, England, VII 963
  Brasilia, Brazil, VII 980
  Byzantium, III 279
  Canberra, Australia, VII 959
  Carcassonne, III 283
  Charleville, III 290
  Cologne, West Germany, VII 974–5
  Constantinople, III 275
  Corinth, III 275
  Cumbernauld, Scotland, VII 971
  Damascus, III 275
  Delos, III 274
  Dura Europus, III 275
  Ephesus, III 275, 282
  Florence, III 280
  Glasgow, Scotland, VII 972
  Harappa, Indus Valley, III 270
  Henrichemont, III 291
  Hull, III 284
  Kahun, Egypt, III 269–70
  Leghorn, III 287
  London, III 278–9, 295–8; VII 968, 969,
      979
  Los Angeles, U.S.A., VII 972
  Lucca, III 280
  Marzabotto, III 276
  Miletos, III 272–3
  Mohenjo-Daro, Indus Valley, III 270
  Monpazier, III 282–3
  Neubrandenburg, III 285

Towns/cities named (planning) (*cont.*)
New York, U.S.A., VII 966
Olynthos, III 273–4
Ostia, III 280–1
Palma Nuova, Venetia, III 286
Palmyra, III 280
Paris, III 293; VII 966
Piraeus, III 271–2
Pompeii, III 274–5
Preston, England, VII 961
Priene, near Miletos, III 273–4
Richelieu, III 291–3
Rome, III 276, 281–2, 287–9
Selinus, Sicily, III 273
Silchester, III 277
Stockholm, Sweden, VII 976–7
Timgad, Algeria, III 280, 282
Turin, III 280, 286–7
Ur, III 271
Verulamium, III 276–8
Vitry-le-François, III 289–90
Winchelsea, England, III 282
Toxophilus (1545), Roger Ascham, III 351
Toys, 6th century B.C., II 6–7
Trade and commerce
beginnings, I 44, 52, 54–7, 500–1, 704
copper, I 563–4
diffusion of culture, I 78–9; II **753–76**
factors affecting, II 493
glass products (from 750 B.C.), II 321
grain, II 103
Greek trade-routes, II 567–9
incense, I 295
ivory, I 664
pottery, I 401, 406, 408–9
radishes, I 356
Roman, II 573–4
silver and lead, I 583
specialization, III 116
textiles, II 191, 197, 527, 763–5
timber, I 685
tin, I 589, 591
tolls and imposts, II 524–5, 533
tools, I 500–1
tokens, IV plate 48
trade cards, III 225, 623, 634
trade-routes, general, II 772–6
trade-routes, medieval, II 525–7, 537, 759
(map), 760, 769 (map)
wine, II 129–31, 136–7
writing, invention, I 102–3
Trade Unions, VI **93–112**
attitudes change, VI 96–7, 100–4
collective bargaining, VI 100, 102, 108–10
demarcation disputes, VI 14, 98–9, 102, 104,
106, 585–6
education, VI 105

function, organization, and growth, VI 95–8,
103, 107, 108
recruitment, VI 106
war-time position, 20th century, VI 100–1, 108
Whitley Councils, VI 101, 109–10
Trade Unions named
Amalgamated Engineering Union, VI 102, 107
Amalgamated Society of Engineers, VI 98
American Federation of Labor, VI 98, 103
American Federation of Musicians, VI 104
Clerical Workers' Union, VI 107
General and Municipal Workers' Union, VI
107
International Association of Machinists, VI 98
International Typographical Union, VI 98
International Photo-engravers' Union of North
America, VI 105
Miners' Federation of Great Britain, VI 104
National Union of Boot and Shoe Operatives,
VI 104
National Union of Railwaymen, VI 107, 108
National Union of Vehicle Builders, VI 107
Order of Railroad Telegraphers, U.S.A., VI 97
Transport and General Workers' Union, VI 107
Typographical Association, VI 97
United Mineworkers, VI 108
United Road Transport Union, VI 107
Trades Union Congress (TUC), VI 107–9
Tradesmen, *see* Artisans
Tradesmen's tools (*c.*1500–1850), III **110–33**
Traffic signals, VII 872
Trams, IV 77, 85, 89–90; VII 732–3
Tramways, V 230, 232 n. 1, 233, 348
Transmutation, base metals, II 352, 731–2, 736
Transistors, VII 1117–20, 1156
Transport, VII **691–710**
air, VII 697–702, 824–9, 990, 1017–19, 1022–3
ancient world, II **492–536**
beginnings, I 704–5
bridges, I 256, 714–15
camel and dromedary, II 497
canals, VII 912–20
coal, III 84–6; VII 694–5
cost, III 438
driving test, VII 872
electric vehicles, VII 731–5
flag discrimination, VII 696
goods vehicles, VII 722–8
grain, II 103, 498; III 19
growth, VII 1478–9
horse-drawn railways, III 85
inland, I **704–29**; III 84–5; VII 702, 708–10,
986–7
leisure market, VII 701–2
litters and sedan chairs, II 182–3
loads, comparison, III 438
metals and ores, II 568 n.

Ministry of, U.K., VII 871
oil tanker fleets, VII 695–6
pack transport, I 705–7, 728
public service vehicles, VII 706, 728–30
railways, III 84; VII 703–4, 706–7, 708–9,
    762–88
road speed limit, VII 872
roads, I 712–14; II 512–16, 524–7
salt, II 527
sea, III 19
shipping, VII 692–6
sledges, I 205, 206, 210, 479, 481, 516, 707–11
stone blocks, II 401, 545, 660, 662
supersonic aircraft, VII 834–5, 1469
tilting-rails, III 85
traffic census, England (1922), VII 872
traffic signals (1929), VII 872
vehicles in use, U.K. (1904–60), VII 705
water, III 84
wine, II 129–31, 136–7, 527, 545
wool, II 527, 537
Travel, ancient world, II **492–536**
Travertine, II 27, 406
Trawling methods and equipment, VI 343–8
Tread-traps, I 518
Treadles, III 160–1, 223–4, 237
Treadmills, II 16, 591, 602, 636–7, 642, 659–60;
    IV 151, 634–5, 671
Treaties and pacts
    Kellogg Pact, VI 14
    Locarno Pact (1925), VI 13–17
    Nuclear Weapons Tests (1963), VI 283
    Peace of Brest–Litovsk (1918), VI 10
    Versailles Treaty (1919), VI 11, 17
    Washington Naval Treaty (1922), VI 12
Trebuchets, II 643, 644, 724
Trees, cultivation, I 371, 372, 450, 551–4
Triangulation, *see under* Surveying
Tribulum (threshing-board), II 97, 106
'Tricel' (textile fibre), VI 650
Tricycles, *see* Motor-tricycles
Trigonella, I 248
Trigonometria instrument (1617), III 626–7
Trigonometry; *see also* Mathematics
    adoption, 3rd century B.C., III 503
    calculators, III 614–15
    navigation, III 526
    tables, III 542, 551
    'Topographical Instrument', III 542
Trinity House, IV 674
Triple Entente, VI 6
Tripoli powder, III 237
Triquetrum (Ptolemy's rulers), III 589–90
Triticum (wheat), I 363–6, 520, plate 7; II
    103 n.
Trobriand Islanders, I 119–20, 739, 742
Trolleybuses, VII 732

Trompe (water-aspirator), III 52–3
Trowels, III 117
Truman Doctrine, VI 24
Tube Alloys Directorate, VI 271, 281
Tubes; *see also* Pipes
    metal, V 625–9
    optical instruments, III 634–5
    rubber, V 755
    seamless steel, V 64, 629
Tufa building stone, II 406, 410
Tungsten
    alloy steels, V 64–5, 98, 640; VI 453–5; VII
        1037
    carbide steel, VI 453–5; VII 1037
    carbide-tipped rock-drills, VI 421
    extraction, IV 119, 124; V 98, 99
    isolation, 18th century, V 98
    lamp filaments (from 1904), V 99; VII 1082–3
    magneto points, VI 430
Tunnels, V **499–521**; *see also* Coal-mining, Metal-
    mining, Mining, Tunnels named
    engineering techniques, IV 462–4; V 516–21,
        527; VI 200; VII 902–5
    hydro-electric schemes, VI 200–1
    railways, VI 200; VII 899–901
    roads, VII 905
    Roman roads, II 510
    water-supply, I 533; II 667
Tunnels named
    Benelux, Rotterdam, VII 905
    Bons, Grenoble to Vienne road, II 511
    Boston, sewer-outlet, U.S.A., V 521
    Breadalbane, Scotland, VII 900
    Clunie, Scotland, VII 900
    Clyde Traffic, VII 905
    Coen, Amsterdam, VII 905
    Dartford, England, VII 905
    Detroit River Railway Tunnel, VII 905
    Furlo pass, *Via Flaminia* (A.D. 78), II 511
    Hudson River, U.S.A., V 519, plate 31A, B
    IJ, Amsterdam, VII 905
    Kingsway (1971), VII 905
    Languedoc canal, III 467–8
    Lochaber, Scotland, VII 900
    Merseyside canal, England, IV 563, 565
    Monmouth Motorway, VII 901
    Mont-Cenis, Alps, V 516–17, 527
    Nieuwe Maas, Rotterdam, VII 905
    Potters Bar Railway, VII 902, 904
    Queensway (1934), VII 905
    Rotherhithe, VII 901, 902, 903
    St Gothard, Switzerland, V 517
    St Quentin canal, France, IV 556
    Scheldt, Antwerp, VII 905
    Severn, England, V 516, 518; VI 217
    Simplon, VII 899
    Stockholm, Sweden, V 520

Tunnels named (*cont.*)
  Thames, England, IV 427, 449, 463–4, plate
    32A; V 516; VII 901–2
  Tower Subway (1869), VII 905
  Tronquoy, France, IV 462
  Vienna area, Roman canals, IV 553
  Woodhead, Pennines, VII 899, 900
Turbines
  compressed air, IV 83
  gas, IV 148; VI 438, 443; VII 760, 998,
    1016–24
  hot-air, IV 160–1
  pit water-wheel (1579), III 328–9
  power recovery, VI 213–14
  steam-, IV 148, 168, 193; V 138–40, 149–52,
    301; VI 444; VII 758–9, 760, **1025–34**,
    1069–71
  turbo-generators, V 137, 139–40; VII
    1069–71
  water-, IV 148, 203; V 347, 528–33, 536; VI
    208–10
  wind, IV 160
Turkey
  parliament, VI 4
  ships, IV 588
  water-supply, IV 498
Turmeric, I 247, 259
Turners/turnery
  Greek craft, II 222–3, 232–3
  Indian method, II 232
  lathes, use, IV 383, 418–19
  Middle Ages, II 249–51, 643–4
  northern regions, 6th century, II 237
  plate, IV 28
  wood-work, II 232–3, 234, 249–51, 643–4
  workshop, 18th century, IV 417–18
Turnips, cultivation, III 15
Turnpikes, IV 536–7; *see also* Roads
Turpentine, III 687; VI 595–6
Tutankhamen's tomb (*c.*1350 B.C.), I table F
  baskets, I 422, 450, plate 10B
  bow-drill, I 224
  camp-bed, II 235–6
  chariot, I 722, 728
  coffin, gold, I 660, frontispiece, plate 22
  cosmetic jar, I plate 3
  dagger of iron, I 618, 619
  fine metal-work, I 658; II 460
  furnishing, glass, II 320
  furniture, I 686, 691–6, 699, 701, plates 23–5;
    II 222
  gold, rose coloured, II 484
  harness, chariots, II 163
  headrests, I 654; II 319 *and* n., 336
  natron, I 269
  pigment discovery, I 239
  pottery wine-jar, I 283

  textiles, I 440, 448, plate 16
  writing materials, I plate 32
Tutty (zinc oxide), II 54, 352
Tweezers, III 394; *see also* Pincers, Pliers
Twill (cloth), III 188, 195–6, 198–9
Type-metal, III 29, 43–4, 385–6, 393; IV 120
Typewriters, V 639, 687, 689–90; VII 1269–71
Typha angusta (basket reed), I 420
Tyres; *see also* Tire-bender
  iron, IV 522; V 422–3, 430
  pneumatic, IV 523; V 418, 424–5, 431, 436,
    771–3; VI 339, 563
  solid rubber, IV 523, 541; V 422, 423, 430,
    433, 771–3

Umbrella frames, V 624
Unemployment, VI 16, 102–3
Union of Soviet Socialist Republics (USSR)
  agriculture, V 16, 304; VI 328–9
  aircraft industry, VII 823, 830–4
  atomic energy research, VI 242–4, 279–80
  blast-furnaces operated by coke, V 70
  canals and waterways, IV 553–4
  Central Aerodynamic Institute, VII 822
  chemical industry, V 243; VI 507–8
  coal, VI 178–9, 180
  copper production, V 76 n. 1
  education, VI 164–6
  food, V 16, 304
  fossil fuels, VI 172, 174, 185–7, 191
  flax, III 156
  geodesy, V 442–3
  gold, IV 141; V 95
  Gregorian calendar, adoption, III 579
  hemp, III 158
  industrial growth, VI 4, 14, 411
  industrialization, V 828
  Iron Age burial structures, I 324–5
  iron and steel, V 67; VI 498
  medical technology, VII 1322–3
  metal-work, I 646–7; II 470; V 632
  navy, V 189
  oil industry, V 111, 113–15, 117, 121–2
  paddle-steamers, V 143
  Palaeolithic settlements, I 301–2, 307, 309
  platinum metals, IV 144
  potash (pearl-ash), III 229
  railways, V 61, 324, 329–30, 342
  Revolution, VI 10, 101, 180; VII 1322
  roads, IV 529, 545, plate 33C
  shipping, IV 558, 588
  space technology, VII 865–70, 990
  surveying equipment, V 449
  surveys, IV 602, 610
  telegraphy, IV 656
  tobacco, III 8
  trade-routes, III 380

water-supply, V 554
wheeled vehicles, I 716, 717, 719
windmills, II 624; III 89, 92–3, 98–9, 103, 106
Yukaghir peoples, I 179
United Kingdom, *see* Britain
United Nations Educational, Scientific and
    Cultural Organization (UNESCO), VI 26
United Nations Organization, VI 25, 49, 50
United States of America; *see also* America, North
    America
  aeronautics, V 391–2, 404, 405–7, 409–12
  agriculture, IV 7, 8, 12, 23; V 3–4, 24, 27, 30,
    33–6, 42–4, 46, 48–9; VI 319–27, 329–36
  aircraft, VII 789–92, 797–9, 800–7, 810–16,
    822–32, 840–52
  aircraft engines, VII 1009–10, 1012, 1014–15,
    1018–22
  aluminium, V 91–2, 93, 249
  atomic energy research, VI 225–40, 252–4,
    257–8, 260–2, 271–9
  blast furnaces, IV 122–3
  bridges and tunnels, IV 458–9, 486; V 62,
    500–2, 508–11, 513, 515
  building, V 466–7, 470–1, 477–9, 481–2, 484–6,
    491; VII 931–49, 954
  canals and waterways, IV 548–9, 551, 555, 562
  cartography, IV 621–2; V 440–2, 444, 449, 453
  ceramics, V 659–64, 666, 668, 670; VI 622,
    625, 629–35
  chemical industry, V 243, 248–52, 255, 299,
    301, 312; VI 500–2, 506, 508–11, 515–23,
    528–30, 541–3, 548–50
  cinematography, V 734, 737, 739, 742–4, 746–8
  civil engineering, VII 873, 880–1, 884–7, 890–1,
    896–7, 913–17
  computers, VII 1154–8, 1164–6, 1168–71,
    1179–1203
  copper, V 77, 79, 85
  cotton, III 159
  discovery, III 548
  domestic appliances, VII 1126, 1128, 1131–5,
    1140–1, 1146–9
  dredging, IV 641
  dyestuffs, V 258, 261, 265–6
  electrical communications, VII 1220–30,
    1233–41, 1244–57, 1259–62
  electricity, V 133–4, 187, 191, 199–201, 203,
    213–14, 216–18, 232; VI 284, 289
  electronic engineering, VII 1092–1125
  engineering, scientific basis, VII 984–6
  engines, V 126, 131, 133–4, 137–8, 140, 142–5,
    147, 149, 152, 155, 160–1; *see also* railways
  explosives, V 292
  fish industry, IV 51; VI 343–5, 349, 362–6
  flax, III 156
  fossil fuels, VI 173–84, 187, 189–92, 362–6,
    376–408

glass, V 675, 678–9; VI 574–6, 578–89
gold refining, IV 142–3; V 96
houses, weather-boarded, III 265
hydraulic engineering, V 529, 532–3, 537, 540
hydro-electric engineering, VI 204, 206, 213–16,
    221
hydrography and oceanography, IV 619–20
iron and steel, V 53–4, 57, 60–70, 606, 612,
    629; VI 458, 463–5, 468–73, 475–81, 488,
    490–8
iron industry, III 32; IV 105, 110–11
machine-tools, IV 417, 435–41; V 638–40,
    645–8, 652–6
McMahon Act (1964), VI 281
management, scientific, VII 1046–58
mercury, IV 135
metals, IV 66, 69
military technology, VI 123; VII 848–9
name, III 535 n.
National Labor Relations Act (1933), VI 108
oil and petroleum, V 105–6, 108–10, 112–13,
    115–17, 119–22
paints, VI 590, 597, 599–606
photography, V 721–2, 725, 729–30; VII
    1284–8, 1294–5
plastics, VI 552–7, 560–2
population changes (1800–1950), V 26, 823
printing, V 683–5, 687–90, 692–4, 697–701,
    704, 707, 709–10, 712–14; VII 1271–9
railways, V 62, 323–4, 326–7, 329–32, 336–7,
    339, 351–6, 348; VII 702–3, 766–72, 774–6,
    778–87
road transport, VII 704–5, 707–8
road vehicles, VII 711–26, 728–35
roads and streets, IV 536–41, 543–5
rubber, V 763–8, 773
sewage disposal, IV 518
ships and ship-building, IV 586, 589–92; V
    117, 383; VII 694–7, 699–700, 740–3, 750–4
silver refining, IV 142–3
sodium cyanide, V 249
spinning-wheels, III 161
steam-engines, IV 188–90, 195
steam navigation, IV 548
steam-power, IV 166
steel, *see* iron and steel
surveying, V 440–2, 444, 449, 453
telegraphy, IV 647, 655, 659–61; V 220, 225–6
telephones, V 226
textile-machinery, IV 282, 286–7, 291, 295, 298,
    305
textiles, V 571–3, 575, 577, 579–82, 584, 586–
    92, 599–600, 602; VI 639, 641–6, 650–1,
    658–60, 663, 676–7, 688–9
tinplate (1890), V 617
tunnels, *see* bridges and tunnels
vehicles, V 414, 425, 433

United States of America (*cont.*)
  water-supply, IV 501; V 554, 557–60, 564–6
  whaling, IV 58, 62
  windmills, III 89, 95, 106
  wire manufacture, V 621, 623, 625
Universities and colleges; *see also* Education
  Basle, VI 552
  Birmingham, VI 147, 150, 161, 235
  California, VI 83, 189, 241; VII 1407
  Cambridge, V 780, 785, 792, 793; VI 141, 150,
    154, 161, 162, 301, 305, 327; VII 1018, 1069,
    1151, 1154, 1188, 1198–9, 1202, 1403
  Chicago, VI 83, 85, 236, 252, 271; VII 1133,
    1223, 1230, 1277, 1318, 1324, 1328–9, 1331
  City and Guilds Institute, London, V 794–5
  Clark, U.S.A., VII 861–2
  Columbia, U.S.A., VI 141, 233
  Copenhagen, Denmark, IV 654
  Cornell, U.S.A., VI 221; VII 1021
  Dartmouth College, England, VI 83
  Durham, England, V 789
  École Central, Paris, V 787
  Edinburgh, Scotland, IV 444
  Giessen, W. Germany, V 788
  Glasgow, Scotland, IV 445; V 780–1; VII 1407
  Gottingen, W. Germany, V 788; VI 44
  Johns Hopkins, U.S.A., VII 1346
  Harvard School of Business, U.S.A., VI 83,
    236, 560; VII 1186
  Heidelberg, W. Germany, V 788
  Leeds, England, VI 141, 147, 150, 161
  Liège, E. Belgium, VI 139
  Liverpool, England, V 778; VI 147, 150
  London, England, IV 445, 657; V 781, 793;
    VII 762, 984
  Manchester, England, V 777–8, 798; VI 147,
    150, 161, 580; VII 1196, 1199, 1202
  Massachusetts Institute of Technology (1865),
    V 789; VI 236; VII 1187, 1195, 1197, 1198
  mechanics institutes, U.K., V 777–80, 783–4
  Mines, Government School, England (1851), V
    782
  Munich, W. Germany, V 788
  New York, U.S.A., VI 83
  Newcastle, England, VI 150
  Oxford, England, V 780, 785, 792; VI 154, 157,
    162
  Pennsylvania, U.S.A., VI 83; VII 1189
  polytechnics, V 787–8, 790, 795
  Princeton, U.S.A., VII 1197, 1198
  Queen Elizabeth College, London, VII 1407
  Queen's Colleges, Ireland, V 783, 792
  Royal College of Science, London, VII 1402
  Sheffield, England, VI 147, 161
  Strathclyde, Scotland, VII 1407
  technical colleges, V 795–6
  Toronto, Canada, VII 919

Université Libre, Belgium, VII 1351
Virginia, U.S.A., VI 230
Westminster, U.S.A., VI 221
Würzburg, W. Germany, VII 1299
Yale, U.S.A., VII 1358
Zürich, Switzerland, VII 984
Upper Palaeolithic culture, *see* Cultures
Urali Kurumbars of Wynad, India, I 410 (14)
Uranium, IV 119, 221, 375; VI 225–7, 228–37,
  256–67
Urtica, I 355–6, 449
Utrecht Psalter, II 95–6, 651

Vaccination, VI 315
Vaccinium (bilberry), I 353
Vacuum; *see also* Aeolipiles
  atmospheric pressure discovery, IV 168–9
  distillation method (1855), V 115
  early discoveries, III 324, 344, 432
  pumps, III 324, 332, 637, plate 27A; IV 169
  steam-engine inventions, IV 168–71, 176
Vacuum-cleaners, VII 1129–31
Valves
  aircraft engines, VI 441, 442–3
  hydraulic engineering, V 536
  I.C. engines, V 165–6, 169
  steam engines, 18th century, IV 175, 185–6,
    188
  suction pumps, 16th century, III 330–2
  thermionic, VII 1091–8, 1100–3, 1237–8
  turbines, VI 202–3
Vanadium, V 640
Varnishes, I 243–5; II 248, 249, 362; III 687,
  695–6; *see also* Lacquer industry
Vasari Society, mud-mill (1589), IV 639
Vats, paper pulp, III 412–13
Vegetable Staticks (1727) Stephen Hales, IV 216,
  258
Vegetables, *see* Agriculture, Food
Vehicles; *see also* Chariots, Motor-entries,
  Transport, Wheeled vehicles
  agricultural, III 134, 139–48
  barges, III 438–9
  caretta, II 541
  classical times, II 541–5
  coach-tools, III 113, 115
  coaches (from 1450), II 547–8; IV 522–3,
    530–1, plate 33B
  early civilizations, II 537–62
  electric, V 418–20
  Greek, II 498, 544–5, 549–50
  hamaxa carts, II 498, 539
  land transport, I 704–29
  load capacity, 18th century, III 438
  medieval times, II 546–8, 552, plate 37A
  motor-cars, 19th century, IV 539

road-, V 414–37; VII 711–35
Roman, II 545, 550
steam, V 420–6, plate 18B
traffic, London Bridge (1850), IV 546
tumbrils, III 143–4
wains, III 144
water transport, early man, I 730–43
Vellum, II 180, 188
Velocity, water-wheels, IV 205, 210–11
Velvet, III 164, 179, 199, 203–5, plate 15C
Veneering, II 227, 240
Venezuela
  natural gas, VI 191
  oil, VI 186
Ventilation
  bearings, air lubrication, VII 991
  coal-mines, III 75; IV 86, 91, 92–4; VI 373
  machinery, VII 989
  metal mines, II 3, 15–16, 17, 19, 23; III 30;
    IV 69, 671; VI 421–2
  rotary fan, Chinese, II 770
  sewers, IV 515–16
  tunnels, V 517–19
Verjuice, II 138 and n., 357
Vermilion pigment, II 5–6, 361, 747
Vernier, IV 380, 385, 397, 400, 402–3, 603
'Verulamium plan' (c.1600), III 680–1
Veterinary science, IV 14, 32, 35
  animal nutrition, IV 14, 41–2
Viaducts named; see also Bridge entries
  Adam, Wigan, VII 892
  Fades Railway, France, VII 881–2
  Goktek, Burma, VII 880
Vibration research, VII 987
Vices, III 110, 114, 120, 337–8
Vicia, I 361, 362
Vikings, II 237, 474, 479, 557, 562, 579–81, 720;
    IV 560
Villanovan culture, II 269, table I
Vinegar; see also Acetic acid
  adulterants, Middle Ages, II 128
  ancient use, I 284, 285
  fermentation product, I 276
  fire-setting myth, II 8 n.
  glazing process, ceramics, II 357
  green pigment process, II 362
  manufacture, V 308, 310
  value (c.1501–8), II 747
  white lead process, II 360 and n.
  wine-making process, II 133
Viola arvensis, I 370
Virgula divina, see Divining-rods
Vision (visual faculty)
  higher primates, I 8
  racial groups, I 35–6
  skilled behaviour, I 6, 7–8, 33
  sustained attention, man, I 12

Vitamins, VI 300–1, 310
Viticulture
  ancient civilizations, I 282–4
  Australian (c.1900), V 9
  copper-based fungicide, V 78
  Greek, II 128–33
  medieval, II 137–9
  Italian (1860s), V 20
  Roman, II 133–7
  Spanish, II 134
Vitis (grape-vine), I 281
Voile (worsted fabric), III 179

Wainscot oak panelling, II 244
Wales; see also Britain, Europe
  anthracite, IV 110–11
  bridges, III 434–6
  canals and waterways, IV 564–5, 572
  coal-mining, IV 87, 93
  copper, IV 126
  farm vehicles, III 140, 144
  iron industry, III 31; IV 114–15
  lacquer manufacture, III 688
  lead industry, IV 101, 123
  surveys, IV 607
  water-supply, IV 490, 497
  wool, III 152
Walls; see also Fortifications
  buildings, III 247, 261, 266–7
  papering, III 415
  towns, III 270–1, 278, 280–7, 291
Warfare, III 347–76; see also Military technology,
    Wars named
  atomic, VI 23–4
  biological and pesticides, VI 322
  Blitzkrieg technique, VI 18–19
  bombing attacks, VII 795–6, 844, 845, 847
  cavalry, I 728–9; II 703, 704, 705–6; III
    348–9, 353
  chariots, I 209–10, 713, 718, 724–8
  coal-industry, VI 178–84, 359–66; VII 759
  galley, development, I 741–3
  Greek and Macedonian period, II 695–703
  illuminants, shortage (c.1800), IV 265
  iron industry, III 31
  mathematics and science, III 371–5, 622
  medicine, III 373–4
  medieval period (c.400–1400), II 720–8
  naval, III 350–68; VI 9–10
  Roman period (c.200 B.C.–A.D. 400), II 703–20
  strategy and tactics, III 347–50
  trench-, VI 8–9
  water-supply, I 526–7, 530–1, 554–5; III 318
Warping, II 209–10; see also Textiles
Wars named; see also Battles, Warfare
  Abyssinian campaign (1868), V 555

Wars named (*cont.*)
  American Civil (1861–65), IV 589; V 3–4, 10,
    35, 51, 226, 373, 397, 569 n., 646, 808, 819,
    827
  Balkan (1912–13), VII 793
  Boer (1890–1902), V 51, 397, 631
  Chinese incident (1931–41), VII 811
  Crimean (1854–57), IV 587–8; V 10
  English Civil War (1642–49), III 369
  Franco–Prussian (1870–71), V 10
  Hundred Years War (1338–1453), III 349
  Italian, 16th century, III 349
  Italo-Turkish (1911–12), VII 793
  Japanese, 20th century, VI 15–16, 23
  Napoleonic (1803–15), IV 582, 586; V 713
  Peninsular (1808–14), V 570
  Religious (*c.*1540–1648), III 349
  Spanish Civil (1936–39), VII 811
  Sudan Campaigns (1883, 1889), V 630, 750
  Thirty Years War, Germany (1618–48), III
    349, 622
  twentieth century, VI 1–26
  Wars of the Roses (1455–85), III 349
  World War I (1914–18), V 399, 403, 803, 833;
    VI 8–11; VII 811
  World War II (1939–45), V 51, 412; VI 18–23;
    VII 811–12
Washington Treaty (1922), VII 752
Waste disposal, IV **504–19**
Watches; *see also* Clocks
  drum-shaped, III 656
  jewelled bearings, III 670–1
  'pendulum', III 669
  pocket, III 636
  repeating, III 669
Water, alchemical element, IV 214
Water-carriers, II 187, 689
Water-carts, dust laying, IV 541
Water-clocks, *see under* Clocks
Water engineering, II **663–94**; III **300–23**; *see*
  also Hydraulic engineering, Water headings
  aqueducts, III 289, 446, 466–7
  dikes and polders, II 681–9
  drainage, II 617, 637, 670, 678–85; III 270,
    276, 281, 291
  installations, Rome, II 671–4
  irrigation, II 617, 637, 670, 678–85; III 286
  pipes and sewers, II 418–19, 527–31, 664–6
  pumping, IV 168–79, 491, 493
  Roman period, II 405, 418–19
  roof drainage, II 404, 444–5
  sewers, III 281
  town-planning, III 285, 286, 293–5
  water-lifting, II 675–8
  water-wheels, *see* Water-wheels
  wells and cisterns, II 663–4, 666, 689
Water-level (surveying), III 511–12, 542–3, 609–11

Water Pollution Research Laboratory, VII 1389,
  1396
Water-power; *see also* Hydro-electric power,
  Natural power
  blast-furnaces, III 30, 32, plate 3
  canal-lifts, IV 549–50, 568–71
  ceramic industry, IV 348
  chemical works, V 301
  corn grinding, II 349
  corn-mill (1588), III 19
  engines, V 527–35
  flour bolting, mechanization, III 18–19
  forge-hammers, IV 104
  fulling-mills, II 216, 349, 609
  glass-engraving machinery, III 223–4; IV 377
  hydraulic hammers, II 69, 73, 75
  industrial revolution, IV 150–1
  iron and steel industry, III 32
  machine-tools, IV 441
  metal-working mills, II 74; III 45, 342–3, 351,
    367; IV 130
  metallurgy, II 62, 68–9, 73–4, 75, 78, 348
  military weapons, III 351, 359, 367
  mine machinery, III 82–3, 327–8; IV 76, 78
  ore crusher, III 327–8
  paper-making machinery, III 412, 415
  pumps, IV 330–3; V 526–7
  steam competition, IV 151–6, 161–2
  textiles machinery, III 162, 168–9; IV 278–9,
    283, 287
  twisting-mills, II 207
  -wheels, *see* Water-wheels
Water-supply, IV **489–503**; V **552–68**; VII
  1363–98; *see also* Irrigation
  ancient civilizations, I **520–57**; II **663–94**
  coal-mines, Low Countries, III 74
  compensation-water, IV 495–6
  Constantinople system, III 279–80
  consumption (1849–1900), V 552
  Corinthian system, III 275
  dams and reservoirs, IV 490, 495–8, 501–2
  distribution, V 559–60; VII 1371–3
  Greek houses, II 404
  House of Commons Select Committee (1840),
    IV 490
  hydraulic engineering, IV 449; VII 1363,
    1364–5
  London, III 464
  mains and pipes, IV 497, 499–500, 502
  management, V 552–4; VII 1363–5
  medieval, II 689–92
  meters, IV 501–2
  organization, V 552–4
  Paris, III 462
  pollution, V 561–2, 567
  purification, II 674; IV 493–5, 498, 501–3
  Roman, II 671–4; III 276, 282

smelting works, Siegen area, II 74
softening, 19th century, IV 503
sources, V 555–8; VII 1366–71
treatment, V 561–7, plate 32; VII 1373–80
valves, IV 501
water-budgets/skins, II 184, 187
waterworks, IV 451, 489–96, 500–3; V 127–9
wells, IV 492–3, 495, 497–9
Water-wheels; *see also* Turbines, Water-power,
    Watermills
  bellows, furnace, II 68
  London Bridge installation, III 420–1; IV
    491–2
  Marly-la-Machine, Versailles, IV 154–5
  medieval, II 681
  metal construction, IV 209, plate 11A
  mines, II 7, 10, 20–1, 23, 66, 78
  overshot types, II 23, 68; III 19; IV 204–5
  prime-movers, II 605–6; III 170, 327, 328, 342
  scoop-wheels, IV 156
  Smeaton's experiments, IV 152–3, 203–9
  steam-assisted, IV 178, 181
  tub/pit-wheel design, III 328–9
  types, IV 201–9
  undershot type, II 20–1, 23, 609, 612–13; III
    331; IV 202–3
  water-lifting, II 675
  wire drawing machinery, II 74
Watermills, II 775; IV 199–213; *see also* Water-
    power, Water-wheels
  breast-wheels, IV 201–2, 204, 206–9
  Brook Mill, Deptford, IV 209
  Chinese use, II 770
  Coldron Mill, Spelsbury, Oxon, IV 212, plate
    11B
  floating mills, II 607–8, 647
  flour-milling, 17th century, III 15, 19
  gunpowder-mill (c.1676), II 381
  industrial uses, IV 199–202, 348, 419, 582
  iron and steel industry, II 69
  medieval period, II 609–11
  Middle Ages, II 608–14
  Norse mills, II 590, 593–5; IV 200–1
  overshot-wheels, IV 200–1, 203–5, 207
  Roman, II 600–6
  Stratford Mill, Somerset, England, IV 209–10
  tide-mills, II 610, 614, 648
  undershot-wheels, IV 200–4, 210
  vertical-, angled blades (1579), III 328
  Vitruvian, II 595–600; IV 200
  wooden construction, II 395
Waterways, III 438–70; *see also* Canals
Wax, metal density standard, III 57
Weapons, I 600–22; III 347–76; *see also* Artillery,
    Firearms, Tool-making
  bayonets, III 348
  foraging/hunting, I 154–68

halberds, III 351
harpoons, I 181–2
lances, III 348, 351
ornamented, II 451, 453, 455–8
pikes, III 348, 350–1, 353
pyrotechnic, II 374–82, 771
spears, *see* Spears
steel, II 63, 456
swords, III 35, 350–1
Weaving, II 210–14; III 164–9; V 579–87; *see
    also* Spinning, Textile industries
  automatic looms, III 168–9; VI 661–6
  basketry, I 416–18, 422–3
  Bronze Age, I 442–3
  carpet weaves, V 581
  cloth types, III 178–9
  draw-looms, III 164–7
  Dutch loom-engine (bar loom), III 167
  figured fabrics, III **187–205**
  flying shuttle, invention, III 169
  Greek and Roman, II 211–12
  Greek vase (c.560 B.C.), I 444
  hand-loom (from 1760), IV 299
  knitting/knitted fabrics, III **181–6**; VI 666–71
  linen loom, Egypt (c.1500 B.C.), I 437
  looms (c.3000 B.C.), I 425–8
  medieval period, II 212–14
  origins, I 413–15
  plain looms, III 164
  power-looms, III 167–9; IV 299–304
  primitive techniques (c.2500 B.C.), I 420–1
  Rameses III, girdle, I 441, plates 14, 15
  silk fabrics, II 198; IV 316–19
  swivel loom, III 168
  twill weaves, I 430, 443, 445
  types, I 428–31, plates 9–17
Wedges
  furniture, Middle Ages, II 242 n., 252
  mines, to 17th century, II 3, 8, 12, 15
  quarrying tool, II 30, 36
  Roman period, II 630, 712
  tradesmen's tools, III 110
  wedge-presses, fruit pulp, II 117–18
Weights, I **774–84**; *see also* Balances,
    Measurement
  ancient balances, I 779, 781, plate 31B
  assaying, III 61, 66–8
  balances, precision, IV 403–8
  capacity, I 781–3
  metallurgical, 11th–12th century, III 57
  Middle Ages, II 128
  standards (c.5000 B.C.–1500 B.C.), I 779–81
  standardization, I 110, 774–5, 783–4
Weirs, III 440–1, 448–9, 456, 458, 462, 464, 467
Welding, II 58; VII 986; *see also* Metals,
    Soldering
Welfare work, VI 84

Wells and cisterns, II 663–4, 666, 689
Whales and whaling, IV **55–63**, plate 4; VI 356–8
 gear, IV 59
 harpoons, IV 58–62
 hunting, IV 57–62
 oil, IV 56, 58, 61–2, 264, 272
 whale-meat, IV 62
 'whalebone' (baleen), IV 55–6, 58
 whalebone spoke-set gauge, III 124–5
 whaling-fleet, IV 468
Wheat, I 363–7, 522, plate 7; II 103–4
Wheelbarrows, II 17, 546, 547, 641, 642, 770
Wheeled vehicles, I **716–29**; *see also* Vehicles
 carts and wagons, I 716–19
 chariots, I 209–10, 700–1, 713, 718, 720, 724–8
 draught animals, I 719–24
 origins, I 209–11, 704
 roads, 17th–18th centuries, IV 522–4
 solid wheels, I 204–9, 716
 spoked wheels, I 211–14
Wheelwright's trade, notes, III **123–8**
Wheels; *see also* Tyres, Water-wheels
 carts, II 545, 548–50; III 142–4, 146–7
 chariots, II 539, 542, 549, plate 5D
 cog-wheels, II 635, 648, 654, 675–6; III 654–5
 crown, III 345
 diffusion, I 203–4, 207, 209–11, 716
 'dished', II 552; III 112, 124, 146
 dividing-engine, IV 392–3
 early wheels, dimensions, I 209
 engraving, III 222–4
 evolution, I 71–2, 80, 187, 194–5, 206–7, 716
 fly-, III 223–4, 326, 423
 gear-, clocks, III 641
 glass-making, III 210, 226, 239
 knurled, for mezzotints, III 407
 lantern-, III 18
 -making (craft), III **123–8**, plate 6
 origin, I 194–5; II 630
 paddle-wheels, II 606, 607, 647–8, 651, plate 37B; V 141–6
 ploughs, II 88–9
 polishing, armour, III 351
 potters', *see* Potters' wheels
 pulleys, II 630
 rotary motion, I **187–215**
 scoop-wheels, II 685, 688; III 303–6, 321–2
 solid, II 545, 548, 550
 spinning-wheels, *see under* Spinning
 spoked development, I 211–14, 721; II 548–52
 tires, II 163, 549, 551–2; *see also* Tyres
 wheel-dredgers, IV 637–8, plate 45A
Whitley Committee (1917–18), VI 101
Wicker-work, II 237, 253–4
Winches, II 630, 642, 646, 685; III 97–9, 177, 441

Wind; *see also* Windmills
 charts, III 555
 -organ, II 614–15
 -power, III 83, 412, 415; VI 214–16
 -rhumbs (points), III 518, 524–7
 -rose, III 517, 523–7
 turbines, *see under* Turbines
 velocity, IV 157–9
Windlasses
 ancient cranes, II 658–9
 cannon-boring operation, III 367
 draw-plate device, III 47
 horse-whims, II 18
 invention, I 214
 lock-gates, canals, II 686; IV 439, 442
 mines, I 566; II 14, 16
 pile-driver, 16th century, III 423
 ships, 17th century, III 483
 'Spanish windlasses', ships, I 735
Windmills, II **623–8**; III **89–109**
 acid cleansing tower, conversion, IV 241
 auxiliary machines, III 18, 106
 books, III 89, 96, 99, 108
 brakes, III 103–4
 development, 14th century, II 349, 395, 396, 775
 development, Europe, II 617–20, 625–8
 drainage-mills, polders, II 688–9
 Dutch types, IV 157–60
 energy output, 18th century, IV 157–9
 fantails, III 94, 95, 96, 98–9, 102, 107
 flour production, III 15, 18, 90–2
 hoists, III 18, 105
 industrial revolution period, IV 151, 203
 land drainage, III 303, 307–8, 317, 321–2, plate 20
 marsh-mills, II 689; IV 156, 157, 159
 mechanical aids, III 97–8, 106
 millstones, III 104–5
 origin, II 614–17
 ownership, 16th century, IV 201
 post-mills, II 536, 623–4, 649, 689; III 89–95, 97–9, 103, 105, plates 4, 5
 prime-movers, IV 156–7, 199
 rollers, III 96–7
 roofs, III 93, 97
 rotating caps, III 303
 round-houses, III 94
 sail-cloths, III 99–100
 sails, III 93–4, 96–103, 106–7
 smock-mills, III 95, 106–7
 survivals, III 89–90, 96
 textile industry, III 170
 tools, III 107–8
 tower-mills, III 89–90, 95–7, 99, 102, 107
 turret, III 330
 -turbines, II 160
 weatherboarding, III 93–5

winches, III 97–9
wind engines, 20th century, VI 214–16
wind shafts, III 102–3
Windows; *see also under* Building construction, Glass
 glass, II 416, 443; III 237–40; IV 365–7
 Gothic period, II 437–8, 443
 Greek period, II 403
 lead strips, III 41–2, 44
 Roman period, from 1st century B.C., II 416
 sashes, III 267
 stained-glass, II 429; III 32, 240–2; IV 373
Wine; *see also* Viticulture
 adulteration, II 128
 ancient civilizations, I 281–5
 date-wine, early Egyptian, I 276–7
 dregs, II 354–5
 grape-presses, II 106, 112–18
 Greeks and Romans, II 119
 kotyle, Syrian measure (*c.*1540 B.C.), I 782
 -making, I 282, 284
 manufacture, II 131–3, 135–6
 origins, I 275–6, 281–2
 palm-wine, I 277
 trade practices, II 129–31, 134, 137
 vessels (amphorae), II 129
Wine-skins, II 160, 161, 167
'Winged Victory' (statue), II 566–7, plate 35
Wire
 barbed-, V 625
 coiled drawn, III 55
 copper, III 59
 -drawing, I 655; II 74, 75; III 32, 342, 414
 filigree-work, I 655–7
 galvanized, V 624–5
 glass core-winding process, II 336
 gold, II 481–3
 hydraulic drawing machine, II 655
 iron, II 74, 75
 manufacture, from 19th century, V 620–4
 origins, I 655
 paper watermarks process, III 413–14
 silver, II 452–3; III 48, 59
 swaging method (before 2500 B.C.), I 655
Woad (dye), I 247, 249; II 349, 361, 365; V 261–3; *see also Isatis tinctoria*
Women
 basket-makers, I 173, 413–14
 beasts of burden, I 705
 food-gatherers, I 154–5, 177
 potters, I 49, 407, 408
 spinners and weavers, I 437, 438, 443–4
Wood; *see also* Carpentry, Charcoal, Timber
 ancient use, I 141–3
 ashes, II 354, 358; III 206–7, 210
 barrels and casks, II 133, 136–7, 255, 354
 bending, II 234

building construction, I 300–25, 371–2; II 397–8, 413–15, 440–3; V 466–70
 carvings, II 245–7
 chemicals source, V 308–11
 clocks, III 672; IV 412
 coal, replacement, III 78–81, 220–1, 685–6
 coopering, III 128–33
 distillation, III 686; V 308–9
 doors and locks, II 415–16, 425, 429–30, 443
 fine wood-work, I 684–703
 fire-making, I 220–6
 flooring-boards, IV 437
 fuel, I 229, 230, 576; II 347–8, 369; IV 234–5
 furniture, *see* Furniture
 machines, wood-working, III 326, 329, 332; IV 382–3, 426–7, 437
 paving, roads, IV 545
 ploughs, II 85, 89
 potters' wheels, I 196
 -pulp, paper, III 411, 415
 sculpturing, I 689–90, 699, 702
 seasoning, II 233
 shields, II 169
 ship-building, V 350, 353, 355 n. 1, 357, 365
 structural methods, I 143, 300–25, 371–2, 469, 472, plate 4
 substitutes, III 685–8; IV 234–5, 371
 supplies, I 684–7; III 77
 tar, III 12; IV 234
 tools and weapons, I 30, 32–3, 70, 141–3, 517–18; III 115
 torches, I 233–4
 vessels, I 517–18, 685; IV 100, 127–8
 wall-linings and panels, II 243–5, 247–8, 363, 443
 water-pipes, II 664; IV 499–500
 wheels, I 204, 206, 207–8; II 549–52; III 123, 126
Woodcuts, III 404–6, plate 2 n.
Woodhenge structure, England, I 314–15, 493
Wool; *see also under* Livestock, Textile industries
 bleaching, Roman period, II 218
 bowing, II 195; III 151–2, 154
 carding, II 193–4; III 153–4; IV 298–9; V 571–3
 cloth, finishing, III 169–70, 172–7
 cloths, list of types, III 178–9
 combing, II 194–5; III 154–6; IV 295–9; V 574–6
 composition, I 451; IV 295
 dye industry (from 2000 B.C.), I 248–9
 figured fabrics, III 196, 197
 production, England, III 152; VI 640–1
 reclamation, used wool, V 569–70
 Roman Middle Ages, II 192–5
 Scandinavian textiles, I 442
 seventeenth century, III 15, 151

Wool (*cont.*)
  spinning machinery, IV 282, 299; V 577–8
  transport, II 527
  weaving, IV 299–305; V 579–87
  yarn, dressing, III 163, 164
Wool Industry Research Association, VI 652
Workshop recipes, III 27
World setting (1900–60), VI 1–26
Worsted, *see under* Textile industries
Wrenches, III 113–14
Writing and scripts, I **744–73**; *see also* Writing-
    materials
  alphabet, origin, I 762–4
  Arabic alphabet, I 767
  Brahmi, Sanskrit and Prakrit scripts, I 771–2
  Chinese scripts, I 772
  cuneiform script, I 83, 103, 750–2, 757, 763,
    plates 29B, 30B
  Cypriote, I 761
  Egyptian hieroglyphs, I 80, 192, 222–3, 237,
    263, 281, 439–40, 747, 752–7, 785
  Elamite, I 760–1
  Etruscan, I 769–70
  Greek, I 767–8, 769–71
  hieratic script, I 754, 759
  Hittite hieroglyphs, I 757–8, 761
  Indus script, I 760, 771
  invention, I 49, 53–4, 80, 101–3, 744–7, 785
  Italic scripts, I 769–70
  Latin scripts, I 770–1
  Maya hieroglyphs, I 83
  Minoan scripts, I 758–9
  Nabataean, I 766, 767
  number-signs, I 785–6
  Phoenician alphabet, I 763–8
  pictographic forms, I 744–7
  Sumarian hieroglyphs, I 80, 101–2, 205–6, 252,
    280, 745–7, 761
  syllabic signs, I 747–50
  Ugaritic, I 762–3
Writing-materials
  Egypt, I 245, 753–4, 757, plate 32
  ink, II 359–60; *see also* Inks, Pigments
  leather, II 164
  Mesopotamia, I 750–1
  paper, II 189, 771

  papyrus, II 188; 197
  parchment, II 164, 180, 187–90
  vellum, II 180, 188

X-rays
  dental industry, VII 1324, 1325
  discovery (1897), VI 223; VII 1092
  electronic emission, VII 1092
  metal structural tests, VI 461; VII 986
  radiology, VII 1299–1300, 1324–5, 1327–34,
    1360

Yami, Botel-Tobago Island, I 118–19
Yokuts, central California, I 173–4
Yukaghir peoples, Siberia, I 179

Zaffre (cobalt blue), III 678, 698, 702–3
Ziggurats, I 467–9
Zinc
  alloys (silver/nickel), IV 146; V 608
  brass alloys, II 53–5, 67; III 37–8; V 607–8
  by-product, lead-smelting, III 30 *and* n.
  coinage, III 46
  die-casting, alloys, VI 434
  extraction, II 54; IV 119, 130–3; V 72, 76;
    VI 434
  galvanized iron (from 1836), V 624–5
  map, sources U.K. (1851), IV 65
  metallurgy, medieval times, II 67
  mining, II 2, 67; IV 66; VI 410, 423
  Munz-metal, V 607–8
  oxide, III 51
  pewter ingredient, III 30, 43
  production, III 690–2; V 73; VI 428
  ship hull sheathing, IV 591
  silver purification process, IV 138
  smelting, III 51; V 78
  speculum-metal ingredient, III 41
  Volta's pile (battery, 1800), V 178
  world production, V 73
Zincography, IV 626
Zingiber officinale, I 288
Zircon extraction, VI 424
Zirconium, IV 221
Zizyphus lotus (lotus tree), I 271
Z. spina (sidder), I 685